THIRD EDITION

INVESTIGATIONS
IN ENVIRONMENTAL GEOLOGY

Duncan Foley
Pacific Lutheran University

Garry D. McKenzie
The Ohio State University

Russell O. Utgard
The Ohio State University

Prentice Hall
is an imprint of

Upper Saddle River, New Jersey 07458

Library of Congress Cataloging-in-Publication Data
Foley, Duncan.
    Investigations in environmental geology / Duncan Foley, Garry D. McKenzie, Russell O. Utgard. — 3rd ed.
       p. cm.
       ISBN-13: 978-0-13-142064-9
       ISBN-10: 0-13-142064-X
1. Environmental geology—Textbooks. I. McKenzie, Garry D.
II. Utgard, Russell O. III. Title.
QE38.F66 2009
550—dc22

                                                                                    2008031528

Publisher, Geosciences: *Daniel Kaveney*
Acquisitions Editor: *Dru Peters*
Editor-in-Chief, Science: *Nicole Folchetti*
Project Manager: *Crissy Dudonis*
Assistant Editor: *Sean Hale*
Media Editor: *Andrew Sobel*
Marketing Manager: *Amy Porubsky*
Editorial Assistant: *Kristen Sanchez*
Managing Editor, Science: *Gina M. Cheselka*
In-House Production Liaison: *Ed Thomas*
Full Service Vendor/Production Editor: *Pine Tree Composition/Patty Donovan*
Senior Operations Supervisor: *Alan Fischer*
Image Permission Coordinator: *Elaine Soares*
Composition: *Laserwords, Inc.*
Art Editor: *Connie Long*
Art Studio: Precision Graphics

ISBN-10   0-13-142064-X
ISBN-13 978-0-13-142064-9

Printed in the United States of America
10 9 8 7 6 5 4 3 2 1

Pearson Education Ltd., *London*
Pearson Education Australia Pty., Limited, *Sydney*
Pearson Education *Singapore,* Pte. Ltd
Pearson Education North Asia, Ltd., *Hong Kong*
Pearson Educación de Mexico, S.A. de C.V.
Pearson Education–Japan, *Tokyo*
Pearson Education Canada, Ltd., *Toronto*
Pearson Education Malaysia, Pte. Ltd.

Prentice Hall
is an imprint of

www.pearsonhighered.com

Plate No.	Figure No.	Description
1	2.9	Topographic map symbols
2	2.10	Bloomington, IN 7.5 minute quadrangle
3	2.11	Geologic map of west-central Great Lakes region of canada and USA
4	2.14	High altitude color photograph of Boston, MA
5	2.16	Color stereopair aerial photos of Little Cottonwood Canyon, Salt Lake City, UT
6	II.3	Presidential disaster declarations: December 24, 1964 to March 3, 2007
7	II.4	Population density for counties: July 1, 2004
8	4.9	Mt. Rainier, WA map
9	5.5	Geologic map of deposits and features of the 1980 eruptions of Mount St. Helens
10	6.11	Draper, UT 7.5 minute quadrangle
11	6.12	Color oblique aerial photo, Little Cottonwood Canyon, Salt Lake City, UT
12	8.13	Calaveras Reservoir, CA (San Jose) 7.5 minute quadrangle
13	8.23	Athens, OH 7.5 minute quadrangle (1975 and 1995)
14	8.29	Avalanche areas in the Aspen, CO 7.5 minute quadrangle
15	10.5	Zanesville West, OH 7.5 minute quadrangle
16	17.3	Geologic map of Waco West, TX quadrangle

PREFACE

Our world is changing rapidly. Population growth in the United States since 1990 has added more than 50 million people, while population growth globally since 1990 has added more than one billion. Many of these people live in geologically active areas, which are subject to potentially hazardous events such as floods, earthquakes or landslides. Decisions that people make, about where and in what they live, greatly influence whether geologic events become disasters. For example, human choices about resource development, such as logging slopes and changing river courses, can increase the number and severity of impacts from geologic events such as landslides and floods. Impacts from the Indian Ocean tsunami after the Sumatra earthquake of December, 2004, which lead to the tragic loss of nearly 200,000 lives, were probably made worse by development choices along coasts. When population is concentrated in the coastal zone by building hotels and condominiums, risks to humans increase. A director of the World Health Organization has been quoted as saying that he doesn't like to use the term natural disasters; human factors are also important in creating disasters.

In addition to more losses from disasters, population increase also has resulted in greater demand for geologic resources such as fresh water, industrial rocks and minerals, metals, and fossil fuels. As we move beyond the peak in global oil production, major changes in energy sources, energy uses, our quality of life, our organization of cities and transportation will continue to occur.

With population increase, human impact on the Earth system also is forcing changes in diversity of the biosphere, the chemistry of the atmosphere, and global climate. The quest for sustainability, complicated and threatened by changes in population, pollution, energy use, environment, economics and political objectives, is seen by some as the ultimate objective for humanity and one in which environmental geology should play a key role.

The main goal of this book is to help students learn how to make wise choices for sustainability in a finite, changing and geologically active world. No one can promise a high-quality life, safe from natural and human-related risks, but education about observation of past events, interpretation of clues from landscapes in potentially hazardous areas, and choices of actions can help mitigate some of the risks from hazardous events. Understanding the processes, materials, landscapes and history of the Earth and the role of humans in using and changing the Earth system, will provide the basis for meeting the sustainability challenge.

ENVIRONMENTAL GEOLOGY AS A DISCIPLINE

The need for environmental geology as a distinct discipline arose in the 1960s in response to obvious changes in environmental quality. Degradation of the environment was blamed on a variety of factors, including affluence in a throwaway society, increasing industrialization, urbanization, rapidly growing population, and lack of a land ethic. Geologists, who typically have long-term and systems-oriented approaches to problems, understand the human colony's impact on and interaction with the geosystem. Environmental geology is basically the application of geology to the sustainability challenges facing humans in the earth system. The scope of environmental geology includes hazards, resources, pollution, regional planning, global change, long-range planning and sustainability, all within the context of an expanding human population and declining fossil fuels.

TEACHING APPROACHES

In this book we provide basic ideas, concepts, and techniques of environmental geology for understanding how the world works primarily in an active learning, problem-based format. Exercises in this manual are designed for use in undergraduate courses in environmental geology, applied geology, and environmental science. They can be used to provide an environmental focus in physical geology, general geology, physical geography, earth systems, and sustainability courses. Although we have created these exercises to provide a comprehensive experience in environmental geology, many of the exercises are also useful in other disciplines such as landscape architecture, natural resources, regional planning, engineering geology, environmental studies, geography, and interdisciplinary courses on global change. Based on our experience and conversations with colleagues, we have selected both classic and recent case histories that will meet the needs of most instructors and students. As with the previous edition of this book, we expect some use in advanced courses at the pre-college level. Generally, the level of difficulty is set for a first or second course in the geosciences or related fields. This book has been written primarily for use during formal laboratory or recitation sessions; however, we also use the exercises for in-class activities and for assigned homework.

Our experience is that students learn best in a hands-on problem-solving mode. We have used this approach as a guide in preparing these updated and new exercises. In most settings this book will provide

the opportunity to explore topics and examples presented in lectures, and thus will supplement the student's lecture notes and any assigned readings from on-line and print resources such as scientific journals, newspapers, reliable websites, network news, podcasts and textbooks. Some instructors may choose to use this book, *Investigations in Environmental Geology*, as their only required book for their active learning course. One option is to break up lectures with brief exercises drawn from this book, followed by homework, recitation, or lab assignments. Another approach is to spend more than 50% of each class period working, alone or in groups, on exercises in this book, interrupted by on-demand mini lectures and discussions. Given the wide availability of materials on-line, the growing preference for active learning, and the unused content of many textbooks, one of the above active learning options may be of interest to many students and instructors. We also have found that supplementing these exercises with direct observations during local field trips and through additional questions using local examples helps students apply the material to their own lives. It should be noted that the exercises used in this book have been developed using data from real cases or research reports. Although very rare, some data have been adjusted for student comprehension but never in a way that would change the interpretation of the geologic event or problem. With the variety of topics and examples included here, students should learn that there are different ways to approach the scientific study of environmental problems and that some problems defy clear-cut solutions.

EXPECTATIONS FOR STUDENTS

In addition to the standard requirements for completing a course, we expect students to keep up on current affairs. Many of the topics considered in the exercises will be covered in newspapers, public and network TV and radio, and internet news, sometimes on the same day as the topic is scheduled to be covered in your class. We also expect students to search the internet, use the library, participate in both formal and informal discussion of issues, and seek out additional sources of information. Students will have opportunities to develop skills of project organization, lab report preparation and presentation, and peer review in group learning sessions.

FORMAT OF THIS TEXT

This book has 18 exercises that are introduced in four sections. The exercises cover many current issues in environmental geology. Exercises 1 through 3 provide students with background information about earth materials, geologic time, geologic processes, the use and interpretation of maps, aerial photographs and remotely sensed images, and fundamental quantitative skills that are used in the exercises that follow. Exercises 1 through 3 complement the basic topics covered in introductory courses in geology. For more experienced students these three exercises will provide a review; for others they might be used as a quick study of a topic needed in later exercises. Exercises 4 through 11 investigate the nature of geologic events, including volcanoes, earthquakes, landslides and snow avalanches, subsidence, river floods and coastal processes. Exercises 12 through 16 investigate water-related resource and pollution issues. The last section of the book explores the topic of sustainability in two exercises. Exercise 17 looks at land use planning from an environmental geology perspective. Exercise 18 and the Section Introduction investigate evidence of global change (CO_2 fluctuations, ice-core paleotemperatures, glacier retreat, ozone losses, and population growth), resource types and availability, Hubbert's Curve, and forecasting through the development of a long-range scenario.

ACKNOWLEDGMENTS

We thank those students, staff, and colleagues who have helped in the development and testing of these exercises. Because many of these exercises are based on the work of others, we are indebted to them for use of their materials. We appreciate efforts of the following in providing reviews, data, and suggestions for development of the 1st, 2nd or 3rd editions. We are particularly indebted to Walter Arabasz, K. Bower, John Carpenter, Robert Carson, Charles Carter, Gary Christenson, Carolyn Dreidger, Jane Forsyth, John Hanchar, David Hirst, Christina Heliker, James Knox, Grahame Larson, Michael May, David McConnell, Hugh Mills, Carol Moody, Bobbie Myers, Gerald Newsom, Ellen Mosley-Thompson, Linda Noson, Mark Peterson, Pat Pringle, David Ramsey, Kenneth Rukstales, Geoffrey Smith, Donald Swanson, Steven Thacker, Lonnie Thompson, Robert Tilling, William Wayne, Cheryl Wilder and others who helped anonymously. Our own students provided suggestions for improvement including some tighter questions for the more difficult topics. They include: Jessica Albrecht, Gerald Allen, Mohammad Asgharzadeh, Elizabeth Birkos, Philip Borkow, Amanda Cavin, Benjamin Chenoweth, Bradley Cramer, Carina Dalton-Sorrell, Elizabeth Demyanick, Sarah Fortner, Steven Goldsmith, Natalie Kehrwald, Giehyeon Lee, Tim Leftwich, Veronica McCann, Gilbert O'Connor, Thomas Schumann, Jonathon Stark, David Urmann, Rebecca Witherow, Seth Young, and Jeffrey Ziga. A special thanks to graduate students at the University of Akron who provided formal revision reviews. The assistance of Susie Shipley, Karen Tyler, Betty Heath and others in OSU's School of Earth Sciences has also been important. We also thank our students who have used the first or second editions and provided important feedback, especially Beth Barnett, Terra McMahon, Jeff Nicoll and Mark Vinciguerra.

Wayne Pettyjohn, who developed some of the hydrogeology exercises in the original manual and encouraged us to develop an environmental geology course and materials in the early 1970s has been our early mentor. Finally, we thank our families for their understanding during this project.

THE CHALLENGE

At its current rate of growth, world population may grow from 6 billion to more than 9 billion people in the next few decades, which could severely impact Earth systems because of increased *human activity*, i.e., human consumption of energy, mineral, water and food resources and production of waste. Unfortunately, the cumulative impact of increased population is not fully appreciated. The impact of population growth over the past few decades, combined with anticipated future growth, could be compared to an asteroid or a bolide (large meteorite) 1 km in diameter striking the Earth. An asteroid of this size would do considerable damage to the immediate area and might impact the environment on a global scale with dust and debris ejected into the atmosphere. An even larger asteroid (9 km in diameter) would be needed to represent the baggage of geological materials that this mass of humans will need in its life-support system during its lifetime. The impact from a still larger asteroid (22 km in diameter) would be needed to represent the food energy this mass of humanity would consume in 60 years at the rate of 2000 calories per day per person! Such an impact would devastate most of the planet. It would be an event similar to that proposed for extinction of some of the dinosaurs. A cartoon sketch of the human asteroid that could impact the Earth is presented on the cover of this book. Think of it when you hear of proposed programs to track and deflect killer asteroids that might collide with the Earth. We have a human asteroid on the way now. Will impacts from this asteroid-sized human growth lead to extinction? Could we be approaching population limits (imposed by a finite Earth) such as fossil fuel energy, soil exhaustion and food supply, contamination of the biosphere (including humans) with related loss of biodiversity, and rapid global change that impacts infrastructure, water resources, and geologic hazards? Have we been propelled above the long-term carrying capacity of the Earth for humans by high-quality and inexpensive buried sunlight?

The Earth is a spaceship for which there are no operating manuals or wiring diagrams. As humans, we are trying to understand interconnected Earth systems—geological, biological, chemical, and social—so that we will be able to survive on this spaceship by making the right choices about human activity. Most scenarios point to a very exciting future for those on board and for those about to join us. There is not much time to figure out how the world works; we believe that this book will help students develop a new understanding of environmental geology and some of the Earth systems. If we are to address seriously the problems presented to us by global change, those in the human sciences—political, social, medical, and economic— must join those in the natural sciences and engineering. The human asteroid, exponential growth in resource use, peak oil, the geoquality of life, and the impact of natural disasters are topics that must be part of discussions about our future.

Duncan Foley
Pacific Lutheran University
Tacoma, Washington 98447-0003
foleyd@plu.edu

Garry McKenzie, Russell Utgard,
The Ohio State University
Columbus, Ohio 43210-1398
mckenzie.4@osu.edu, and utgard.1@osu.edu

Environmental Statement

This book is carefully crafted to minimize environmental impact. Pearson Prentice Hall is proud to report that the materials used to manufacture this book originated from sources committed to responsible forestry practices. The paper is FSC certified. In order to earn this certification, the forestry practices used to harvest trees must meet 10 stringent standards that ensure environmental responsibility—that these forest resources and associated lands are managed so that they can continue to meet the social, economic, cultural, and spiritual needs of present and future generations. Additionally, each step in the supply chain from forest to consumer must be FSC-certified and adhere to FSC standards in manufacturing and selling FSC-certified products. The binding, cover, and paper come from facilities that minimize waste, energy usage, and the use of harmful chemicals.

Equally important, Pearson Prentice Hall closes the loop by recycling every out-of-date text returned to our warehouse. We pulp the books, and the pulp is used to produce other items such as paper coffee cups or shopping bags.

Additional information from the Forest Stewardship Council is available on their Website, http://www.fsc.org. The future holds great promise for reducing our impact on Earth and its atmosphere, and Pearson Prentice Hall is proud to be leading the way in this initiative. From production of the book to putting a copy in your hands, we strive to publish the best books with the most up-to-date and accurate content, and to do so in ways that minimize our impact on Earth.

Mixed Sources

Product group from well-managed forests and other controlled sources
www.fsc.org Cert no. SCS-COC-00648
© 1996 Forest Stewardship Council

The FSC-certified printer SCS-COC-00648 printed the text of this book.

Prentice Hall
is an imprint of

I. Introduction to Geology

INTRODUCTION

"Our physical environment is fundamentally interesting. To know it is a pleasure, to understand it is a joy, and to solve its puzzles is creative work of the highest type."

—BRETZ, 1940

"The science of geology has long concerned itself with the real-world natural experience of the planet we inhabit. Its methodology more directly accords with the common sense reasoning familiar to all human beings. Because its study focuses on the concrete particulars of nature rather than on abstract generalizations, its results are also more attuned to the perceptions that compel people to take action, and to the needs of decision makers who must implement this action."

—BAKER, 1996

IMPORTANCE OF GEOLOGY

In addition to seeking fundamental knowledge in the Earth sciences to help us better understand our planet, geologists help ensure our health by understanding toxins in the environment, enhance our wealth by studying water and other geologic resources, and improve our security by helping us avoid geologic hazards and prepare for global climate change (Thorleifson, 2003). Your daily life and safety depend on geology. From the economic wealth of society to the energy you use, the water you drink, and the food you eat, even to the atoms that make up your body—you depend on geology. Earthquakes, volcanic eruptions, floods, and climate change are among the natural events that can dramatically alter your lives. This book looks at geologic processes and products that are important to understand for your safety and well-being. But what is geology and how does it work?

GEOLOGY IS A DISTINCT SCIENCE

Geological science is making observations, asking questions, creating ideas about how the world works, gathering information and testing the ideas, and communicating with others. Questions geologists seek to answer come from observations made in the field, observations that then are often extended through work in scientific laboratories. To geologists, "the field" is broadly defined. It may be the rocks on the surface of the Earth, it may be the core of the Earth, or it may even be, as space probes demonstrate, other planets.

Environmental geology applies geological methods to questions that arise from the interaction of humans with the earth. Environmental geology seeks knowledge about how we live with geological events, how we use the Earth for resources, how we use the land to live and work, and what we may be doing, as humans, to impact the future of Earth.

Successful geological studies depend on good observations, good descriptions, asking good questions, and integrating all available knowledge into testable models of how the earth works. One tool of geology is math; in this manual, the math required is straightforward.

Geological systems are complex. Where other sciences may be able to isolate individual actions and investigate them, ultimate knowledge in geology comes from integrating many often diverse observations and using information from many different fields. We may learn, for instance, about the movement of groundwater near a landslide from the distribution of plants on the surface of the land.

Time is important to geologists. What we see today is the result of 4.6 billion years of Earth history. In environmental geology we are worried about how the accumulated history of the Earth impacts human occupation and use of land now and into the future. A basic concept in geology is "the present is the key to the past." But geologists often state that "the past is the key to the present (and the future)." By understanding past geologic processes, events, and products, we

become more aware of the importance of and potential impacts from current geologic issues.

Geologists must often accept that answers to questions may be uncertain. As a field-based science, geologists are often limited by nature in the ability to observe complex processes directly. We can not, for instance, cut through an active volcano and observe directly how magma is working at depth. Nor can geologists, despite fervent wishes, travel through time to see the past directly. Limits of nature, however, mean that geologists must be very careful about observations that can be made, so the maximum amount of knowledge can be gained.

Many apparently permanent geological features are actually moving slowly, often in interconnected cycles. Continents move on large plates; rocks are created and erode. These processes form new rocks and illustrate the continual tug of war between constructive and destructive Earth processes. The hydrologic cycle links circulation of waters in the atmosphere, on land, underground and in the ocean. It has been said that one can never cross the same river twice, which means that it is not possible to regather all the same water molecules moving in exactly the same way. Through geology, however, humans are able to view the processes of the river, and to understand, for instance, that floods are a normal part of a river's cycle, and that a river is a normal part of the hydrologic cycle.

OBSERVATIONS AND DESCRIPTIONS

A key skill in geology is careful observation and description of geologic products and processes. Features of rocks can be interpreted to reveal a great deal about their history. Deposits at the surface of the Earth will be different if they form from a river, from a landslide, or from a volcanic eruption.

Where direct observation is not possible, geologists use other methods to gather data. In environmental geology, these methods may include seismographs, in order to measure earthquakes and interpret subsurface structures of the Earth. Another method is satellite imagery. Satellite images are able to show us parts of the world that we can not see in other ways. Airplanes flying over sites take photographs or gather radar images.

Maps are a very powerful tool in geologic studies. Patterns made by contours on U.S. Geological Survey topographic maps are very useful in interpreting zones with geologic hazards. Hummocky topography of landslides, rivers meandering on floodplains and distinctive shapes of active volcanoes are typical features that easily can be identified on topographic maps.

This manual introduces you to many of the tools and techniques that geologists use to gather data and answer questions. Doing well on these exercises will help you become a more aware, and safer, citizen.

OVERVIEW OF EXERCISES 1–3

The three exercises in this section provide basic concepts, tools, and techniques of the geosciences. For some this material will be a review, for others an introduction, and for all a handy reference while answering questions in this manual.

In *Exercise 1*, we explore minerals; common igneous, sedimentary, and metamorphic rocks; regolith (the engineer's soil); and geologic time and geologic processes in the Earth system. Understanding minerals is important for our use of geologic resources and building on sediments and rocks. Sedimentary rock, because of its widespread extent, is most likely to be the bedrock encountered in environmental investigations. But more important is a related component of the rock cycle, the surficial sediment or regolith that overlies the rocks. With geologic time, we focus on the last few hundred thousand years and present a geologic time chart that emphasizes this. At the same time we realize that understanding paleoenvironments and changes over deeper time is important. In this exercise we ask you to explore the Earth system, the connections that exist between the processes (Figure 1.3), and the rates of these processes. In seeking to understand and address environmental problems, an Earth systems science approach which looks at frequency, rates, connectivity and controls on events is essential.

In *Exercise 2* we gain increased familiarity with maps and images that provide the spatial framework for understanding environmental problems. A review of topographic maps is followed by interpretation of geologic maps and an introduction to other useful images including aerial photographs and satellite and LIDAR images. Fortunately for the environmental geologist, as our impact on the planet has increased exponentially, the technology to observe the nature, extent, and rates of change has grown in a similar fashion.

Scientific measurements and notation, useful calculations and conversions, display of data on graphs and tables, and simple statistics are in *Exercise 3*. Additional useful information for problem-solving exercises in this manual is in the Appendices.

Bibliography

Baker, V. R., 1996, The geological approach to understanding the environment: *GSA Today*, v. 6, no. 3, p. 41–43.

Bretz, J H., 1940, *Earth sciences*: New York, Wiley and Sons, 260 p.

Frodeman, R., 1995, Geological reasoning: Geology as an interpretive and historical science: *Geological Society of America Bulletin*, v. 107, p. 960–968.

Thorleifson, H., 2003, Why we do geology: *Geology*, v. 32, no. 4, p. 1 and 4.

Earth Materials, Geologic Time, and Geologic Processes

INTRODUCTION

One of the necessary phases of geologic studies is gaining knowledge of the materials that make up the Earth's crust. The ability to identify these earth materials (rocks and minerals) provides one with an appreciation of the natural environment and supplies earth scientists with a tool that may aid in geologic and environmental studies of an area.

The minerals (Part A) and rocks (Part B) studied in this exercise are the most commonly occurring types and those of interest because of their importance in environmental considerations. Over geologic time (Part C), these mineral and rock materials are modified by geologic processes (Part D) acting on and within the Earth.

PART A. MINERALS

A mineral is a naturally occurring inorganic substance that has an orderly internal structure and characteristic chemical composition, crystal form, and physical properties. Minerals possess many fundamental characteristics that are external evidence of an orderly internal arrangement of atoms. The physical properties of minerals reflect their internal structure and provide clues to their identity. Many chemical properties also aid in mineral identification. For instance, the carbonate minerals calcite and dolomite (the main constituents of limestone and dolostone) effervesce with application of dilute (5 percent) hydrochloric acid.

Physical Properties

CRYSTAL FORM Crystal form refers to the orderly geometric arrangement of external planes or surfaces that are controlled by the orderly internal arrangement of atoms and/or molecules that make up a mineral. Minerals may exhibit many different characteristic external forms, for example, sheets (mica), cubes (halite and galena), rhombohedrons (calcite), and hexagons (rock crystal quartz).

HARDNESS The resistance that a mineral offers to abrasion is its hardness. It is determined by scratching the surface of a mineral with another mineral or material of known hardness. The Mohs Scale of Hardness consists of 10 minerals ranked in ascending order with diamond, the hardest known mineral, assigned the number 10. This scale, together with several common objects of known hardness, is shown below and should be referred to in determining the hardness of unknown minerals. Because weathering may affect hardness, it is important to make tests on fresh surfaces.

MOHS SCALE OF HARDNESS	COMMON OBJECTS
1. talc	
2. gypsum	2-2.5 fingernail
3. calcite	3-3.5 copper penny
4. fluorite	
5. apatite	5-5.5 knife blade,
6. feldspar	5.5-6 glass plate
7. quartz	6.5 steel file
8. topaz	
9. corundum	
10. diamond	

CLEAVAGE AND FRACTURE The bonds that hold atoms together in a crystalline structure are not necessarily equal in all directions. If definite planes of weakness exist, a mineral will tend to break along these cleavage planes. Cleavage is described with reference to the number of cleavage planes and their angles of intersection, and may be further qualified as

perfect when well developed, or indistinct when poorly developed. Some examples of different types of cleavage are:

Basal (1 plane or direction)	mica and selenite gypsum
Cubic (3 planes at right angles)	halite and galena
Rhombic (3 planes not at right angles)	calcite

The breaking of a mineral along directions other than cleavage planes is termed *fracture*. Some common types of fracture are conchoidal (quartz), irregular (pyrite), and fibrous (asbestos).

LUSTER   The luster of a mineral is the appearance of a fresh surface in reflected light. There are two major types of luster: metallic and nonmetallic. Galena and pyrite have a metallic luster; nonmetallic lusters include glassy or vitreous (quartz), resinous (sphalerite), pearly (talc), and dull or earthy (kaolinite).

COLOR   Color is one of the most obvious properties of a mineral but in general it is of limited diagnostic value. For example, quartz may vary from colorless to white (milky), gray, rose, purple (amethyst), green, and black. In contrast, galena is a distinctive gray and the micas can often be separated on the basis of color, which results from different chemical compositions.

STREAK   The color of a mineral's powder is its streak. It is obtained by rubbing a mineral specimen across a streak plate (a piece of unglazed porcelain). Streak is commonly a diagnostic characteristic of metallic minerals. Color may vary greatly within one mineral species but the streak is generally constant (e.g., hematite has a reddish streak).

RELATIVE DENSITY OR SPECIFIC GRAVITY   Specific gravity (s.g.) is the ratio between the weight of a mineral and the weight of an equal volume of water. Although s.g. can be determined accurately, it is sufficient for general laboratory and field purposes to observe the "heft" of a mineral simply by handling it. Galena (7.5) and pyrite (5) are heavy; the common rock-forming minerals (calcite, quartz, and the feldspars) are of medium weight (2.5 to 2.8); and halite is light (2.1).

TENACITY   Tenacity is the resistance of a mineral to being bent or broken. It should not be confused with hardness. Some of the terms used to describe tenacity are:

Elastic	will bend and return to original position when force is released	mica
Flexible	will bend but does not return to original position when force is released	selenite gypsum
Malleable	can be hammered into thin sheets without breaking	native copper
Brittle	does not bend but shatters when sufficient pressure is applied	quartz

Other Mineral Properties

Magnetism	Magnetite is the only common mineral to show obvious magnetic attraction.
Taste	Halite has a salty taste, which is a definite property of this mineral.
Odor	The clay minerals (kaolinite, etc.) smell earthy, especially when breathed upon.
Feel	Some minerals have a distinctive feel (e.g., talc feels soapy).
Chemical Reaction	Calcite will effervesce (fizz) when cold, dilute hydrochloric acid (HCl) is applied. Dolomite reacts with dilute HCl if it is powdered.

QUESTIONS 1, PART A

1. Examine each mineral specimen that your instructor selects for this exercise. Determine and record in Table 1.1 all of the properties you test and/or observe. This includes determination of luster, color, cleavage, fracture, and specific gravity; and testing for hardness, streak, and other properties.

2. After you have recorded the observed physical properties, with the aid of earlier information and the descriptive information about each mineral found in Table 1.2, determine the names of the minerals. Record the chemical composition of each mineral and note the information regarding their geologic, environmental, and economic significance. As you will see, some of these minerals are of particular importance because of the way they influence the environment when not properly used or when their presence is not considered.

PART B. COMMON ROCKS

A rock is an accumulation or aggregate of one or more minerals, although some nonmineral substances such as coal or volcanic glass are also considered rocks. Rocks are generally classified into three major categories: igneous, sedimentary, and metamorphic. The interrelations among the three groups of rocks are shown by the *rock system* (Figure 1.1). Note that you need to add another arrow from metamorphic rocks.

TABLE 1.1 Mineral Identification

Sample Number	Color	Streak	Luster	Hardness	Specific Gravity	Fracture or Cleavage	Other Properties	Name of Mineral	Chemical Composition

Igneous Rocks

Igneous rocks are formed by the crystallization of a molten silicate-rich liquid known as magma. Magma that cools relatively slowly beneath the Earth's surface forms *plutonic* or *intrusive* igneous rocks, whereas magma that crystallizes at or near the Earth's surface as lava forms *volcanic* or *extrusive* igneous rocks. The rate at which magma cools determines the texture of igneous rocks. The composition of the magma determines the mineral composition of the rock. These two properties, texture and composition, are the basis for classifying and identifying igneous rocks.

Igneous rock texture

Coarse (Phaneritic)	Interlocking mineral grains that can be distinguished with an unaided eye
Fine (Aphanitic)	Individual mineral grains are small and cannot be distinguished with an unaided eye
Porphyritic	Two sizes of minerals, with large crystals (phenocrysts) imbedded in fine- or coarse-textured groundmass
Vesicular	Rocks contain many cavities or voids, giving a texture similar to a hard sponge
Pyroclastic	Rocks made from volcanic materials that have flown through the air, such as ash (tephra)
Glassy	Texture that resembles glass

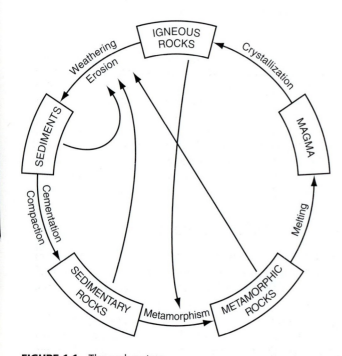

FIGURE 1.1   The rock system.

TABLE 1.2 Properties of Common Minerals

Mineral Name	Chemical Composition	Color	Streak	Luster	Hardness	Specific Gravity	Fracture or Cleavage	Other Properties	Geologic, Environmental, and Economic Significance
Augite (pyroxene)	Ca, Mg, Fe, Al silicate	Dark green to black	Greenish gray	Glassy on fresh surfaces	5–6	3.2–3.4	Two cleavages at nearly 90°, often poor and yielding splintery surfaces		Important igneous rock-forming mineral
Calcite	$CaCO_3$	Usually colorless or white	White	Glassy	3	2.7	Perfect rhombohedral, three directions at 75°	Effervesces with application of dilute hydrochloric acid	Major constituent of limestone; contributes to hard water
Clay Minerals Kaolinite, Illite, Montmorillonite	Hydrous aluminum silicates	White		Dull, earthy	2–2.5	2.6	Irregular fracture	Plastic; montmorillonite has unique capacity for absorbing water and expanding	Derived from weathering of feldspars; used in manufacture of ceramics. Montmorillonite can expand when wet, causing damage to structures
Dolomite	$CaMg(CO_3)_2$	White, pink, gray, brown	White	Glassy to pearly	3.5–4	2.85	May cleave to rhombohedral crystals	Effervesces with application of dilute HCl when powdered	Major constituent of dolostone and dolomitic limestone
Feldspars Orthoclase Plagioclase	$KAISi_3O_8$	Pink, white, reddish	White	Glassy	6	2.57	Good in two directions at or near 90°		Important igneous rock-forming mineral
	Ca, Na, Al silicates	White, gray, green, black	White	Glassy	6	2.62–2.76	Good in two directions at or near 90°	Striations frequently visible; continuous series from soda-rich to calcium-rich	Important igneous rock-forming mineral
Fluorite	CaF_2	Clear, white, blue, green, purple, yellow	White	Glassy	4	3.2	Good in four directions parallel to the faces of an octahedron	Cubic crystals	Used as a flux in steel and aluminum manufacturing and in making hydrofluoric acid
Galena	PbS	Metallic gray	Gray	Metallic	2.5	7.5	Good cubic	Cubic crystals	Principal ore of lead
Gypsum	$CaSO_4 \cdot 2H_2O$	Colorless, white	White	Glassy, pearly, silky	2	2.3	Selenite has good cleavage in one direction; cleaves to thin folia	Selenite is flexible; satin spar is fibrous; alabaster is massive, fine-grained	Principal ingredient of plaster of Paris, major use is wallboard for building construction

TABLE 1.2 Properties of Common Minerals (continued)

Mineral Name	Chemical Composition	Color	Streak	Luster	Hardness	Specific Gravity	Fracture or Cleavage	Other Properties	Geologic, Environmental, and Economic Significance
Halite	NaCl	Colorless	White	Glassy, pearly, silky	2.5	2.16	Perfect cubic	Salty taste	Used as table salt and road salt and in chemical industries; occurs in oil field brines
Hematite	Fe_2O_3	Shades of reddish brown to black	Reddish brown	Metallic to dull	5.5–6.5	5.3	Irregular fracture	Coloring agent responsible for most red rocks	Major ore of iron
Hornblende (amphibole)	Complex ferromagnesium silicate	Dark-green to black	Colorless to gray-green	Glassy on fresh surface	5–6	3.2	Good in two directions at 56° and 124°		Important igneous rock-forming mineral
Limonite	Hydrous iron oxides	Yellow to brown and black	Yellow-brown	Dull to glassy	1–5.5	3.6–4	Irregular fracture	Amorphous, yellow and brown rock coloring agent	
Micas Biotite (black mica)	Complex ferromagnesium silicate	Black, greenish black	White to gray	Pearly, glassy	2.5–3	2.8–3.2	Perfect in one direction	Thin sheets, elastic	Common rock-forming mineral in igneous and metamorphic rocks
Muscovite (white mica)	Complex potassium aluminum silicate	Colorless to light yellow and light green	White	Glassy, pearly	2–2.5	2.76–3.1	Perfect in one direction	Transparent in thin sheets	Common mineral in rocks; used as an insulating material
Olivine	Ferromagnesium silicate	Olive to graygreen	Colorless to white; pale green	Glassy	6.5–7	3.2–4.3	Conchoidal fracture; often breaks around grains	Usually granular	Important igneous rock-forming mineral
Pyrite	FeS_2	Brassy yellow	Greenish black to black	Metallic	6–6.5	5.02	Irregular fracture, may show striations	Cubic crystals, fool's gold	Source of much of sulfide pollution and mine-acid drainage
Quartz	SiO_2	Colorless white, rose, gray	Colorless	Glassy	7	2.65	Usually imperfect cleavage, may be conchoidal	Hexagonal crystals	Common mineral in rocks
Sphalerite	ZnS	Yellow brown to black	Pale yellow, white, brown	Resinous	3.5–4	3.9–4.1	Good cleavage in six directions		Ore of zinc

COMPOSITION Generally, the composition of igneous rocks can be divided into two major groups: (1) those in which the light-colored minerals quartz and orthoclase feldspar predominate (acidic igneous rocks); and (2) those in which the dark-colored minerals olivine, augite (pyroxene), and hornblende predominate (basic igneous rocks). Plagioclase feldspar is usually present in the darker igneous rocks.

SIGNIFICANCE Igneous rocks are economically important as most metallic minerals are found in or associated with them. Many other economic products, including *building* stone, aggregate, gemstones, and ceramic and glass materials are derived from igneous rocks. Geothermal energy, a source of which is heat from igneous rocks, is a potentially important, although limited, future energy source. Since igneous rocks are relatively nonporous and impermeable, the groundwater obtainable from them is generally limited in quantity, although usually of good quality. However, some igneous rocks may be jointed or fractured, and some volcanic rocks may have lava tunnels so that large voids exist that provide a reservoir of abundant water. When these rocks are used for waste disposal, groundwater may readily become polluted.

Important engineering properties of igneous rocks include their strength and stability; that is, they provide a sound foundation for construction projects. Commonly, soils that have developed on igneous rocks are thin and on slopes such soils may slide when saturated with water. Although igneous rocks usually provide a firm foundation, the benefits may be overcome by adverse factors such as excavation difficulty, thin soils, and lack of groundwater.

QUESTIONS 1, PART B1

1. Your instructor will provide specimens of igneous rocks for you to identify. Determine the texture, composition, and any other important characteristics of each specimen and record your observations in Table 1.4. Using the information you have compiled and Table 1.3, determine the name of each rock.

2. After you have made a record of the names and characteristics of the igneous rocks, note information regarding the geologic, environmental, and economic significance of these rocks in the proper column in Table 1.4.

Sedimentary Rocks

Sedimentary rocks are formed from the products of the weathering (breakdown) of preexisting rocks. When these products (sediments) are *compacted* by pressure of overlying material or *cemented* together by precipitation of mineral matter between the particles, new rock is formed. Sedimentary rocks may also be formed from the accumulation of organic materials and from chemical precipitation of mineral matter from water. The common sedimentary rocks and their characteristics are listed in Table 1.5.

SIGNIFICANCE Since sedimentary rocks cover approximately 75 percent of the Earth's surface, human interaction with earth materials is more likely to be with these rocks than with any other type. When we seek a water supply, dispose of waste materials, engage in construction projects, or extract mineral resources, the rock encountered is likely to be sedimentary.

TABLE 1.3 Classification of Igneous Rocks

Texture \ Composition	Quartz, Orthoclase, Biotite	Na-rich plagioclase, Hornblende, Biotite	Ca-rich plagioclase, Augite, Olivine
Phaneritic	Granite	Diorite	Gabbro
Phaneritic (with phenocrysts)	Porphyritic granite		
Aphanitic	Rhyolite (Felsite)	Andesite	Basalt
Aphanitic (with phenocrysts)	Porphyritic rhyolite	Porphyritic andesite	Porphyritic basalt
Glassy	Obsidian		
Vesicular	Pumice		Scoria
Pyroclastic	Tuff and Volcanic breccia		

TABLE 1.4 Igneous Rock Identification

Sample Number	Texture	Mineral Composition	Other Identifying Characteristics	Name of Rock	Geologic, Environmental, and Economic Significance

TABLE 1.5 Sedimentary Rocks: Detrital (by granular texture*) and Chemical/Biochemical (by composition)

Conglomerate*	Rounded gravel-sized (>2 mm) rock and mineral fragments cemented together by silica (SiO_2), calcite, or iron oxides.
Sandstone*	Sand-sized (1/16 to 2 mm) particles of chiefly quartz cemented together by silica, calcite, or iron oxides.
Siltstone*	Silt-sized (1/256 to 1/16 mm) particles of rock fragments, quartz, and other minerals held together by calcite, silica, or iron oxides or by a clay matrix.
Shale (mudstone)*	Clay-sized (<1/256 mm) particles that are chiefly the product of weathering of feldspars and are lithified by compaction. Mineral composition is chiefly clay minerals.
Limestone	A calcium carbonate rock, $CaCO_3$, which may be formed by accumulation of fossil remains or chemical precipitation from solution upon evaporation, it is composed of calcite and occurs in several varieties including oolitic, fossiliferous, coquina, and chalk.
Dolostone	A calcium-magnesium carbonate rock, $CaMg(CO_3)_2$, composed largely of the mineral dolomite. Texture and appearance are similar to limestone. Can usually be distinguished from limestone by its slow reaction to HCl.
Coal	An accumulation of vegetation that through compaction loses its water and volatile matter to become relatively pure carbon. Peat and lignite are early stages of coal formation; bituminous coal represents a later stage; anthracite is further transformed by heat and pressure and may be considered a metamorphic rock.
Rock gypsum	A chemical precipitate composed mainly of the mineral gypsum, $CaSO_4 \cdot 2H_2O$, which originates from evaporation of lakes and seas in restricted bays and thus is an indicator of an arid climate at the time of formation.
Rock salt (halite)	A chemical precipitate composed of the mineral halite, NaCl whose origin is similar to that of gypsum. A significant physical property of rock salt is that it flows at relatively low temperature and pressure.

The fact that shale is the most abundant sedimentary rock is somewhat unfortunate, as it has several undesirable properties. These properties are sometimes ignored in construction and engineering projects. Some shales absorb large quantities of water and expand greatly; others are plastic and flow easily. Shales typically weather rapidly, are weak and unable to support great weight, and, when lubricated by water, slide readily. The impermeability of shales makes them a poor source of groundwater.

The coarser clastic (made of fragments) rocks, conglomerate and sandstone, are generally porous and permeable and as a result are the source of abundant groundwater, usually of high quality. These rocks generally possess good foundation-bearing qualities or support strength.

Limestones may be highly soluble and hence unsuitable sites for many construction projects. Solution channels, caves, and sinkholes which develop in limestone may result in subsidence problems and make unsuitable sites for surface water reservoirs and for waste disposal.

Sedimentary rocks are of great economic significance. Building materials, aggregate, glass, ceramic products, and the fossil fuels, as well as some metallic mineral deposits are the products of sedimentary rocks.

QUESTIONS 1, PART B2

1. Your instructor will provide specimens of sedimentary rocks for you to identify. Using the information presented in Table 1.5, determine the composition and texture of each specimen. On this basis draw your conclusions about the origin of the sediments, the process of lithification, and the name of the rock. Record this information in Table 1.6.

2. After you have identified the rocks, determine some of the important uses and geologic and environmental significance of each rock and record this in the last column in Table 1.6.

Metamorphic Rocks

Metamorphic rocks are formed from preexisting rocks that have undergone substantial alteration by pressure, heat, and chemically active fluids. Metamorphic changes include (1) recrystallization of minerals, normally into larger grains; (2) chemical recombination and growth of new minerals; and (3) deformation and reorientation of constituent minerals. Classification of metamorphic rocks is based primarily on texture and composition. Two general categories result: foliated and nonfoliated. Foliated texture is characterized by an alignment of platy minerals such as mica, giving the rock a banded or layered appearance. Nonfoliated

TABLE 1.6 Sedimentary Rock Identification

Sample Number	Texture	Mineral Composition	Other Characteristics (clastic/non-clastic)	Name of Rock	Geologic, Environmental, and Economic Significance

TABLE 1.7 Common Metamorphic Rocks

Texture	Diagnostic Characteristics	Rock Name
Foliated (coarse)	Alternating bands of unlike minerals, such as feldspar and quartz, with lesser amounts of mica, and ferromagnesian minerals	Gneiss
Foliated (medium to fine)	Foliation due to parallel arrangement of platy minerals; mica, chlorite, talc, hornblende, and garnet are common minerals	Schist
Foliated (very fine)	Quartz, muscovite, and chlorite are common minerals but are not visible with the unaided eye; may resemble shale but has no earthy odor; compact and brittle; varicolored. Generally represents metamorphosed shale.	Slate
Nonfoliated	Principally quartz; varicolored; very hard, massive. Original rock is generally quartz rich sandstone.	Quartzite
	Principally calcite or dolomite; crystals commonly large and interlocking; varicolored. Original rocks are limestone and dolostone.	Marble

texture is characterized by an absence of an alignment of minerals due to the random orientation or equidimensional nature of constituent minerals. The common metamorphic rock characteristics are summarized in Table 1.7.

SIGNIFICANCE The environmental significance of metamorphic rocks is similar to that of igneous rocks in that they are generally nonporous and impermeable; they are, therefore, a poor source of groundwater, and provide a firm foundation for construction projects. However, one characteristic of foliated metamorphic rocks that needs special consideration is the direction of orientation of the foliation or planes of schistocity. These may represent planes of weakness that must be considered in foundation support, excavations, and in any situation where sliding may take place along these planes.

In some cases, metamorphic rocks retain properties of the original rock and, therefore, respond in the same way as the original rock. This is especially true of marble, which should be subject to the same considerations as limestone because of its high solubility.

Metamorphic rocks provide many economic products including building stone, monument and decorative stone, aggregate, and gemstones, as well as fossil fuel (anthracite).

QUESTIONS 1, PART B3

1. Your instructor will provide specimens of metamorphic rocks for you to identify. Using the information presented in Table 1.7, determine the texture, composition, and other properties of each specimen to identify the metamorphic rock and probable original rock. Record your observations in Table 1.8.

TABLE 1.8 Metamorphic Rock Identification

Sample Number	Texture	Mineral Composition	Other Characteristics (original rock)	Name of Rock	Geologic, Environmental, and Economic Significance

2. After you have identified the metamorphic rocks, note information regarding the geologic, environmental, and economic significance of these rocks in the proper column in Table 1.8.

Regolith—Unconsolidated Material on Bedrock

One of the most important geologic materials with which the human colony interacts at the surface of the Earth is regolith. This unconsolidated material that overlies the bedrock is often loose and readily excavated. There are four basic processes of formation: (1) in place by *weathering* of underlying bedrock; (2) by *mass-wasting* of bedrock and sediments moving downslope under the force of gravity; (3) by *accumulation of organics* such as peat; or (4) by *fluvial, glacial, or eolian transport* of eroded material, which is deposited as sediment. The general term used by geologists for all these materials is **regolith**; however, for the same unconsolidated materials over bedrock engineers apply the term **soil**. Geologists, biologists, and soil scientists generally use the term soil for those weathered horizons developed at the top of the transported regolith or directly from weathering of the bedrock. The biologist's soil is considered to be able to support life.

Most regolith is described primarily on the basis of grain size (and sorting) in much the same way as are detrital sedimentary rocks (see Table 1.5). Grain sizes (or textures) of unconsolidated materials are clay, silt, sand, and gravel, which correspond to the sedimentary rock materials of shale, siltstone, sandstone, and conglomerate, respectively. Sorting of sedimentary material into these grain sizes occurs by wind or moving water. Unsorted sediments are a mixture of many of these grain sizes and include such materials as those deposited directly from glaciers (till) and from landslides (colluvium). See Table 1.9.

SIGNIFICANCE Many of the environmental problems encountered in use of the surface of the Earth occur in or on the regolith. Failure or movement of this material often occurs during hazardous geologic processes and may result in destruction of buildings and highways. The regolith is where solid and liquid waste disposal and recharge, protection, storage, and extraction of groundwater can occur. In this material we also find gravel and sand resources for construction and agricultural soils for crops. And in glacial and other deposits, including those in rivers, lakes, and bogs, we are able to decipher Quaternary history, which shows us how climate, geologic processes, and landscapes have changed over the last 2 million years. Thus the regolith provides essential resources for humanity and provides scientific data that helps us understand what the future might hold. Where does the regolith fit in the rock cycle? It is basically unconsolidated sediment that overlies the rocks.

TABLE 1.9 Common Regolith (Unconsolidated Sedimentary) Materials

Sample Number	Texture (or grain size—gravel, sand, silt, clay) (sizes in Table 1.5)	Sorting (well sorted, poorly sorted, unsorted)	Sediment Type (detrital, organic, residual)*	Possible Origin (e.g., river, beach, humans, wetland, dune, glacier)	Geologic Significance (good aquifer, mineral resource, or building site, etc.)

* Detrital, loose grains of broken rocks and minerals (e.g., sand, etc.); organic, peat or loose plant matter; residual, chemically weathered residual regolith with loose silicate fragments, clay minerals, and concentrations of iron oxide and aluminum oxide.

The properties of surficial materials are determined largely by grain size, sorting, and moisture content. Generally fine-grained sediments (silts and clays) are less suitable for construction substrates because of loss of strength and resulting deformation that occurs, particularly with high moisture content. But such fine-grained materials also provide benefits as they are more suitable for trapping subsurface pollutants and restricting groundwater flow. Coarse-grained sediment such as gravel is generally a good substrate for construction because of its strength and well-drained character. Gravel units often are good groundwater reservoirs. The strength of till (an unsorted component of the regolith) depends on the range of grain sizes and the dominant grain size. In glaciated regions of North America, much of the infrastructure is built on glacial till or gravel, so understanding their properties is important. If clayey and unfractured, till serves as a barrier to movement of toxic materials and polluted groundwater. Clay mineral type is also important in understanding the behavior of regolith (e.g., some clays expand more than 50 percent when wet). More information on the engineering aspects of the materials on bedrock is available in engineering geology books (e.g., Rahn, 1996; West, 1995).

To minimize the structural losses from hazards and maximize availability of the resources in the regolith, land-use planners use information from two basic sources: surficial geology maps (often known as glacial geology maps in the northern states and Canada) and soil surveys. The surficial maps provide information on geologic materials at the surface, stratigraphy of the regolith, depth to bedrock (which might range from less than 1m to more than 200m), and landforms. The soil surveys typically focus on the upper meter of the regolith, but also give us a clue to the substrate beneath. Soil surveys also contain site-specific aerial photographs marked with soil types and important tables that contain resource and engineering limitations. Groundwater resource maps provide additional information on the nature of the regolith and groundwater. By combining these information sources in what are known as stack maps, it is possible to obtain a 3-D picture of the regolith that aids us in planning land use. For more information on surficial materials consult Graf (1987), Easterbrook (2003), Walker and Cohen (2006), and state soil surveys (many on the Web).

QUESTIONS 1, PART B4

1. Your instructor will provide samples of sedimentary materials for you to identify. Use information in the description of regolith types above, and descriptions in the sedimentary rocks section of the manual, to help identify the samples, their possible origin, and their geologic significance. For example, a dry, compact mass of well-sorted, very fine-grained sediment that can be abraded with a knife might be a clay, formed in a lake bed, and have low strength, slow movement of water through it, and thus not a good aquifer (but useful for containment of wastes, e.g., a potential landfill site). Coarser-grained sorted sediments reflect higher energy environments. See Table 1.5 for grain-size information.

2. Define each of the following terms and any others assigned by the instructor.
   Regolith
   Soil (engineering)
   Soil (biological or soil science)
   Till

3. List four possible environments of formation or origins of regolith. (Hint: See the introduction to the section "Regolith.")

4. Think of the types of materials that make up the regolith and determine which material or materials would be most useful for:
   a. a groundwater reservoir (aquifer)

   b. protecting an aquifer from contamination from a gasoline spill on the land surface

5. If the local surficial geology map and county soil survey reports are available in lab or online, determine:
   a. the material and origin of the regolith beneath your building or campus

   b. the soil type beneath your building or campus. (Soil maps are available for most counties; they contain a wealth of data on a site, often including aerial photos.)

PART C. GEOLOGIC TIME

One of the basic concepts underlying geologic studies is that of geologic time, a term which implies an exceedingly long duration of Earth history. Geologic time encompasses the total age of the Earth, about 4.6 billion years, from the time of formation of the primitive Earth to the present. The immensity of geologic time is difficult to comprehend because we must consider time spans that far exceed our common experiences.

Geologic processes acting on the Earth since its beginning have produced tremendous changes that are possible only because of almost unlimited time.

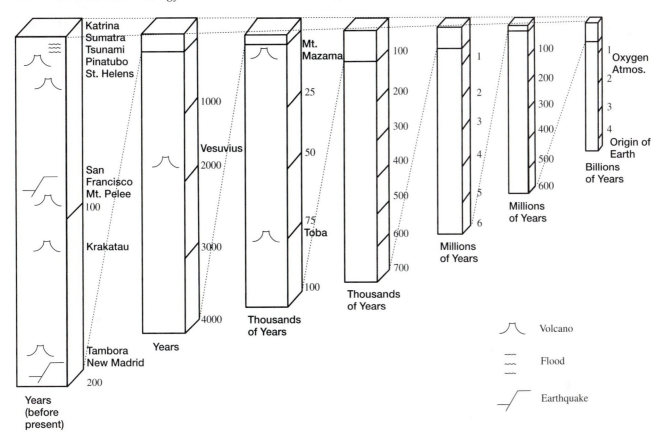

FIGURE 1.2   Geoenvironmental time scale. See Appendix II for a more detailed time scale. (Modified from McKenzie and others, 1996)

From the human perspective these natural processes often act so slowly that change may not be visible over an entire lifetime. Conversely, humans as agents of geologic change have had an enormous impact on the Earth's surface, an impact that is greatly out of proportion to the amount of time humans have occupied the Earth.

A necessary outgrowth of years of geologic studies has been the development of a calendar of geologic time, known as the geologic time scale. In the early days of collection of fossil and rock specimens or the timing of geologic events, a standard chronologic sequence of relative time was developed based on the way strata occur one on top of another and on stages of evolutionary progress shown by fossil organisms. In the last half century radiometric dating methods have been developed that allow precise determination of absolute time in terms of actual years before the present (Appendix II).

Global environmental changes in the last 50,000 years have been particularly rapid and significant for the human colony; however, the standard geologic time scale does not adequately represent most recent time. The Geoenvironmental Time Scale (McKenzie et al., 1996) encompasses the standard time scale but emphasizes, with greater detail, recent time and includes historical events (Figure 1.2). It is designed to place environmental events in a geologic perspective. Use it and the more detailed traditional time scale showing eras, periods and epochs from Appendix II for answering the questions below.

QUESTIONS 1, PART C

1. Which subdivision of geologic time represents the greatest length of time (era, period, epoch)?

2. What is the length in millions of years of the:
Cenozoic Era?
Tertiary Period?
Quaternary Period?
Eocene Epoch?
Pleistocene Epoch?

3. How does the Pleistocene Epoch compare in length to the Ordovician Period?

4. Approximately how many times longer is Precambrian time than the Cenozoic Era?

5. What is the age in million years (Ma) of the boundary between the Cretaceous and the Tertiary (K/T boundary)?

6. Approximately how many years ago did the following geologic events occur?
Mt. St. Helens eruption
New Madrid, Mo., earthquake
Vesuvius eruption
Toba eruption

PART D. EARTH SYSTEMS AND GEOLOGIC PROCESSES

The Earth's surface is not static; it changes in response to processes that operate at the surface and within the Earth. The results of these interactions are the landforms and landscapes on which the human colony lives.

External geologic processes are generated primarily by solar energy and gravity. Precipitation and other components of the hydrologic cycle work on the materials of the Earth to shape the land. The processes include: weathering of rocks and sediments; mass wasting (commonly described as landsliding); and erosion and sedimentation by rivers, lakes, oceans, wind and glaciers. Biologic processes also produce landforms, for example, organic reefs. In fact, humans are the prime geomorphic agent. We are reshaping the Earth's surface directly with anthropogenic (human-produced) landforms and indirectly through modification of the rates and scales of natural geologic processes. The processes working on the Earth are summarized in Figure 1.3. In addition, humans are also modifying climate, which controls many geologic processes. The Earth has been both colder and warmer in the last 130,000 years, but our current understanding of the Earth system incorporates a mechanism for increases in global temperatures because of increased concentrations of greenhouse gases. Water vapor accounts for 80 percent of the greenhouse effect, which is an essential process for the maintenance of life-supporting temperatures on Earth (Jacobsen and Price, 1990). Anthropogenic increases in natural trace gases such as carbon dioxide (CO_2), methane (CH_4), and tropospheric ozone (O_3) and manufactured gases chlorofluorocarbons (CFCs) and halons are now causing significant changes in temperatures, climate, and geologic processes. Excess quantities of greenhouse gases are caused primarily by the burning of fossil fuels or "buried sunlight" (CO_2, CH_4 and N_2O), but agricultural activity (livestock management, rice paddies, biomass burning, nitrogenous fertilizers, and deforestation) and industrialization (chlorofluorocarbons and halons) also contribute to these gases. Industrial gases also cause stratospheric ozone depletion, which leads to increases in ultraviolet (UV) radiation and impacts human health, plant productivity, and climate (Jacobson and Price, 1990).

Environmentalists commonly speak of "Spaceship Earth" to describe the life-support characteristics of the Earth. Note that the Earth is not a closed system. For example, infall of meteoritic material is generally a minor process; however, on very rare occasions (in geologic time) large-scale meteorite impact occurs and can cause significant changes in the Earth's landscape, climate, and life. Although not always obvious on Earth, cratering is the most important geomorphic process in the solar system.

Internal geologic processes are driven by geothermal heat, produced primarily by the radioactive decay of elements within the Earth. As indicated in Figure 1.3, the rotational energy of the solar system also acts on the Earth, producing earth and ocean tides. The result of all these processes is deformation of the crust, primarily through the mechanism of plate tectonics, in which large and small segments of the crust and upper mantle known as plates collide, separate, deform, grow, and disappear. Due to these plate tectonic processes and to phase changes in minerals in the Earth, the surface of the Earth rises and falls. Although much of this deformation is very slow (on the order of a few centimeters or less per year), some is very rapid. These seismic events or earthquakes not only rapidly displace portions of the crust, but cause shaking and secondary processes that can seriously impact human structures designed for a generally stable surface. In addition to these tectonic processes, there are volcanic processes that affect the surface and near surface areas of the Earth. The rates at which many of the geologic processes that shape the Earth occur are summarized in Figure 1.4.

The interaction of the internal and external processes working on earth materials comprises the "geosystem" and produces our landscapes and the substrate for human structures. As the surface changes on short- and long-term bases, humans must adjust to it. The climate component of the geosystem, which is responsible for the relative importance of different external geomorphic processes, is a major factor in the rate and scale of changes on the Earth. With the widespread recognition of global warming, both natural and human-induced (IPCC, 2001, 2007), our need for improved understanding of the geosystem has grown. The impacts of global warming will not all be negative; some regions may benefit (Ausubel, 1991). Global change is now a focus of much educational, research, and political activity. Given the expected growth in the human component of this system, we can expect humans to become a more important geomorphic agent and to have a more difficult time adjusting their activities to this changing system. In partial recognition of this development, the 1990s were designated the International Decade of Natural Disaster Reduction (IDNDR). When humans have the potential of being overwhelmed by natural geologic processes, these processes are referred to as geologic hazards. If the events actually produce losses in property and life, they are then described as geologic disasters.

Much of this book explores the realm of geologic hazards and disasters as they are a major component of environmental geology. The last section explores global climate change, possibly the most important geologic hazard, in the context of sustainability.

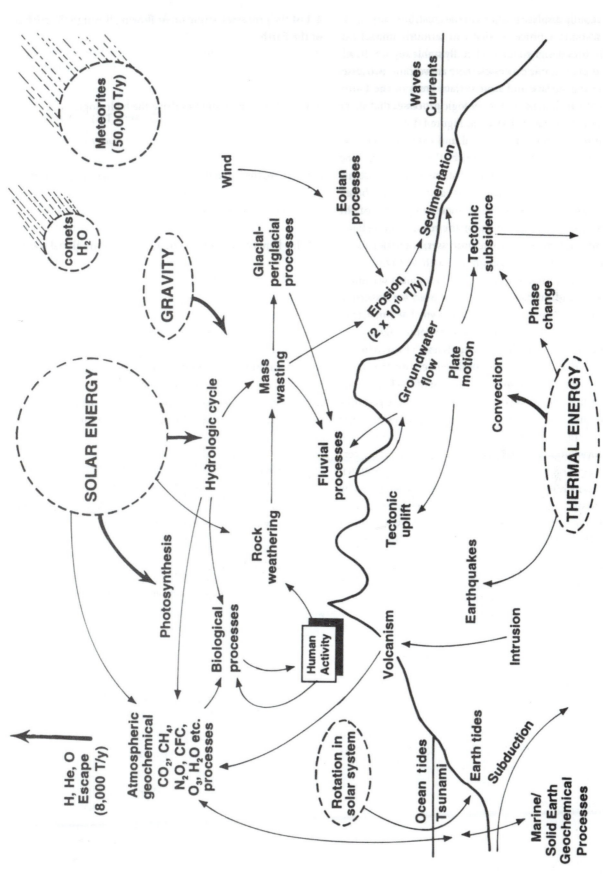

FIGURE 1.3 The geosystem showing some of the links between the primary internal and external geologic processes operating on the Earth. For examples of subsystems, see also the rock system (Figure 1.1) and a diagram of the hydrologic cycle. Human activity is impacted by most processes shown; more arrows can be drawn from human activity. (After Bloom, 1998 and Selby, 1985)

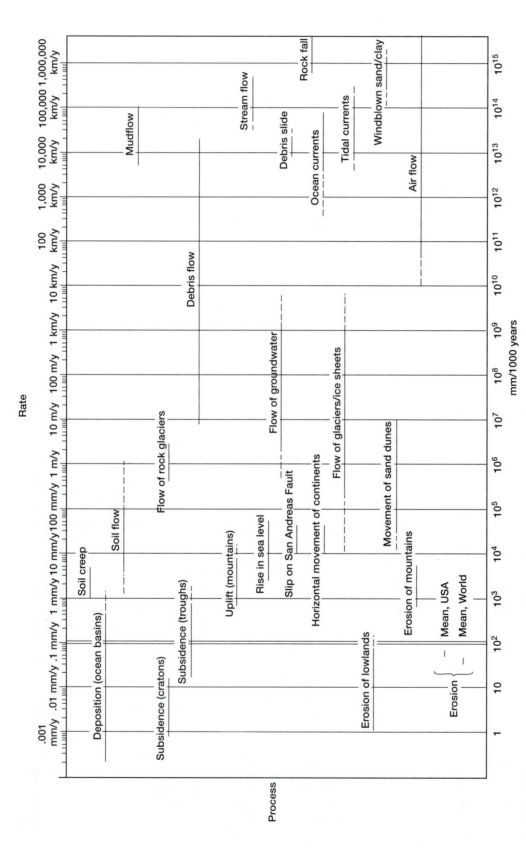

FIGURE 1.4 Rates of selected geologic processes. (After Fischer, 1969; Bloom, 1991; and Selby, 1985)

QUESTIONS 1, PART D

1. Figure 1.3 identifies many components of the geosystem, but does not show linkages between all of them. Sketch on this figure links that you think might exist between various geosystem components. If you are uncertain, it may help to refer to your text. Add to your links a few key words that explain why you think a particular link exists. Add soils to the diagram, and indicate important links in their formation and erosion.

2. Is the Earth a closed system? Explain.

3. According to this figure, what is the net annual gain in the mass of the Earth?

4. List the processes acting on or flowing through the surface of the Earth.

5. What two energy sources drive the hydrologic cycle?

6. What are the "greenhouse gases" and what do they do?

7. What benefits for humans might accompany global warming?

The following questions refer to Figure 1.4.
8. If you can run a four-minute mile (only slightly slower than the world record pace), which geologic processes will be able to catch you? (Hint: Use the conversion factors given in Appendix 1 and refer, if needed, to exercise 3 for a discussion of unit conversion.)

9. If you are driving a car at 65 miles per hour, which geologic processes will be able to catch you?

10. If fingernails typically grow at a rate of about 15 mm per year, what geologic processes are occurring at the same rate?

11. Soil profiles are about one meter thick in many parts of the world. How long, at an average rate of erosion, will it be before all the soil is removed (assuming that it is not replenished)?

Bibliography

Ausubel, J. H., 1991, A second look at the impacts of climatic change: *American Scientist*, v. 79, no. 3, p. 210–221.

Bloom, A. L., 1998, *Geomorphology: A systematic analysis of Late Cenozoic landforms* (3rd ed.): Englewood Cliffs, NJ, Prentice-Hall, 482 p.

Busch, R. M. ed., 1990, *Laboratory* manual *in physical geology* (2nd ed.): Columbus, OH: Merrill, 222 p.

Easterbrook, D. (ed.), 2003, *Quaternary geology of the United States: INQUA 2003 field guide volume*: Reno, NV, Nevada Desert Research Institute, 438 p.

Fischer, A. G., 1969, Geologic time-distance rates: the Bubnoff Unit: *Geological Society of America Bulletin*, v. 80, p. 549–551.

Graf, W. L. (ed.), 1987, *Geomorphic systems of North America*: Boulder, CO, Geological Society of America, 643 p.

IPCC, 2001, *Climate Change 2001: Synthesis report*. A Contribution of Working Groups I, II, and III to the Third Assessment Report of the Integovernmental Panel on Climate Change [Watson, R. T. and the Core Writing Team (eds.)]. Cambridge, United Kingdom, and New York, Cambridge University Press, 398 p. See also: http://grida.no/climate/ipcc_tar/vol4/english/pdf/spm.pdf

IPCC, 2007, *Climate Change 2007: The physical basis*. Summary for Policy Makers. Contribution of Working Group I to the Fourth Assessment Report of the Intergovernmental Panel on Climate Change [Drafting Authors and Draft Contributing Authors]. Retrieved April 28, 2007, from: http://www.ipcc.ch/SPM2feb07.pdf

Jacobsen, H. K. and Price, M. F., 1990, A framework for research on the human dimensions of global environmental change. *International Social Science Council/UNESCO*: ISSC/UNESCO Series 3, 71 p.

McKenzie, G. D., Foley, D., and Utgard, R. O., 1996, Geoenvironmental time scale: *Geological Society of America Abstracts with Programs*, v. 27, p. 477

Miller, E. C., and Westerback, M. E., 1989, *Interpretation of topographic maps*: New York, Merrill Publishing Company (now Prentice-Hall), 340 p.

Okulitch, A. V., 1988, Proposals for time classification and correlation of Precambrian rocks and events in Canada and adjacent areas of the Canadian Shield, Part 3, A Precambrian time chart for the Geological Atlas of Canada: Geological Survey of Canada, Paper 87–23, 20 p.

Palmer, A. R., 1983, The decade of North American geology 1983 geologic time scale: *Geology*, v. 11, p. 503–504.

Rahn, P. H., 1996, *Engineering Geology*: Englewood Cliffs, NJ, Prentice Hall, 657 p.

Selby, M. J., 1985, *Earth's changing surface: An introduction to geomorphology*: Oxford, England, Clarendon Press, 607 p.

Walker, J.D., and Cohen, H.A. (Compilers), 2006, *The geoscience handbook* (AGI Datasheets, 4th ed.): Alexandria, VA, American Geological Institute, 300 p.

West, T. R., 1995, *Geology applied to engineering*: Englewood Cliffs, NJ, Prentice Hall, 560 p.

Maps, Aerial Photographs, and Satellite Images

INTRODUCTION

Maps and aerial photographs are basic tools of earth scientists. They are scale models of a portion of the Earth's surface that show details of size, shape, and spatial relations between features. There are many different kinds of maps, several of which will be used in this manual. Therefore, it is very important to have a basic understanding of how to interpret maps, aerial photographs and other images.

PART A. TOPOGRAPHIC MAPS

A topographic map is a representation of the configuration of the Earth's surface. In addition to showing the position, relation, sizes, and shapes of the physical features of an area, it typically shows cultural features. Topographic maps are of great value to earth and environmental scientists for observation and analysis of the effects of geologic processes that are constantly changing the face of the Earth. They are also used as base maps for recording and interpreting geologic and environmental information that may be used in analyzing and solving environmental problems. The U.S. Geological Survey, which has been actively engaged in making topographic maps since 1882, has produced maps covering practically all of the United States. Appendix I contains useful conversion factors for working with maps using data in both feet and miles and meters and kilometers.

Interpretation of topographic maps is a great deal easier if one has an understanding of the following terminology:

Topography is the configuration of the surface of the land.

Relief refers to difference in elevation between the tops of hills and the bottoms of valleys within a given area.

The datum plane is the plane of zero elevation, which for nearly all topographic maps is mean sea-level.

Elevation is the vertical distance between a given point and the datum plane.

Height is the difference in elevation between a topographic feature and its immediate surroundings.

A bench mark is a point of known elevation. It is usually designated on a map by the letters BM, with the elevation given to the nearest foot or meter.

Contour lines are lines connecting points of equal elevation. Contour lines show the vertical or third dimension on a map and, in addition, the size and shape of physical features.

Contour interval is the vertical distance between adjacent contour lines.

Hachured contour lines are used to indicate closed depressions. They resemble ordinary contour lines except for the short lines (hachures) on one side, which point toward the center of the depression.

Scale refers to the ratio of the distance between two points on the ground and the same two points on the map. The scale is commonly shown at the center of the bottom margin of the map sheet and may be expressed in three ways: (1) a fractional or ratio scale, such as 1:24,000, meaning that one unit of measurement on the map represents 24,000 of the same units on the ground; (2) a graphic scale, which is simply a line or bar that is subdivided to show how many units of distance on the ground are equivalent to a given distance on the map; and (3) a verbal scale is a convenient way of stating the scale, such as "one inch equals one mile."

In the Questions for Part A, the "Tour of a Topo" section explains many of the additional types of information that are on a topographic map.

Map Symbols

Conventional symbols, used to indicate natural and manmade features, make it possible to put a vast amount of detailed information on topographic maps. In general, cultural (man-made) features are shown in black and red, all water features are blue, and relief is shown by contours in brown. Green overprint is used on some maps to indicate forest areas. Maps that have been photorevised use a purple or gray overprint to show cultural features constructed since the map was originally published.

Topographic map symbols used by the U.S. Geological Survey are inside the front cover. Information on how to order maps, a description of the national mapping program, and map scales are in Appendix IV.

Methods of Describing Locations

There are many ways to locate places or areas on maps. One way that is commonly used is a description in terms of *latitude and longitude*. Lines of latitude (parallels) and lines of longitude (meridians), form a coordinate system superimposed on the Earth's surface. Parallels are east–west lines that parallel the equator. Meridians are north–south lines that converge toward the poles. By this method location is expressed in terms of the number of degrees (°), minutes ('), and seconds (") of latitude north or south of the equator and longitude east or west of the prime meridian. The prime meridian, which passes through Greenwich, England, is the reference meridian (0°00'00") from which longitude is measured. The equator is the reference parallel (0°00'00"). The most commonly used topographic maps are either $7\frac{1}{2}'$ or 15' quadrangles.

A second method by which areas or places may be located is by using the *township and range system* (TRS). This system, established by congress as part of the "Land Ordinance of 1785," uses selected principal meridians and base lines as reference points (Figure 2.1). Starting from these meridians and base lines, the land was surveyed into congressional townships, six miles square, making the problem of locating smaller tracts of land simpler. Townships were further subdivided into 36 sections, each one mile square (Figure 2.1).

One section of land contains 640 acres, a half-section 320 acres, and so on. As a matter of convenience, sections are divided into halves, quarters, halves and quarters of quarters, and so on, called fractions, which are named or located by points of the compass; such as NW 1/4, SE 1/4, S 1/2, or NE 1/4. In finding the location of a described tract of land, the description is analyzed by first determining the location of the township by range and township number, then the section by its number. The exact fraction is determined by beginning with the last (which is the largest) and working back: T. 2N, R. 2E, Sec. 10, NW 1/4, NE 1/4 (Figure 2.1).

A third way for locating places is the *Universal Transverse Mercator (UTM) grid*. The UTM system is becoming more popular with increasing use of Global Positioning Systems (GPS). This is in part due to the fact that the UTM system locates features using meters, rather than degrees, minutes, and seconds. It is far easier to imagine what the distance between two features is if the distance is reported in meters. It is difficult at best to imagine distances that are given in minutes and seconds.

In the UTM system, the world is divided into 60 zones of 6 degrees longitude. As shown in Figure 2.2, the conterminous United States is located in zones 10–19.

Locations in the US are reported in meters north of the equator (called northings) and meters east of an arbitrary zero line that lies 500,000 meters west of

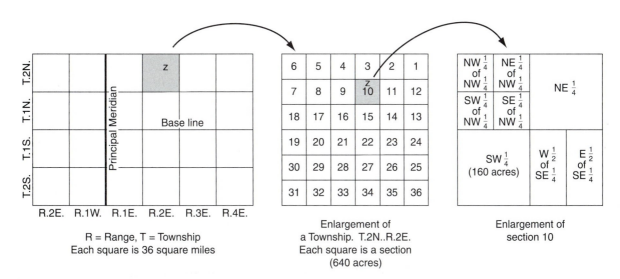

FIGURE 2.1 Subdivision of land by township, range, and section. A township consists of 36 square miles; a section, 1 square mile (640 acres). Point Z, shown on the maps above, is located in the NE1/4 of the NW1/4 of Sec. 10, T. 2N, R.2E.

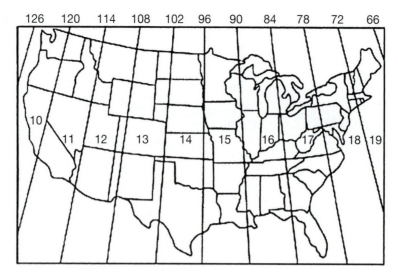

FIGURE 2.2 The Universal Transverse Mercator grid that covers the conterminous 48 United States comprises 10 zones—from Zone 10 on the west coast through Zone 19 in New England. Numbers at top of figure are degrees of longitude. (USGS)

the central N–S meridian in any zone. The central line in any zone has a value of 500,000 meters to ensure that all UTM eastings in a specific zone are reported as positive numbers (Figure 2.3). Since there are 10 UTM zones in the United States, it is important that the specific zone always be indicated. In the United States, for instance, each zone may have identical values for northing and easting; specifying the zone will allow someone to determine if you are in Washington, Minnesota, or Maine. Topographic maps indicate UTM values along their margins shown as blue tic marks.

Contour Lines

A distinctive characteristic of topographic maps is that they portray the shape and elevation of the terrain through the use of contour lines. The manner in which contour lines express topography is illustrated in Figure 2.4. Since contours are not ordinary lines, they must meet certain criteria to satisfy their definition. The following principles govern the use and interpretation of contour lines.

1. All points on a contour line have the same elevation.

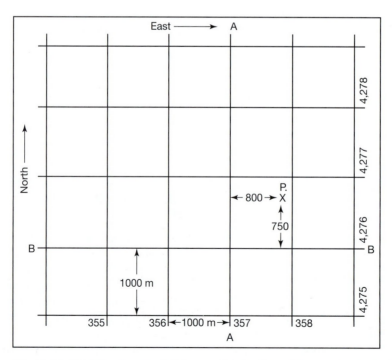

FIGURE 2.3 The grid value of line A-A is 357,000 meters east (of the zero line for that UTM zone). The grid value of line B-B is 4,276,000 meters north (of the equator), Point P is 800 meters east and 750 meters north of the grid lines; therefore, the grid coordinates of point P are north 4,276,750 and east 357,800. (USGS)

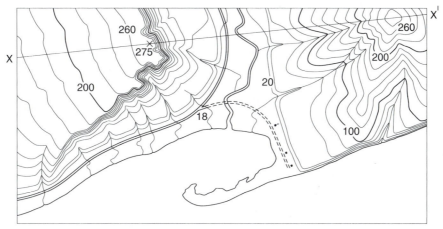

FIGURE 2.4 Topography expressed by contour lines. (USGS)

THE USE OF SYMBOLS IN MAPPING

These illustrations show how various features are depicted on a topographic map. The upper illustration is a perspective view of a river valley and the adjoining hills. The river flows into a bay which is partly enclosed by a hooked sandbar. On either side of the valley are terraces through which streams have cut gullies. The hill on the right has a smoothly eroded form and gradual slopes, whereas the one on the left rises abruptly in a sharp precipice from which it slopes gently, and forms an inclined tableland traversed by a few shallow gullies. A road provides access to a church and two houses situated across the river from a highway which follows the seacoast and curves up the river valley.

The lower illustration shows the same features represented by symbols on a topographic map. The contour interval (the vertical distance between adjacent contours) is 20 feet.

2. Contours separate all points of higher elevation than the contour from all points of lower elevation.

3. The elevation represented by a contour line is always a simple multiple of the contour interval.

4. Usually, every contour line that is a multiple of five times the contour interval is printed as a darker line than the others.

5. Contour lines never intersect or divide, they may, however, merge at a vertical or overhanging cliff.

6. Every contour closes on itself either within or beyond the limits of the map.

7. Contours that close within a relatively small area on a map represent hills or knobs.

8. Steep slopes are shown by closely spaced contours, gentle slopes by widely spaced contours.

9. Uniformly spaced contour lines represent a uniform slope.

10. Minimum valley and maximum ridge contour lines must be in pairs. That is, no single lower contour can lie between two higher ones, and vice versa.

Topographic Profiles

Topographic maps represent a view of the landscape from above, and even though contour lines show the relief of this landscape, it is often desirable to obtain a better picture of the actual shape of the land in an area. This can be achieved by constructing a topographic profile, that is, a cross section of the Earth's surface along a given line. Profiles may be constructed quickly and accurately from topographic maps using graph paper. The procedure for constructing a profile along a selected line on the map is as follows:

1. Examine the line of profile on the map and note the elevations of the highest and lowest contour lines crossed by it. Select a vertical scale that fits the graph paper and note how many inches or centimeters on the graph paper correspond to a selected vertical elevation interval. For instance, one inch on the graph paper might equal 10, 100, or 1,000 feet (or 120, 1,200, or 12,000 inches) of relief in your cross section. Label on the graph paper equally spaced horizontal lines that correspond to the elevation of each contour line

CONTENTS

Preface   v

I Introduction to Geology   1

Exercise 1.   Earth Materials, Geologic Time, and Geologic Processes   3
Exercise 2.   Maps, Aerial Photographs, and Satellite Images   19
Exercise 3.   Measurements, Basic Calculations and Conversions, and Graphs   36

II Introduction to Geologic Hazards   45

Exercise 4.   Volcanoes and Volcanic Hazards   52
Exercise 5.   Hazards of Mount St. Helens   69
Exercise 6.   Earthquake Epicenters, Intensities, Risks, Faults, Nonstructural Hazards and Preparation   82
Exercise 7.   The Loma Prieta Earthquake of 1989   102
Exercise 8.   Landslides and Avalanches   114
Exercise 9.   Subsidence   141
Exercise 10.   River Floods   154
Exercise 11.   Coastal Hazards   174

III Water Resources and Contamination   193

Exercise 12.   Groundwater Hydrology   198
Exercise 13.   Water Quality Data and Pollution Sources   206
Exercise 14.   Lake and River Contamination from Industrial Waste   217
Exercise 15.   Groundwater and Surface Water Contamination from Resource Extraction   227
Exercise 16.   Groundwater Overdraft and Saltwater Intrusion   239

IV Sustainability: Resource Planning and Global Change   251

Exercise 17.   Geology and Regional Planning   257
Exercise 18.   Global Change and Sustainability   268

Appendices

Units and Conversions   281
Geologic Time Scale   283
Population Data   284
Topographic Map Scales   285
Color Plates   286

(use only alternate or heavy contour lines if they are closely spaced).

2. Place the edge of the graph paper along the profile line. Opposite each intersection of a contour line with the profile line, place a short pencil mark on the graph paper and indicate the elevation of the contour. Also mark the intersections of all streams and depressions and note, where appropriate, depressions or hills that do not quite reach the next contour line.

3. Project each marked elevation perpendicularly to the horizontal line of the same elevation on the graph paper.

4. Connect the points with smooth line and label prominent features.

5. State the vertical exaggeration. To avoid distortion, profiles are drawn with equal vertical and horizontal scales. However, it is usually necessary to use a vertical scale several times larger than the horizontal scale.

This is done to bring out the topographic details in areas of low relief. The vertical exaggeration is calculated by dividing the horizontal scale by the vertical scale. Thus a profile with a horizontal scale of $1'' = 2000' (1:24,000)$ and a vertical scale of $1'' = 400'$ $(1:4800)$ would have a vertical exaggeration of five times, that is $2000/400 = 5x (24,000/4800 = 5x)$. Note that the units of measurement must be the same for both the numerator (24,000) and denominator (4800) to calculate a correct vertical exaggeration.

QUESTIONS 2, PART A

1. *Construction of a Contour Map.* The elevation of points shown on the map in Figure 2.5 was determined by survey. Abiding by the principles governing contour lines, you are to draw in contour lines to make a topographic map. Use a 20-foot contour interval and assume that the land slopes uniformly between any two points of different elevation.

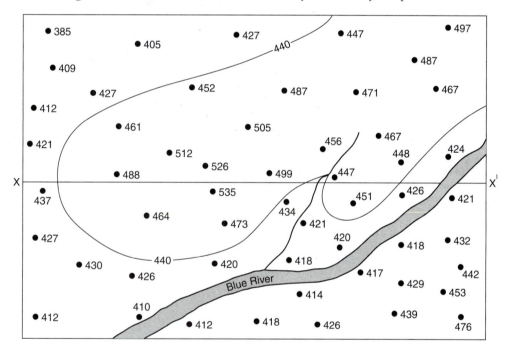

FIGURE 2.5   Elevations (in feet) for drawing a contour map. (Horizontal scale: $1'' = 1$ mile $= 5280'$)

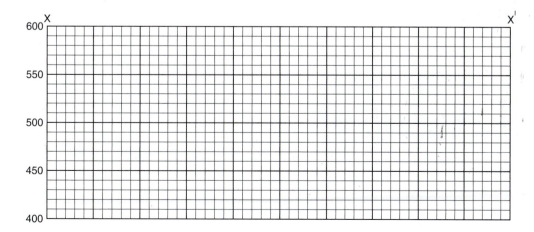

FIGURE 2.6   Profile along line X-X' in Figure 2.5. Vertical scale: $1'' = $ _____ ft. Vertical exaggeration _____ x.

2. *Construction of a Topographic Profile.* Using the contour map you have just completed, draw a topographic profile along line X-X'. This profile may be constructed by "dropping" perpendicular dotted lines from each intersection of a contour line with the profile line, to the same elevation on the graph paper (Figure 2.6). Then connect the points with a smooth line. State the vertical scale and vertical exaggeration in the spaces provided.

3. Use either Figure 2.7 or a map supplied by your instructor to answer the following questions.

    a. What year was the map made? _____ Revised? _____

    b. What year were the photographs taken that were used to create the map? _____ Were later photographs also taken? If so, when? _____

    c. What are the UTM values for the southeast corner of the map? _____ N, _____ E

    d. What are the latitude and longitude values for the southeast corner of the map? _____ degrees, minutes, and seconds N; _____ degrees, minutes, and seconds W

    e. What is the scale of the map?

    f. What is the contour interval of the map?

    g. What is the quadrangle immediately southwest of this quadrangle?

    h. How many degrees difference are there between true north and magnetic north?

4. True North (TN) and Magnetic North seldom are exactly the same and magnetic north changes continually. Since we use a compass to determine our direction of travel and sightings, knowing the difference between TN and Magnetic North becomes essential, especially for mapping. In the SW corner of most maps, *the difference in degrees and direction* between these two north indicators is given for the year of publication of the map. This is known as the **Magnetic Declination** from True North.

    a. What is the mean magnetic declination on the Zanesville West, Ohio, map (Figure 10.5)?

b. Would you expect the magnetic declination to be the same the following year? _____

c. Explain why or why not.

5. Use Figure 2.8 (inside front cover), the description of symbols used on U.S. Geological Survey topographic maps, and Figure 2.10, the Bloomington, Indiana, $7\frac{1}{2}$ minute quadrangle (in the color map section at the back of this book) to answer this question.

    a. What is the fractional or ratio scale?

    b. What is the contour interval?

    c. What is the elevation of the filtration plant below Griffy Reservoir?

    d. Using the township and range system, locate Payne Cemetery to the nearest 40 acres (quarter of a quarter section).

    _____ 1/4, _____ 1/4, Sec._____, T. 9N., R. 1W.

    e. What is the elevation of the major contour adjacent to the cemetery?

6. Using a topographic map of your local area (provided by your instructor), fill in the following information:

    a. Name of the map?

    b. Contour interval?

    c. Datum plane?

    d. Year area first surveyed?

    e. Fractional or ratio scale?

    f. E–W distance covered by map?

    g. N–S distance covered by map?

FIGURE 2.7 Topographic map of the Kalispell Quadrangle, Montana.

h. Approximate area covered by the map (in both mi^2 and km^2)?

i. Latitude and longitude of the northeast, northwest, southeast, and southwest corners of the map?

(NE)

(NW)

(SE)

(SW)

j. Names of quadrangles located north, east, south, west, northeast, southeast, southwest, and northwest of this quadrangle?

k. Note the variety of symbols used on the map. Find examples of each of the following and give its location by either latitude and longitude coordinates or by township, range, section, and quarter-section locations. Also note the relationship to major features or to the general area covered.

A bench mark.

A spot elevation other than a bench mark.

A school or church.

Features shown in:

blue:

brown:

black:

red:

green:

PART B. GEOLOGIC MAPS

Geologic maps show the distribution of rock units at the earth's surface, as though the regolith (loose unconsolidated materials) were stripped from the bedrock. The basic rock unit mapped is the geologic formation. Some geologic maps combine formations and show only units that include all rocks of a given age. Soil and regolith are shown as map units only when they are unusually thick or are directly associated with recent geologic processes, such as floodplain and landslide deposits, or glacial drift. In addition to earth materials, faults and other geologic structures are shown on geologic maps and the relative age of these materials and their arrangement beneath the surface are indicated.

By convention, formations are identified on geologic maps by color or pattern (see Figure 2.10, in map section at the back of the book), and are labeled with a letter abbreviation, which gives the geological age or name of the formation. Usually the color or pattern and the letter symbols for all ages and formations used on a map are printed in an explanation along the map margin. The oldest formation is listed at the bottom. In addition to color, line, and letter symbols for formations, geologic maps may also contain a wide variety of other symbols, depicting rock structure (strike and dip), economic rock or mineral deposits, and other geologic features.

A good geologic map contains a great amount of information. When used in conjunction with topographic maps or aerial photographs, they can provide much of the basic information needed for studies connected with mineral and mineral fuel exploration, construction projects, urban planning, and surface and subsurface water problems. They also tell us much about the geologic history of an area.

Study the geologic map of central North America Figure 2.10 (in the maps section) to help answer the questions on geologic maps below.

QUESTIONS 2, PART B

1. Using Table 2.1, the explanation for Figure 2.10, give the words for the following symbols:

p€m

D$_1$

K

2. What is the age in millions of years of the boundary between the Cambrian and the Precambrian (Hint: Consult the geologic time scale in Appendix II)?

TABLE 2.1 Explanation for Figure 2.11 (Note: U, Upper; L, Lower; und, Undivided)

Period	Color/Pattern	Symbol
Cretaceous	Light green	K
U Pennsylvanian	Blue	IP_2
L Pennsylvanian	Blue diagonal	IP_1
Mississippian	Light purple diagonal	M_1
Devonian (und)	Brown	D
U Devonian	Brown	D_2
L Devonian	Brownish gray	D_1
Silurian	Light purple	S
Ordovician (und, L, U)	Reddish purple	O, O_1, O_2
Cambrian	Brownish orange	€
Era (and rock type)		
U Precambrian	Yellowish tan	p€ u
Keweenawan sed rocks	Reddish brown	p€ u_2
Keweenawan volc rocks	Light brown	p€ u_1
Basic Intrusives	Brownish green	p€ b
Middle Precambrian	Light brown	p€ m
Granite and Granite Gneiss	Pink	p€ i
Lower Precambrian	Olive green	p€ l

3. Almost all of the Precambrian rocks are metamorphic or _____? Precambrian rocks form the basement rock of continents and are known as "shields" where they outcrop. They extend under the younger Phanerozic Eon rocks, which in the map area are sedimentary rocks, and present different environmental challenges to land use planning.

4. The geologic cross section (Figure 2.11) for the bedrock geology map of central North America (Figure 2.10) shows the stratigraphy of part of the Michigan Basin between **A**, northwest of Milwaukee, Wisconsin and A', northeast of London, Ontario, Canada. In the cross section the Silurian system is labeled (S). Label other geologic units shown.

5. The circular geologic structure centered on the blue area in the lower peninsula of Michigan is a basin; however, in the cross section it appears as a (circle one)
       anticline   syncline.

6. If you drilled a hole at the "D" north of Milwaukee, what geologic systems would you pass through before you reached the Precambrian igneous or metamorphic basement rocks? List the sequence of rocks from youngest to oldest.

7. Although this bedrock map shows the distribution of rocks of different ages (and in the Precambrian some of the rock types), in most areas of the map these rocks are buried beneath unconsolidated sediments ranging from a few to as much as 700 feet thick. Which of the following cities is underlain by the oldest bedrock? (check one)

Chicago _____, Detroit _____, Duluth _____

Note: Geologic maps of larger scale also show details on rock type that are important in seeking resources (water, metallic minerals, gravel, stone, etc.) and meeting environmental challenges.

PART C. AERIAL PHOTOGRAPHS AND REMOTELY SENSED IMAGES

Aerial Photographs

An aerial photograph is a picture taken from a camera position located somewhere above the earth's surface. Such pictures may be taken obliquely (for example, see Figure 5.2, which shows oblique photographs of Mt. St. Helens in Washington). In geology, it is common to use aerial photographs that have been taken vertically looking down to the ground. These photographs are often overlapped

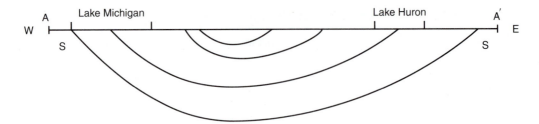

FIGURE 2.11 West–east cross section of the Michigan Basin from NW of Milwaukee, Wisconsin (A), to NE of London, Ontario, Canada (A'). See color map, Figure 2.11, for the line of the cross section.

and combined to create stereopairs of photos, in which relief of the ground surface can easily be seen. Aerial photos are valuable tools for use by earth scientists in studying geologic and environmental phenomena. Aerial photos have proven invaluable in making topographic, geologic, and soil maps, in aiding military operations, in regional planning, and in monitoring the environment.

Federal, state, and other governmental and private agencies have photographed practically all of the United States and Canada from aircraft flying at altitudes as great as 60,000 feet.

Satellite Imagery

Beginning in July 1972, with the launching of ERTS-1 (Earth Resources Technology Satellite), the United States started collecting, on a repetitive basis, satellite images of the earth's surface. Several similar Land Satellites (Landsat) have subsequently been launched in near-polar orbits at an altitude of about 570 miles to collect a variety of images of large areas of the earth's surface. Landsat images can be photographed in different wavelengths and may be produced in false color or enhanced by computer to make available a tremendous amount of data. The satellite passes over the same area at regular intervals permitting monitoring of surface changes in addition to providing information for mapping and detecting changes in vegetation.

Many satellites from different national programs and private organizations are now providing improved understanding of Earth processes, landforms, materials, and change in the Earth system. Exceptional image resolution, availability on the Web (e.g., Google Earth, TerraServer, etc.) and ease of use are enticing the public to explore and understand as never before how components of the Earth work.

LIDAR

LIDAR (*LIght Detection And Ranging*), also known as Airborne Laser Scanning (ALS), is a relatively new remote sensing technique that transmits laser light from an airplane and records the light that bounces back to the plane. About 30,000 points per second are recorded as the airplane flies over the land surface. These data are then compiled into an image for interpretation of natural and human-related features on the ground. LIDAR is particularly powerful in terrain that has tree cover, as the data allow "virtual deforestation" or removal of vegetation from the image. Thus LIDAR images show the surface of the ground, not the tops of the trees. Fault identification, landslide recognition, floodplain delineation, and basic geologic mapping are among the applications in environmental geology. A LIDAR image analysis is included in Exercise 6 where it is used to locate new faults that had escaped traditional field and aerial photographic techniques.

Interpretation of Photos and Images

The interpretation of either single aerial photos or stereopairs requires practice. One of the greatest difficulties experienced in viewing an air photo for the first time is recognition of features on the photographs, such as roads and buildings that are very familiar when seen on the ground. However, someone unfamiliar with aerial photos can comprehend them amazingly well with a little practice and the guidance of a few basic principles of recognition of common features on black and white photos. Some aspects of features to consider when looking at an aerial photograph are listed in Table 2.2.

On natural color aerial photographs, colors are similar to those seen on the ground. Some geological studies use infrared photographs, in which healthy vegetation appears as shades of red. Satellite images are manipulated in computers to use colors to emphasize different aspects of geology or vegetation. It is important to pay close attention to color keys when working with infrared photographs or satellite images.

Determination of Scale

The most commonly used aerial photographs are vertical photos that accurately illustrate the land in the

TABLE 2.2 Aspects of Features That Aid in Their Recognition on Aerial Photographs (modified after Avery & Berlin, 1992)

Aspects of Features on Photographs	Characteristics of Features Helpful for Their Identification
Shape	Natural features, such as rivers, faults, coastlines, or volcanoes, tend to be irregular in shape. Human constructed features, such as roads or buildings, tend to have regular geometric shapes. Agricultural fields are typically rectangular. Circular crop patterns occur where center-pivot irrigation systems are used.
Size	Some features, such as homes or highways, have approximate sizes familiar to most observers. Where such features can be located on a photograph, they can be used as a basis to infer the sizes of other features, such as rivers or landslides.
Pattern	Some geologic features, such as landslides, have characteristic patterns. For landslides, these patterns commonly include a steep scarp at the top, and a lower area of displaced soil, rocks, vegetation (trees are especially easy to see), or human structures. Refer to introductory material in many exercises for descriptions of typical features related to geologic hazards. Unmodified drainage patterns can reveal much about the underlying geology of an area.
Shadow	Taller objects will cast longer shadows when sun angles are low. A high sun angle, such as directly overhead of the feature being observed, will create only a small shadow. Low-sun-angle photography, with enhanced shadows, is often helpful in identifying geologic features such as linear fault scarps.
Tone (or color)	Water is dark gray or black except where sunlight is reflected. Soil moisture, which is controlled by texture and type of soil, largely determines the difference in appearance of different fields on air photos. Wetter areas generally appear as darker grays; dryer soils as lighter grays. Vegetation accounts for a great many differences in pattern and shades of gray. Heavily forested areas are usually medium to dark gray. Grasslands are light gray. Cultivated fields vary greatly in color tone depending on whether or not just plowed, or if planted, the kind of crop and stage of growth.
Texture	Roughness or smoothness of features on aerial photographs can be helpful in identifying them. The texture of a given feature on a particular photograph depends on the scale of the photograph. For example, a fresh lava flow may appear rough when viewed closely, but will appear smooth when viewed from farther away. Water will often be smooth; surf zones along coasts may, when viewed closely, appear rough.
Association	When looking at some features, other features are often also found. Landslides may cause rivers to bend or, if the landslide is large enough, a lake may be created. Linear faults often are zones with more water and therefore greater vegetation. Active volcanoes may have fresh lava areas where there is no vegetation.
Site	Thinking about typical sites for features can help identify objects on photos. Large shopping malls are not typically found in rural areas. Major airports are located near big cities. Floodplains are a natural part of river systems.

central portion of the photo directly beneath the camera. The basic scale of a vertical air photo, overlooking variations due to distortion near the edges of the photo and due to relief, can be computed if the focal length of the camera (the distance between the lens and the film) and the altitude of the camera above the ground surface are known.

The scale of a photograph can also be determined by measuring distances between objects on the ground and comparing them with distances measured between the same objects on the photo. Section lines, where available, are well suited for this purpose since they are supposed to be one mile apart and are frequently visible on photos as roads or fencelines. Scale is computed using this equation:

$$Sc = Dph/Dgd$$

where Sc = the scale of the photo,
Dph = the distance on the photograph, and
Dgd = the distance on the ground.

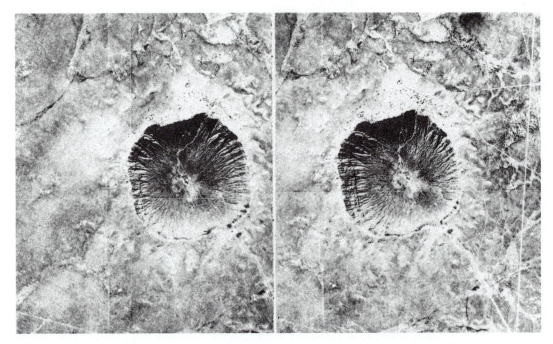

FIGURE 2.12   Stereopair aerial photos of Meteor Crater, Arizona. The scale of these photographs is about 3,500 ft/in. The crater is approximately 600 ft deep. (Reprinted by permission from Avery, 1968. Photograph courtesy of U.S. Department of Agriculture. Copyright © 1968 T. Eugene Avery)

Remember that scale is described as a ratio (such as 1:24,000), and that the units of measure must be the same on the photo and the ground (for example, inches on both photo and ground, not inches on the photo and feet or miles on the ground).

Stereoscopic Viewing

Air photos are usually taken in sequence along a path of flight so that there is considerable overlap, such that any given feature will appear on at least two different photos. Viewing these overlapping photos (a stereopair) through a stereoscope results in a three-dimensional view of the surface. Figure 2.12 is a stereopair that allows the third dimension to be easily seen with a standard stereoscope. Note that features on the ground will appear to be vertically exaggerated on stereo photographs.

QUESTIONS 2, PART C

Meteor Crater

1. Examine Figure 2.12. Given the scale in the caption, what is the approximate diameter of the crater?

2. Using a stereoscope to see the third dimension in Figure 2.12, outline the floor of the crater and mark the lowest point with an "L."

3. On Figure 2.12, label any other visible relief features, vegetation, roads and buildings.

Boston

The following questions refer to Figure 2.13, which is in the color map section at the back of this book. This is a high altitude photograph of the Boston, Massachusetts, area. The Charles River enters from the middle west of the photograph. The Mystic River enters the northern Inner Harbor from the west, and Chelsea Creek enters the Inner Harbor from the east. The channel with the ship docked south of Logan Airport is oriented E–W. The instructor will identify place names and locations of specific sites or you can consult topographic or road maps.

4. Study the color high altitude photograph (Figure 2.13) using a magnifying glass or hand lens if necessary to identify geologic and cultural features. Label the following features on the photo, using the corresponding numbers 1–15.

    1. park
    2. airport runway
    3. airport terminal with planes
    4. tidal flat
    5. beach
    6. star-shaped fort (SE quarter of image)
    7. road
    8. bridge
    9. railway
    10. pond
    11. small boat
    12. large docked ship
    13. racetrack

14. baseball field

15. oil and gasoline storage tanks

5. In which direction is the small boat traveling at the entrance to Boston Inner Harbor?

6. Notice the diffuse white zone in the Mystic River, between the first and second bridges. What might cause such a zone in the river? What is your evidence from this photograph?

7. The Back Bay Fens is a swampy area that drains from the south into the Charles River. A Fen is an alkaline bog. Given this information, locate and mark the baseball field named Fenway Park.

8. The L Street Beach, which is located west of the circular bay in the SE quarter of the image, has structures that extend into the Old Harbor. Give the name for these erosion control structures and the direction of the longshore current in the area.

9. Many elongated hills in this glaciated landscape have influenced street patterns. In the NE quarter of the photograph is Suffolk Downs Racetrack. What is the direction of elongation (which gives the ice-flow direction) and the name of the landforms beneath the communities of Beachmont (1 cm E of the racetrack) and Orient Heights (1 cm W of the racetrack)?

10. If the long axis of Suffolk Downs is 650 m (2100 ft), what is the scale of the photograph?

11. Tall structures can be identified by their shadows. At what time of day (morning, noon, afternoon) in mid-April 1985 was the picture taken?

Salt Lake City, Utah

The following questions refer to Figures 2.14 and 2.15. Figure 2.15 is a black and white stereo pair of aerial photographs taken of the Salt Lake City area in 1958. Figure 2.15 (back of book) shows color aerial stereo photographs of an area south of the black and white photographs. The color stereo photographs were taken in 1992. The area of Figure 2.15 is included on Figure 6.10, the colored topographic map of the Draper, Utah, quadrangle, which also is in the map section at the back of this book.

12. Using a pocket stereoscope, examine the stereo pair in Figure 2.14 to determine the nature of the landscape. The shades of gray indicate vegetation or soil moisture differences. Identify a feature that shows:
   a. darkest shading
   b. lightest shading
   c. intermediate shading

13. Identify the natural or built feature at each of the following sites on Figure 2.14: Most can be identified without the stereoscope.

   A. _____

   B. _____

   C. _____

   D. _____

   E. _____

   F. _____

   G. _____

   H. _____

   I. _____

   J. _____

   K. _____

   L. (Select a Site) _____

   M. _____

14. What land uses are visible in the 1958 stereopair (Figure 2.14)? Are other land uses visible on the 1992 photographs (Figure 2.15)?

15. Use the definitions provided below to help determine in which geomorphic feature site "J" (Figure 2.14) is located. What is your evidence?
   terrace (flat area near a river but above the river in elevation)
   floodplain (low-lying area along a river that is flooded at high water)
   drumlin (rounded hill deposited by a glacier)
   landslide (area where land has slid)

16. a. In Figure 2.14 what is the distance between the middle of the G to the middle of the H? _____ mm

   b. Given that the scale of the photo is 1:14,400, what is the distance on the ground between G and H? _____ km (there are 1,000,000 mm in 1 km)

17. Compare similar features on the black and white stereo photographs (Figure 2.14) with those on the color stereo photographs (Figure 2.15).
   a. Are the two sets of photographs the same or different scales? What is your evidence?

   b. What are some advantages of using color stereo photographs over using black and white stereo photographs?

   c. What are some advantages (if any) of using black and white photographs over color photographs?

FIGURE 2.14 Stereopair aerial photos (1958) of area at edge of Wasatch Front South of Salt Lake City, UT. South Street (W-E) meets Holladay Blvd (N-S) in NW quarter of photo.

San Francisco, California

The following questions refer to Figure 2.16, which is a satellite image of the San Francisco Bay area. The dots and small circles are earthquake epicenters, which are discussed in Exercise 7.

18. Locate examples of the following land uses on the satellite image. Mark these features and a north arrow on the outline map of the San Francisco Bay area (Figure 2.17 (back cover)):

    a. bridges across San Francisco Bay
    b. airport(s)
    c. an urban park
    d. a major highway
    e. elongated lakes along the San Andreas Fault

19. What pattern differences help distinguish natural vegetation areas from vegetation patterns in agricultural areas?

20. Compare natural and human features on this satellite image with those on the black and white and color stereo photographs. What are some of the advantages and disadvantages of using satellite images rather than aerial photographs in environmental studies? Under what conditions do you think that satellite images might be especially useful?

Bainbridge Island near Seattle, Washington.

21. What are the kinds of features that can more easily be seen on the LIDAR image than on the topographic map? Which of these features are natural and which are related to human activity? What is your evidence?

22. What are the kinds of features that can more easily be seen on the topographic map than on the LIDAR image?

FIGURE 2.17 Outline map of the San Francisco Bay area for interpretation of Figure 2.17.

23. How does having a topographic map help interpret human-related features on a LIDAR image?

24. When would the use of each type of map be an advantage? A disadvantage?

FIGURE 2.18 LIDAR map of the northern tip of Bainbridge Island, which is located in Puget Sound near Seattle, Washington.

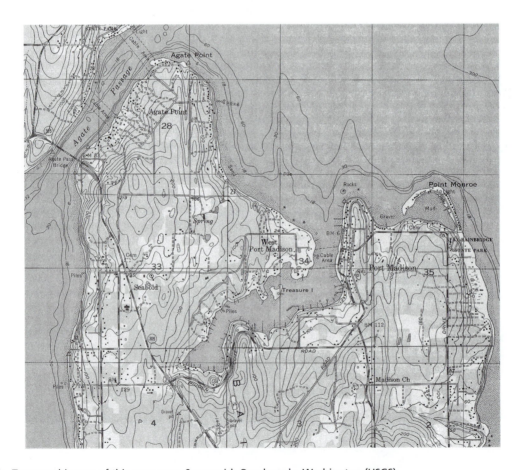

FIGURE 2.19 Topographic map of this same area. Suquamish Quadrangle, Washington (USGS).

Bibliography

Arnold, Robert H., 1997, *Interpretation of airphotos and remotely sensed imagery*: Saddle River, NJ, Prentice Hall, 250 p.

Avery, T. E., 1968, *Interpretation of aerial photographs*: 2nd edition, Minneapolis, MN, Burgess, 324 p.

Avery, T. E., and Berlin, G. L., 1992, *Fundamentals of remote sensing and airphoto interpretation*: 5th edition New York, Macmillan, 472 p.

Miller, V. C., and Westerback, M., 1989, *Interpretation of topographic maps*: New York, Prentice Hall, 339 p.

Sabins, Floyd F., 1987, *Remote sensing principles and interpretation*: 2nd edition, New York, Freeman, 449 p.

U.S. Geological Survey, undated, Topographic map symbols: 4 p.

Measurements, Basic Calculations and Conversions, and Graphs

INTRODUCTION

Scientific study of Earth requires careful observation, description, measurement, and analysis of data that are collected in the field and laboratory. Some of the data are in numerical form and are obtained by measurement of some quantity—such as the size of pebbles on a beach, temperature of geothermal or surface waters, thickness of a sequence of sediments, or location of stakes on a glacier. We then study the patterns and trends in these data using mathematics, tables, graphs, and maps to help us understand the materials, processes, and landforms of the Earth. With this approach we are able to establish the rates of processes, the nature of change in and on the Earth and, if we really understand the system, predict changes. The techniques used can be quite sophisticated; however, there are basic calculations that non-scientists can master that will help them understand how the Earth system operates. Many of these will be useful in day-to-day activities, also. These are reviewed in this chapter; additional related information is in Appendix I.

PART A. SCIENTIFIC MEASUREMENTS AND NOTATION

In measuring a feature on the Earth, such as the width of a stream, geoscientists obtain a quantity (a number) and associated units (such as meters or feet). Standards have been devised for measurements that ensure reliability and uniformity in handling these numbers and units; these standards include the International System (SI) and other units, scientific notation, and significant figures.

International System of Units

The International System of Units is the standard for making scientific measurements. With a standarized format for measurements, the results of science are readily transformed from one nation to another and from one discipline to another. The International System of Units, or SI Units, (Systeme International d'Unites), is similar to the metric (or CGS, centimeter-gram-second) system. The standard base units most often used in the geosciences are: meter (m) for distance, sec (s) for time, and kilogram (kg) for mass. Temperature is measured in degrees Kelvin (K). But often we use centimeters (cm), grams (g), and degrees Celsius (C) because they are more convenient. So-called derived units are formed through a combination of these base units and sometimes other units, such as cm^2, m^3, m^3/sec (or $m^3\,s^{-1}$), and gm/cm^3. Additional units are given in Appendix I.

Unfortunately, the geosciences still use a combination of SI units (actually modified as metric units) and English units to describe the size or dimensions of things we measure in everyday activity. This situation makes it easier for non-scientists to understand some things; however, it means that in the geosciences we must often deal with English units, for instance, on older topographic maps, contours and elevations are in feet. Other than energy units (e.g., Btu), most of the measurements employ modified SI units. Interfacing with old maps and some disciplines that use English units requires that we be able to switch from English to metric or SI units (see the conversion tables in Appendix I). Actually, exercises in this book are in English or metric units. Eventually, the United States will make the transition to metric to compete more effectively in international trade. One additional advantage of SI is the ease of making conversions from one set of units to another as shown below:

1 kilometer (km) = 1,000 meters (m) = 1,000,000 millmeters (mm)

which is easier to work with than:

1 mile = 5,280 feet = 63,360 inches

Scientific Notation

Often numbers in science are very large or very small. Scientists have devised scientific notation as a way of saving space, minimizing the chance of dropping a zero or two, and speeding calculations and communications. The technique employs the exponent (in this case the power of 10), which is tacked on to a 10 (i.e., 10^3 or $10 \times 10 \times 10$ or 1000 are equal; the "3" is the exponent), and is multiplied by another number (usually between 0.1 and 10) to indicate the size of the number. If it is a positive exponent it means we need to increase the number in front of the 10 by the number of zeros equal to the exponent. If it is a negative number, it means we decrease the number by the number of zeros equivalent to the exponent. For example, 4×10^3 actually means $4 \times 10 \times 10 \times 10 = 4 \times 1000 = 4,000$. 2.3×10^9 actually means $2.3 \times 1,000,000,000$, or, when we move the decimal point nine spaces to the right in 2.3, the number is 2,300,000,000 and would be said as two billion 300 million, or more easily as 2.3 billion. In fact when we see 2.3×10^9, in most cases we say 2.3 billion.

We also use this system for small numbers. 2.3×10^{-3} actually means that we move the decimal point over three spaces to the left; the number is 0.0023. We also put a zero to the left of the decimal point to emphasize the decimal. Another advantage of scientific notation is in recognition of significant figures (described below).

Another option is to use prefixes to describe very large and very small numbers. We often deal in millions and billions of years in the geosciences. We use the terms million and billion, but we also use the term mega for million and giga for billion, as in megawatts and gigayears. Actually, the French word annee is used for year in SI; it is abbreviated as (a), but some geologists still use the word year (yr).

The full ranges of these prefixes are in Appendix I. They range from 10^{18}, which is known as exa- (E is the symbol) for 1,000,000,000,000,000,000 (a quintillion, which has 18 zeros after the 1) to 10^{-18}, which is known as atto (a is the symbol) for a very small quantity, that is, 0.000,000,000,000,000,001. Spaces, not commas, are used in the SI system between the sets of zeros. Commas have been included here for ease of reading.

Significant Figures

Significant figures provide an indication of the precision with which a quantity is known. The more precise a number is, the more significant figures it has. For example, the number 2.43 has three significant figures. The actual value of this number may be between 2.425 and 2.435. If there is no decimal point in a number, the number of significant figures is the number of digits from the first nonzero digit on the left to the last nonzero digit on the right. For example, 22,000 has two significant figures. If there is a decimal point in the number, the number of significant figures is the number of digits from the first nonzero digit on the left to the last digit (zero or nonzero) on the right. For example, 22,000.0 has six significant figures. Other examples of significant figures are 0.0022 has two significant figures, 2.2×10^3 has two significant figures, and 2.20×10^3 has three significant figures. The last two examples show the use of scientific notation in determining significant digits. It is important to note that when two numbers (or more) are multiplied or divided, the number of significant figures of the new number is the value of the smallest number of significant figures of the numbers multiplied or divided. For instance, if 17.0 is divided by 7.00, the answer is reported as 2.42 (three significant figures). If 17.0 is divided by 7.0, the number is reported as 2.4 (two significant figures, as 7.0 has only two significant figures). In neither example is the answer reported as 2.428571429, which is the result you may see if you use a calculator. Reporting that number as your answer would indicate 10 significant figures, when, in fact, there were only two or three significant figures (depending on example above).

QUESTIONS 3, PART A

1. What is the SI unit for measuring:
  a. distance?

  b. temperature?

2. Why might geoscientists not use SI Units in working with some maps produced in the United States?

3. Give the value in words and numbers for the following numbers, which are given in scientific notation (the first is given for you).

Scientific Notation	Numbers	Words
a. 1.2×10^6	1,200,000	one million two hundred thousand (or 1.2 million)
b. 4.12×10^3		
c. 5.8×10^9		
d. 2.2×10^{15}		
e. 0.1×10^6		

4. Consult Appendix I to determine the value and name for the following prefixes in the International System of Units.

	Value	Name
a. T	10^{12}	tera
b. G		
c. M		
d. k		
e. c		
f. m		
g. μ		
h. n		

5. Complete the blanks in the columns to provide the name and equivalent values in meters.

		Full Name	in Meters
a. 1 cm	=	1 centimeter	= 0.01 meter
b. 5.5 cm	=		= 5.5×10^{-2} meters
c. 5.5 km	=		
d. 1 mm	=		

6. Write in long form and scientific notation the equivalent of 4.7 Gyr.

7. Give the significant figures for the following:

   a. 2.2×10^2

   b. 1500

   c. 0.028 or 2.8×10^{-2}

   d. 0.0440

   e. 125.00

8. Complete the sentence. The number of significant figures in a number is an indication of the _____; the more significant figures, the more _____ is the value.

PART B. BASIC CALCULATIONS AND CONVERSIONS

Calculations

Percentage. This calculation determines some part of a given amount or quantity. The simplest example for most of us is our score on a test consisting of 100 questions. If we got 70 questions correct we would have 70 out of 100 or 70% correct. We obtain this by taking the 70/100 amount and multiplying it by 100% as follows:

$$\frac{70}{100} \times 100\% = 70\%.$$

In this example we have taken some fraction (70/100) of the 100%. If the total of the points on the test were 106 and the score were the same (70 correct), we would calculate the percentage score by

$$\frac{70}{106} \times 100\% = 66\%.$$

If we were told we got 20 questions wrong of 106, how would we calculate our percentage score? Subtract the 20 from the 106 = 86. Then divide 86 by the total number of questions and multiply by 100% as below:

$$106 - 20 = 86 \text{ (i.e., total questions} - \text{wrong answers} = \text{correct answers)}$$

$$\frac{86}{106} \times 100\% = 81\%$$

The key here is knowing the quantity for which you want to determine the percent.

Often we want to know percentage change in something over a period of time. For example, if the ozone level in the stratosphere over Antarctica in October averaged 200 in 1988 and 180 in 1989, what would the percentage change have been from one year to the next?

We need to compare the change with the original amount. The change was a decrease of 20; that is (200 − 180 = 20). The original amount was 200. Therefore the percentage change is:

$$\frac{20}{200} \times 100\% = 10\%$$

Note that this is not 20/180, which would be dividing by the final, not the original, amount. Dividing by the final amount to determine a percent change is a common error that must be avoided.

As another example, we might ask for a discount of 10% on the price of some quartz crystals that we need for our collection. How would we calculate it?

$$10\% \text{ of } \$200 = (10/100) \times \$200 = \$20$$

which is the amount the price would be lowered, that is, $200 − $20. The new, discounted price is $180.

Remember the following.

1. When determining percent, you are taking a fraction of 100 (per cent, from the Latin, centum [hundred] and means out of 100).

2. When determining percentage increase or decrease in a quantity, use the amount of change to construct the numerator (top) of the fraction and the amount you are comparing to (the original amount) as the denominator (bottom) of the fraction.

Slope or Gradient

The slope of the land surface or a river is often of interest in the study of Earth. It is often inadequate to characterize a slope as "steep" or "gentle," more precision is needed. Slope is calculated by using the difference in elevation between two points (either a rise or a drop if we were traversing from one point to the other point)

and the length or distance between the two points. Both of these quantities can be obtained from a topographic map. Then the relationship of the "rise" divided by the "run" is used to calculate slope. Let's use the data in Figure 3.1 to determine the slope between points A and B, both on contour lines, on a hillslope.

The difference in elevation between A and B is 400 meters obtained from the contour lines (900m − 500m = 400m); the horizontal distance is 2 kilometers, obtained by comparison with the scale for the map. The gradient or slope is a fraction that gives the rate of change from A to B and is determined by:

$$\frac{\text{rise (vertical change)}}{\text{run (horizontal dist.)}} = \frac{400 \text{ meters}}{2 \text{ kilometers}} = \frac{400\text{m}}{2000\text{m}}$$
$$= \frac{1}{5} = 0.2$$

That is, there is a 1m drop for every 5m of horizontal distance or 0.2 m in 1 meter on the map. The units (meters) cancel, making the slope dimensionless. We can think of this as a slope of 1m drop in 5m of horizontal distance. We can convert the ratio of 0.2 to a percent by multiplying as follows (remember that 0.2 = 2/10):

$$0.2 \times 100\% = 2/10 \times 100\% = 20\%$$

Sometimes slope is expressed in the units actually measured on the map. For example, there might be a drop of 150 feet in the elevation of a road between the center of the city and the river's edge over a distance of 3 miles. The slope or gradient could be expressed in feet/mile as follows:

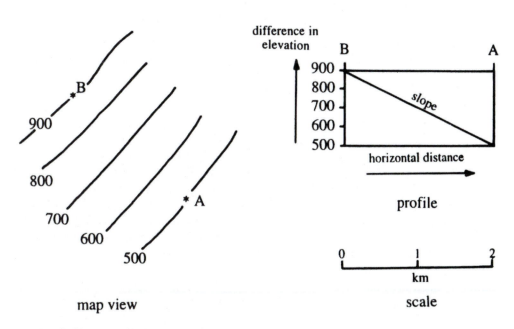

FIGURE 3.1 Map view (left) and profile (right) showing the relative position of Points A and B for gradient determination.

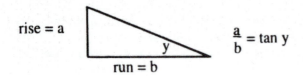

FIGURE 3.2   Tangent (opposite side/adjacent side) of angle y is a/b.

150ft/3 miles = 50 feet/mile or 50 feet per mile.

For those who recall geometry and trigonometry, this relationship of rise over the run is the tangent of the slope angle. In the example in Figure 3.2, it is the angle y.

rise = a; run = b

a/b = tan y

We can convert this fraction for slope between A and B in Figure 3.1 to a slope angle by using the inverse tan function on a calculator, which may be labeled $\tan^{-1}$. Slope = 400/2000 = 0.2 (0.2 is the tangent of the slope angle).

The angle that has a tangent of 0.2 is about 11.3 degrees or 11 degrees 19 minutes.

Rates

Although the determination of slope (above) is really a rate calculation (how does the elevation change with distance), we often think of rates as change over time. The rate of movement of your car is given in miles per hour (mph or miles/hour) or in km/hr. The general relationship is velocity = distance/time. The slash or the "per" indicates one quantity over the other, in this case rate. The velocity of water in a stream or the movement of debris in a landslide is usually described in feet per second, meters per second, or miles per hour. In Figure 1.4, the ranges in rates for many geologic processes are given in m/s, mm/s, km/y, m/y, and mm/y. These could be written as mm/yr or even mm yr^{-1}, where the −1 superscript for the yr indicates 1/yr and means per year.

In addition, the rate of use of a quantity such as miles per gallon or liters per kilometer in transportation is familiar to most of us. In the geosciences we also speak of the volume of material per time in the discharge of river. For example, river flow in English units is measured using cubic feet per second, or cfs, units.

Conversions

When working practical problems, measurements in many different units may need to be combined to arrive at an answer. In so doing we need to be able to convert the units and the amounts of the quantities readily, without becoming confused. Fortunately, scientists and engineers have devised a simple way to keep these numbers and their units straight. It sometimes is described as dimensional analysis because we are measuring or calculating the dimensions of a feature, although that term has other meanings and it is based on the need to arrive at the answer in specific units. To make unit conversions, which are sometimes lengthy and complicated, we set up a dimensional equation, with the given quantity and units on the left and the desired units (the ones we wish to have after the conversion) on the right.

A familiar example of this conversion technique is given by the following problem. If we wish to convert 1 hour into seconds and we know the number of minutes in an hour and the number of seconds in a minute we can set up a relationship to do the conversion. This relationship covers the numbers and the units portions of our data. All except the unit you want (what you will have after the conversion) cancel each other. Here we use seconds, minutes, hours as the units of the quantity we are working with. We use the rule (as in algebra) that zeros and the same units on the top and bottom of the same side of an equation can be cancelled or simplified. Although the answer to this problem can be done in your head, let us see what the relationship would look like for the conversion of hours to seconds.

1. List the unit you are seeking as part of a proportionality on the right side of the page (below it is seconds, given in italics).

2. Set up the relation to convert to this unit by elimination of the other units by cancellation.

3. Carry over and reduce the numerical parts of the relationship where possible.

Note: Although we can only add and subtract like values, we can combine and convert units by multiplying and dividing as in the case below.

1 hour = 1 hr × (60 min/1 hr) × (60 *sec*/1 min)
= 60 × 60 *sec* = 3,600 seconds

Note that hours and minutes both cancel out as units during this conversion; the answer is in seconds.

We could mix dimensions such as distance and time (miles and hours) as we change miles per hour to meters per second:

60 miles/hour = (60 miles/hr) × (1 hr/3600 sec) × (1 km/0.6214 miles) × 1000 m/1 km = 26.8 m/sec

We could change mass and volume dimensions as we seek to determine the weight of a cubic foot of gold, as described below. The density of gold is 19.3

grams/cubic centimeter, which can also be written as gm/cc or gm/cm^3. We want to convert this to pounds per cubic foot (or lbs/ft^3), so we set up these units on the right side of the relationship (in italics). We also need to know the conversions between pounds and kilograms and meters and feet.

$$19.3 \text{ gm/cc} = (19.3 \text{ gm/1cc}) \times (1 \text{ lb/0.4536 kg})$$

$$\times (1 \text{ kg/1000 gm}) \times (1 \text{ cc/0.061 in}^3) \times (1728 \text{ in}^3/1 \text{ ft}^3)$$

$$= 1205 \text{ lbs/ft}^3$$

QUESTIONS 3, PART B

1. You had 23 questions incorrect on an exam that had a total of 96 questions. What was your percentage score? (Show how you set up the relationship.)

2. You had a 10 percent deduction in the price of a book that had a list price of $22.50. What was your cost, before taxes? (Show how you set up the relationship.)

3. Ozone levels increased from 250 to 275; what was the percentage increase? (Show how you set up the relationship.)

4. Ozone levels decreased from 400 to 300; what was the percentage decline? (Show your work.)

5. What is the slope between two points on a map, B and S, given that B has an elevation of 150 feet and S has an elevation of 100 feet above sea level and they are 1 mile apart?

a. in feet per mile?

b. in feet per feet (or without dimensions)?

c. in percent?

d. in degrees?

6. What is the gradient or slope of the following?

a. slope of 5°?

b. slope of 50%?

7. Convert the following, showing your work using a dimensional relationship.
a. 12 miles to feet

b. 15 m to cm

c. 10 km to miles

d. 5 miles to kilometers

e. 12 cubic feet/sec to m^3/sec

f. 60 miles per hour to km/hr

g. gm/cc to lbs/ft^3

PART C. GRAPHS

Graphs provide insight to relationships and patterns between sets of data that would be difficult to discover or describe in tabular or text form. Although calculators and software packages for computers (spreadsheets) have the capability of producing graphs, we explore the construction and interpretation of three simple graphs. We also look at the techniques for taking data from published graphs and guidelines for making titles for graphs.

For most graphs, each point represents a pair of numbers. They are read or plotted from the horizontal (x axis) and vertical (y axis) axes, both of which must have labels and values. The axial scales may be of arithmetic or logarithmic, meaning that the numbered units on the axial scales are uniform or equally spaced (10, 20, and 30) or they are not uniform in size (1, 10, 100), respectively.

Simple Graphs

A *bar chart* or histogram is a series of rectangular blocks, the area of which represents the quantity of an item. Usually, the width of each block is constant, so the height represents the quantity of the item.

An *x-y plot* consists of an array of data points or a line that joins the points. The line may also be a "best-fit" line, which shows the approximate trend of the data points. Each point on an x-y plot has two values, one value that is plotted for the x-axis, and one value that is plotted for the y-axis. An x-y plot is a good way to show the relationship between two quantities and often a best fit line can be used to make estimates where data are missing. Usually, the scales on x-y graphs are arithmetic (also known as linear); however, these graphs can also be constructed using logarithmic scales on one or both axes. Logarithmic scales save space for quantities that have wide variation and also produce straight lines that can improve predictions.

Selecting Scales

In drawing graphs it is important to select a scale that includes the range of point values expected or included in the data. The scale chosen must also be able to be plotted easily on available graph paper. One technique for including a wide range of data on a scale is to break the axis with a zig-zag line and change scales; another is to use a logarithmic scale. Axes scales must be labeled to give the values and the units. Scientific notation may be used to reduce the size of the numbers needed on axis labels.

Logarithmic scales are usually recognized by the variable grid pattern made by the changing size of the units. The scales usually increase by units of 10. Logarithmic scales are often used in plotting quantities such as stream discharge measurements, grain size analyses of sands and sediments, and Richter magnitudes. These are all units that have broad ranges of values.

Writing Captions

All graphs must have a caption that explains what the graph shows, describes and defines symbols used for points and lines, identifies the source(s) of data shown, and indicates when the data were collected. If the graph is based on an earlier graph or graphs, the source(s) of those earlier graphs must be acknowledged in the title. In scientific papers, graphs are usually categorized as figures. Figures are numbered for reference in the text of scientific papers or books. Captions are typically placed below figures but above tables. In some figures the key to the symbols used is included in the figure. Being able to write a clear caption for graphs you construct is an important aspect of understanding the meaning of the graphs.

Data Sources

Gathering data (which is a plural word; the singular word is datum) to plot in graphs and interpret is an important aspect of the scientific method. Data in this book are available from measurements or information that you might make in lab exercises. Some data are provided in text descriptions in the book, and more data are available from textbooks. Data are also available from the Internet, but it is necessary to be very careful using Internet data, as much of it is of poor quality. With computer-based data plotting programs such as Excel, it is possible to select data from many sources and plot them in a wide variety of formats. It is important to finally select the graph style that is most informative in conveying the information you desire.

Plotting Lines

Joining the points on an x-y graph with a line may be appropriate for showing the exact number of sales per month or the calculated variation of mass with volume. In experimental results, particularly when the experiment involves measurement of some aspect of nature, scientific uncertainty often precludes joining the points with an exactly straight line. Because it is reasonable to assume there is a relationship between two quantities that we are plotting, a straight line is often drawn through the array of points, balancing those above with those below. We can do this "best-fit" straight line by eye (that is, estimating what the straight line that comes closest to being aligned with all the points looks like) or the line can be calculated using a hand calculator. If we had joined each point, it would suggest an irregular relationship or no uncertainty in our measurements.

QUESTIONS 3, PART C

1. On the two types of graph paper provided here, plot the number of landslides by year, using the data in Table 3.1. Add captions.

TABLE 3.1 Total number of landslides, by year, in Creepville, for the period from 1991–2001

Number of Landslides	Year
20	1991
6	1992
7	1993
10	1994
8	1995
12	1996
45	1997
29	1998
47	1999
30	2000
10	2001

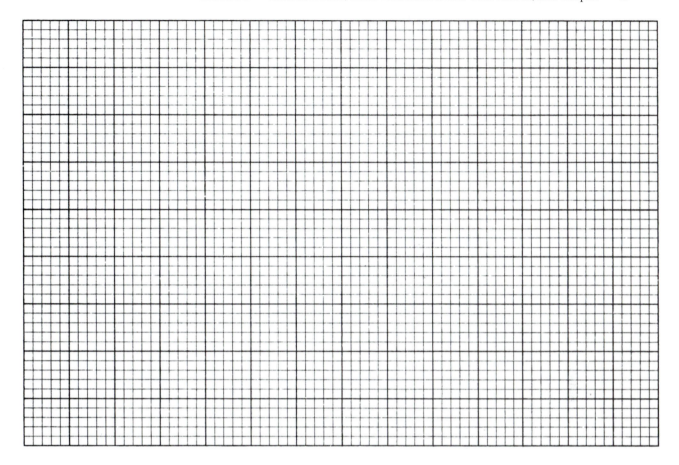

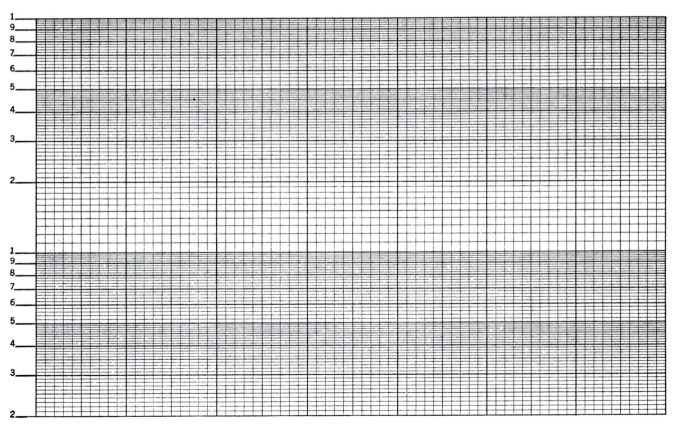

2. From the data in Table 3.1, determine the following for the number of landslides per year.

    a. Mean (the arithmetic average; divide total number of landslides by number of years)

b. Median (the middle value; with equal number of values above and below)

c. Mode (the most common; what value occurred most frequently)

Bibliography

Fenner, P., (ed.), 1972, *Quantitative geology*. Geological Society of America, Special Paper 146, 101 p.

Jones, G. E., 1995, *How to lie with charts*: San Francisco, Sybex, 279 p.

Nelson, R. A., 1995, Guide for metric practice: *Physics Today*, v. 48, no. 8, p. BG15–16

Youden, W. J., 1985, *Experimentation and measurement*: Washington, DC, National Science Teachers Association, 97 p.

II. Introduction to Geologic Hazards

INTRODUCTION

"Civilization exists by geological consent, subject to change without notice."

—WILL DURANT

Humans and our societies are ultimately dependent upon geology. Our food is grown from soils that are the weathering product of rocks, our bodies need water that moves in the hydrologic cycle, and we build using either trees grown on soils, metals and other resources mined from the earth, or plastics that come from oil and other geological products. Global energy resources are primarily oil, natural gas, coal or nuclear; these are all geological. In brief, what we eat, what we wear, where we live, how we work, and how we move about are all ultimately dependent upon geology.

One aspect of living with geology, however, is that there are hazardous geologic events such as earthquakes, volcanoes, or floods that can become human disasters if we choose not to live wisely and be aware of them. A **geologic hazard** is a geologic condition, process or potential event that poses a threat to the human colony (safety, health, structures, functions of society and the economy). A natural hazard (of which geologic hazard is one category) may be defined as a natural geologic or biologic process or condition that has the potential to harm humans or their structures. Section II provides an introduction to the nature, occurrence and mitigation of most of the important geologic hazards.

It is not possible to include all hazards in the exercises selected for this book. Biological hazards (e.g., plagues and disease, pests, and overpopulation by humans) and most weather-related geophysical hazards (e.g., tornadoes, lightning, extreme heat and cold, winter storms, and wildland fires) are omitted. Also omitted are magnetic storms, asteroid impacts, and most technological hazards (e.g., air pollution,

automobiles, nuclear power plant emergencies, and terrorism activities). And there has not been room to include some of the very slow or chronic geologic processes, such as wind and water erosion of soil, permafrost degradation and desertification. We realize that all these hazards are important and interconnected. This is most obvious when considering the impact that global climate change (GCC) has on processes in the Earth system. The current warming of the globe is a geologic and geophysical hazard that impacts surficial geologic processes and the biosphere. GCC also has a technological component because of the human systems that are changing the chemistry of the atmosphere. *Global warming is the ultimate geologic hazard* that impacts everything from local land-use planning to global long-range planning; we include it in the final exercise of the manual. And as we look at planning we realize that many of the natural hazards are not completely natural. They occur because human activity modifies a component of the Earth system (e.g., triggers landslides by overuse of water on slopes) or selects hazardous sites for building because of poor land-use planning and enforcement (Wijkman and Timberlake, 1984). The changing impact of human activity on the Earth system is driven by changes in human population and per capita consumption.

Geography and geologic hazards globally and in the USA

No one is entirely without risk from geologic hazards. They are everywhere but do exhibit geographic variations. On a global scale earthquakes and volcanoes show definite patterns of occurrence (Figures II.1 A and B). At the national scale, geographic variation of natural hazards is well illustrated by expansive soils, those that

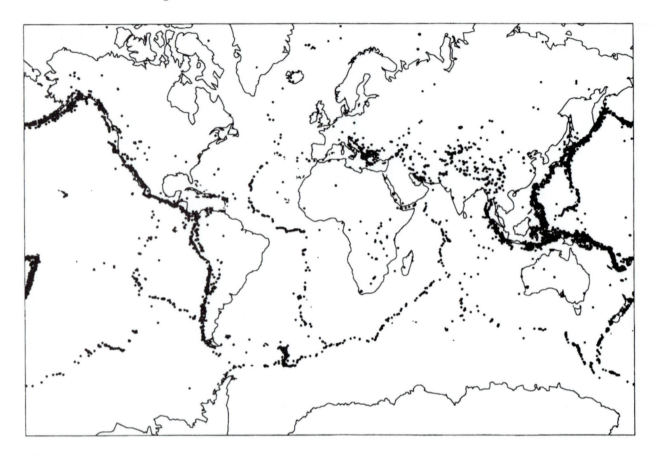

FIGURE II.1A   Map showing global distribution of earthquakes. (from *Yeats, Sieh, and Allen*, 1997)

cause damage to structures by expanding and contracting with available moisture. This soil problem is related to the mineral materials and chemistry of the soils. Soils also exhibit regional trace element differences that can contribute to excesses and deficiencies in plants and animals, thus creating a potential health hazard. Other hazards, such as floods, can occur almost anywhere that people live or work. It is important, therefore, that each citizen be aware of geologic events that exist where they live, and be prepared to cope with these events, should they occur and become hazards.

In this Introduction to Geologic Hazards, we include two sets of questions. The first set is related to the general nature and types of hazards, their occurrence on a global scale, and the key concepts of hazards. The second set of questions explores disasters in the USA: types, locations, and relationships between population, population density and number of disasters (Presidential Disaster Declarations).

Key concepts about geologic hazards

When working the exercises in Section II, it will be helpful to keep the following concepts in mind.

Geologic events are natural. The events become hazards and disasters when people chose to live unwisely in areas that can be impacted. For instance, much flood damage and many coastal problems can be avoided by not living on floodplains (or "in the river" as some geologists suggest) or living directly on the changing coast.

People can avoid or reduce the risks that the hazards present. Understanding why geologic events happen in the way they do, when they are likely to happen, where they are likely to happen and the impacts they can have on humans is a major goal of these exercises.

Geologic events may not be predictable. Geologists distinguish "predictions" (3 pm Friday) from "forecasts" (there is a 20% chance sometime in the next 30 years). Some events, such as some landslides are more likely to occur in wet seasons (i.e., they have rainfall as a precursor) and therefore general statements about the times of their occurrence, or a forecast, may be made. Other events, such as earthquakes, have proven to be very difficult to predict. People need to be prepared, therefore, to cope with events that may occur unexpectedly.

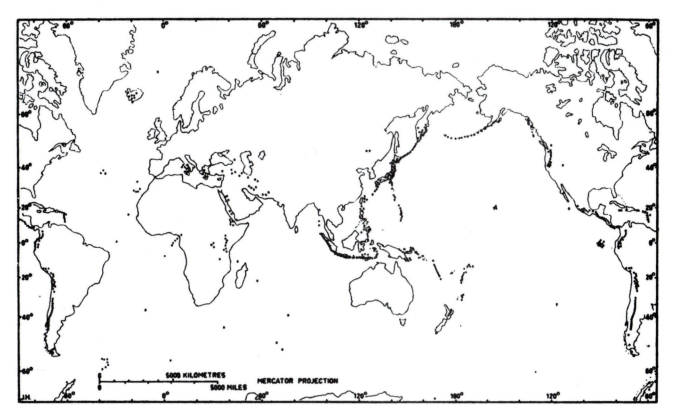

FIGURE II.1B   Map showing global distribution of volcanoes.

(modified from *Ollier*, 1999)

Geology is three-dimensional. Normally humans just look at the surface of the earth, but in order to understand what happens at the surface of the earth, we must also appreciate what is happening below the surface.

Time is very important in geology. Geologists take a long-term view of events. Even though most of the time an earthquake may not be occurring at a particular site, if one has occurred in the past, it is reasonable to expect one again in the future. So too, even though a flood along a particular river may not be every year, unusual events occur and people who are not prepared suffer. Geologic events may have a frequency (how often they occur; how many years between occurrences).

Rare geologic events can cause great changes. Although coasts are dynamic and regularly changing, storms provide dramatic punctuation to the daily equilibrium and in a day or so can bring changes that exceed many years of normal processes. Volcanoes, dormant for long periods of time, may have new eruptions that greatly change the landscape.

Different geologic processes have different rates. Some events, like snow avalanches or earthquakes, may be over in just a few seconds. Other processes, like floods, may take days or weeks. See Figure 1.4.

Geologic setting is important. The more we can learn about the geologic locations, conditions, and environments of past or current geologic events, the better prepared we can be for future occurrences. For instance, many landslides take place at sites of previous landslides. The tragic debris flow that wiped out Armero, Columbia in 1985 provides a useful lesson for people living near volcanoes throughout the world.

Different geologic events have different magnitudes. Some events, like floods, may impact only one area along a river. Even the disastrous floods along the Mississippi River in the US in 1993 did not have a major impact in many towns, once the water flowed downstream and the river channel was large enough to hold the discharge. Other events, like a major volcanic eruption, may have global impacts on climate. Even small events, such as a landslide, may have disastrous impacts on individuals, however.

Geology is different from other sciences. Geologists focus on interpreting the historical record of the earth. Environmental geologists seek to apply our understanding of this record to the needs of human societies by integrating knowledge from many disciplines. It is not always possible for geologists to go to a lab and determine the answer to questions. Geologic questions and answers ultimately come field observations and investigations.

Humans impact geology. Through diverting rivers, undercutting slopes, drying or wetting soils, extracting water, or many other activities, humans have the power to create geologic events where they might not have otherwise occurred. The deep injection of waste fluids at the Rocky Mountain Arsenal near Denver actually triggered earthquakes. Floods in urban areas may be worse because of the type of development that has occurred.

Change is natural and normal. Rivers meander and change their courses. People who live near a river should expect that natural change will continue, and the river might move from its present location. This change is predictable but often unwelcome, especially to landowners near the river. Human activities, however, can also increase the rate of change. Many coastal structures have impacts far beyond the site they are designed to protect.

Population growth places more people at risk. As population increases, more people live and work in areas that are subject to geologic hazards. If hazardous events occur, more people are impacted. In developed countries, the impacts are largely monetary; in developing countries, the impacts are more often measured in loss of lives.

QUESTIONS INTRODUCTION II GEOLOGIC HAZARDS

1. Give the definition of a geologic hazard (in your own words or that provided in the Introduction).

2. List 5 geologic hazards.

3. What types of hazards have been omitted from this book?

4. Can human activity induce a geologic hazard? _____ Explain your answer.

5. a. How many "Key Concepts about Geologic Hazards" are listed in the Introduction?

b. List the three that you think are the most important.

6. a. From Figure II.1, which geologic process or hazards map would be the most useful for delineating tectonic plate boundaries?

b. Explain why there are places where earthquakes occur but no volcanoes are shown.

c. On the most useful map for showing plate boundaries, delineate the boundaries by playing "connect-the-dots". Show as many plates as you can.

7. Briefly explain Will Durant's statement at the beginning of the Introduction.

Geologic events and natural disasters in the US

No state (and indeed, nowhere in the world) is completely free from geologic or other natural hazards. In this section we will look at the types and distributions of some of the more common hazards that are found in the United States. We then compare the types of hazards with their frequency, as measured by Presidential Disaster Declarations. The President may declare disasters for many reasons, including geological events. A disaster declaration is made when the large and severe events occur that have broad impacts, and when these events are beyond the capabilities of state and local government to respond. The tremendous damage from wind and flooding caused by hurricane Katrina is an example of an event that met all the criteria for a disaster declaration.

Figure II.2 is a map of the United States that identifies some of the more common hazardous geologic events in each state.

1. a. What is your home state? _____

b. Find your home state on Figure II.2, and list its common hazardous geologic events.

2. Do you think that you personally live in an area of the state that is subject to these hazards? Why or why not? If so, which hazards?

3. Identify states that have volcanic hazards. If winds are blowing from the west to the east, is your area subject to ash from a major eruption?

4. In 1811 and 1812, a series of major earthquakes in New Madrid, Missouri (near the southeastern corner of Missouri) rang church bells along the east coast.
   a. Use a compass and draw on Figure II.2 a circle centered on southeastern Missouri with a radius that is the distance from New Madrid to Boston.

Selected Geologic Hazards by State

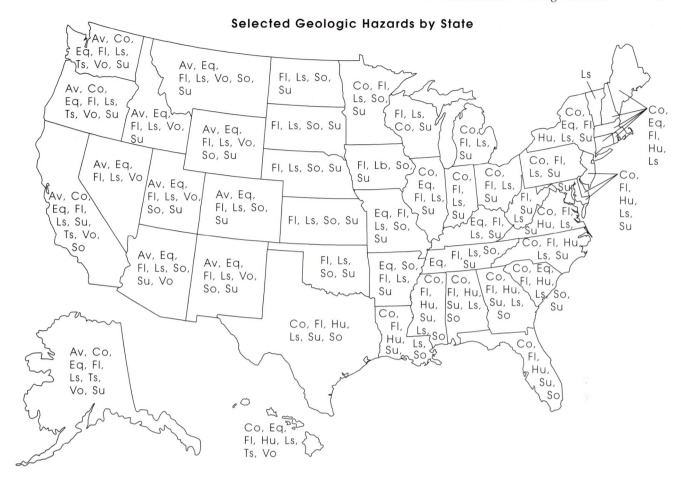

FIGURE II.2 Common geohazards by state: Av, avalanche; Co, coastal; Eq, earthquakes; Fl, floods; Hu, hurricanes; Ls, landslides; Ts, tsunamis; Vo, volcanoes; So, soils; Su, subsidence.

b. If you live in this circle, what kinds of seismic hazards might exist in your state if there were a repeat of the major New Madrid earthquakes of 1811 and 1812?

c. Are earthquakes a major hazard in your state, according to Figure II.2?

Figure II.3 (in color plate section) is a map that shows Presidential Disaster Declarations (PDDs). Find your county on this map. What is its name _____ and what is the name of the state _____?

5. From 1965 through 2000, how many PDDs were there in your county?

6. From your knowledge, does the number of PDDs correspond or not with the number of potentially hazardous geologic events in your county?

7. What kinds of geologic or other events do you think might have led to disaster declarations in your county or state?

8. a. Do you think that there might or might not be a correspondence between the number and types of geologic hazards in your state and the number of PDDs (Figure II.3)?

b. Why?

9. Figure II.4 (also in the colored plates section) is a map of population densities.
   a. Describe what, if any, relationship exists between disaster declarations shown on Figure II.3 and population density shown on Figure II.4.

b. Identify some areas where there are lots of disaster declarations but few people.

c. Identify some areas where there are few disaster declarations but lots of people.

Figure II.5 is a graph that shows annual numbers of Presidential Disaster Declarations. Table II.1 lists US population growth for the past few decades.
10. Describe the trend (increasing, decreasing, or similar terms) of the annual number of Presidential Disaster Declarations between 1953 and 2007.

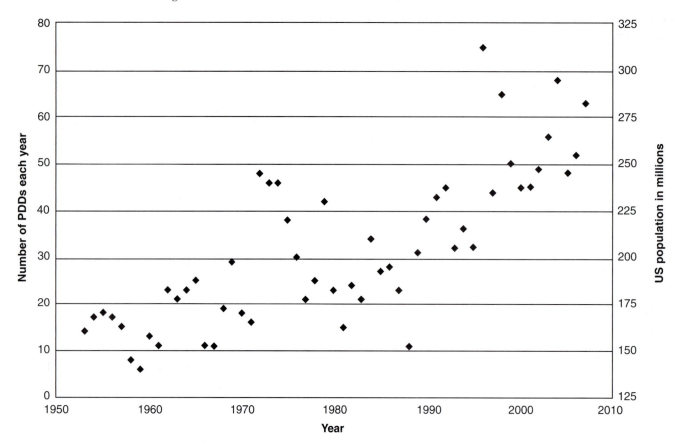

FIGURE II.5 Annual number of Presidential Disaster Declarations (PDDs), 1953-2007. Disaster declaration data from http://www.fema .gov/news/disasters_totals_annual.fema (downloaded August 10, 2008). Population data from http://www.census.gov.

11. Refer to the data in Table II.1. Use the scale on the right side of the graph and plot these points on the graph. Draw a line that connects the points you placed on the graph. What is the correlation between the number of Presidential Disaster Declarations and population growth?

Year	US Population in millions
2007	303
2000	282
1990	250
1980	228
1970	205
1960	181
1950	152

Table II.1, US population in millions of people (data from US Census Bureau).

12. List several strategies that might help decrease the number of Presidential Disaster Declarations. It may be helpful to think again about this question near the end of your course.

Geologic hazards and plate tectonics

13. Refer back to Figure II.2, the state map of hazards, and to your map of plate boundaries that are drawn on Figure II.1.
   a. Where is your state located in relationship to plate boundaries?

   b. What geologic hazards are more typically found near plate boundaries?

   c. What hazards are found typically away from plate boundaries?

14. Refer back to Figure II.3, the map of Presidential Disaster Declarations, and to your map of plate boundaries. Describe and explain the relationship, if any, of PDDs with plate boundaries. For example, is the number of PDDs near plate boundaries greater, about the same, or lower than the number of PDDs away from plate boundaries?

Bibliography

Ackerman, J., 1997, Islands at the edge: National Geographic, Vol. 192, no. 2 (August), p. 2–31

Gerstel, W. J., Brunengo, M. J., Lingley, W. S., Jr., Logan, R. L., Shipman, H., and Walsh, T. J., 1997, Puget Sound bluffs: The where, why, and when of landslides following the holiday 1996/97 storms: Washington Geology, Vol. 25, No. 1, p. 17–31

Keller, E. A., 1996, Environmental geology (7th ed.): Upper Saddle River, NJ, Prentice Hall, 560 p.

Mount, J. F., 1995, California Rivers and Streams, the conflict between fluvial process and land use: University of California Press, Berkeley, CA, 359 p.

Nuhfer, E. B., Proctor, R. J., and Moser, P. H., with 12 others, 1993, The Citizens Guide to Geologic Hazards: American Institute of Professional Geologists, Arvada, CO, 134 p.

Ollier, C., 1988, Volcanoes: Basil Blackwell, Inc., New York, NY, 228 p.

Parfit, M., 1998, Living with Natural Hazards: National Geographic, v194, no.7, p 2–39

Radbruch-Hall, D. H., 1981, Overview map of expansive soils in the conterminous United States, in, Hays, W. W., ed., Facing Geologic and Hydrologic Hazards: U. S. Geological Survey Professional Paper 1240–B, p. B69

Rahn, P. H., 1996, Engineering geology, an environmental approach: Prentice Hall, Upper Saddle River, NJ, 657 p.

Smithsonian Institution/SEAN, 1989, Global volcanism 1975–1985: Prentice-Hall, Englewood Cliffs, NJ, and American Geophysical Union, Washington, DC, 657 p.

Tyler, M. B., 1994, Look before you build–geologic studies for safer land development in the San Francisco Bay area: U. S. Geological Survey Circular 1130, 54 p.

Wijkman, A., and Timberlake, L., 1988, Natural disasters, acts of God or acts of man?: New Society Publishers, Philadelphia, PA, 144 p.

Wijkman, A. and Timberlake, L., 1984, Natural Disasters: Acts of God or Acts of Man: Washington DC, Earthscan Paperback.

Wilshire, H. G., Howard, K. A., Wentworth, C. M., and Gibbons, H., 1996, Geologic processes at the land surface: U.S. Geological Survey Bulletin 2149, 41 p.

Yeats, R. S., Sieh, K., and Allen, C.R., 1997, The geology of earthquakes: Oxford University Press, New York, 568 p.

Volcanoes and Volcanic Hazards

INTRODUCTION

The objective of this exercise is to investigate the different types of volcanoes, volcanic products, and volcanic hazards that exist in the United States. Even though you may not live in an area with active volcanoes, given appropriate wind directions and large enough eruptions, all citizens of the United States live downwind from volcanoes (Wright and Pierson, 1991).

There are two major types of volcanoes that typically present hazards in the United States: basaltic *shield volcanoes,* as found in Hawaii; and *composite volcanoes* (also known as stratovolcanoes), as found in the Cascade Mountains of Washington, Oregon, and California, and in Alaska. It is unusual for there to be more than a few years without an eruption from a Hawaiian volcano (Tilling, Heliker, and Wright, 1987), and the eruption of Pu'u O'o has had more than 50 eruptive episodes over nearly a quarter century. In the Cascade Range, Lassen Peak erupted in 1914–15, and Mount St. Helens had a major eruption in 1980, which was followed by smaller eruptions into 1986. After nearly a 20-year pause, Mount St. Helens rejuvenated in 2004 with an ongoing series of dome-building eruptions. Mount St. Helens is explored in more detail in the next exercise.

Shield volcanoes are relatively flat and composite volcanoes are often conical. The shapes are related to the viscosity of the magma and the eruptive process. Viscosity is a measure of how easily a substance flows; it depends on chemical composition, temperature, and gas content of the magma. A low-viscosity substance will be very fluid, like water, while a high-viscosity substance will be very thick, like molasses or honey.

Most magmas are composed primarily of silica, with lesser amounts of other elements. Basaltic magma has comparatively low silica (approximately 50 percent SiO_2), andesitic magma has an intermediate silica content (approximately 60 percent SiO_2), dacitic magma has a slightly higher silica content (approximately 65 percent SiO_2), and rhyolitic (also know as silicic)

magma has silica concentrations that range up to about 75 percent SiO_2. The relationships among magma composition, magma viscosity, volcano form, and typical eruption products are given in Table 4.1.

In addition to shield and composite volcanoes, there are several other types of volcanoes that erupt less frequently and have potential impacts that range from minor to catastrophic. These include both relatively small volcanic domes and large calderas (volcanoes that collapse as a result of an eruption). Pyroclastic flow eruptions take place typically in volcanoes with dacitic or rhyolitic magma, and can occur from the collapse of small domes or erupt from large calderas, and are composed of volcanic ash (tephra) fragments that are transported in a hot, gas-rich cloud. Pyroclastic flows are the most devastating type of volcanic eruption, as they can travel at speeds of over 100 km per hour with temperatures above 500°C. Exceptionally large but fortunately rare eruptions known as "supervolcanoes" can have worldwide impacts.

Dormant volcanoes can also be hazardous. Landslides and debris flows can be triggered on steep slopes by earthquakes or rocks that have been weakened by chemical actions of hot water and steam can suddenly slide without warning. Weakened ground can suddenly collapse, and volcanoes, even when not erupting, can give off gases.

Figure 4.1 is a simplified sketch of a composite volcano and its associated hazardous phenomena. Table 4.2 describes volcanic products and their hazards in greater detail.

The first part of this exercise explores types of volcanoes using data presented in Tables 4.1 and 4.2, Figure 4.1, and this introduction. Then we look at volcanic activity associated with Yellowstone National Park, which is a large rhyolitic volcanic center; Mount Rainier, which is typical of volcanoes of the andesitic volcanic centers of the Cascade Range; and the active volcanoes of Hawaii, which have basaltic composition.

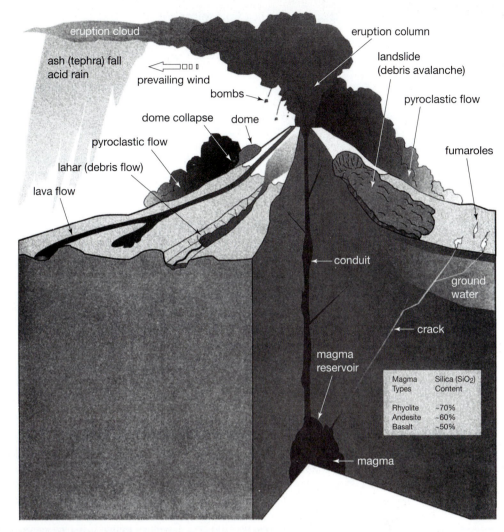

FIGURE 4.1   Diagram of a typical composite volcano, illustrating the locations and extent of selected volcanic hazards. Courtesy of Bobbie Meyers, U.S. Geological Survey

TABLE 4.1 Relationships among Magmatic Composition, Viscosity, Volcano Form, and Typical Volcanic Eruption Products for Different Types of Volcanoes

Magma Composition	Relative Viscosity	Volcano Form	Typical Eruption Products
Rhyolitic (approximately 70–75% silica)	High (sticky) for lavas; gas-charged pyroclastic flows (hot ash clouds) can be very mobile	Smaller domes, local flows, or large calderas from explosive eruptions	Pyroclastic flows, lava flows, lateral blasts, tephra, lava domes, gases, lahars (especially if caldera-filling lakes are suddenly drained)
Dacitic (approximately 65% silica)	Intermediate to high	Composite volcanoes, domes	Intermediate between rhyolitic and andesitic: pyroclastic flows, lava flows, lateral blasts, tephra, lava domes, lahars, gases
Andesitic (approximately 60% silica)	Intermediate	Conical composite volcanoes	Lava flows, tephra, pyroclastic flows, landslides, lahars, gases
Basaltic (approximately 50% silica)	Low (fluid)	Shields and rifts; cones where formed of tephra	Lava flows, tephra, gases

Note that all volcanoes can erupt from central vents or from fracture zones on the flanks of the volcano. Calderas (volcanic collapse areas) can form in all types of volcanoes, with their size controlled by the amount of new magma erupted.

TABLE 4.2 Volcanic Products and Their Hazards (modified from Miller, 1989)

Volcanic Products	Processes that Create Volcanic Products and Characteristics of Volcanic Hazards	Physical Locations of Volcanic Hazards
Volcanic landslides (debris avalanches)	Small rock falls to large failures of volcanic slopes. May be triggered without eruption from steep slopes composed of weak or thermally altered rock. Move at high speed on steep slopes and slow when they flatten out.	Partly topographically controlled[1] on flanks of volcanoes. Large debris avalanches can cover broad areas of low-angle slopes up to 50 km or more from volcanoes and may extend further down valleys.
Pyroclastic flows	Eruption of fragments of hot rock or explosion or collapse of a lava flow or lava dome. Ash fragments transported in gas-rich cloud move away from volcano at tens to >100 km/hr.	Mostly topographically controlled. Effects may extend 40 km downslope and 65 km or more down valleys. Adjacent areas may be affected for many km by hot ash clouds.
Lateral blasts	Explosive ejection of rock fragments from the side of the volcano. Sudden; debris moves at speeds up to hundreds of km/hr.	Distribution of blast controlled by direction of blast. Blasts may impact a 180° sector up to tens of km away from vent.
Lava flows	Eruption of molten lava, which flows downslope slowly. Fronts of flows on low-gradient slopes may move slower than a person can walk.	Most flows follow topographic lows and reach <10 km from the vent. Can pond and fill valleys if enough lava erupts.
Lava domes	Result from nonexplosive, high-viscosity (sticky) lava, which is erupted slowly and accumulates above the vent.	Domes typically limited to above and within a few km of a vent, usually in areas of rhyolitic or dacitic volcanism. May also collapse or explode, creating pyroclastic flows.
Lahars (debris flows)	From eruption onto snow and ice or into rivers; also from eruptive displacement of crater lakes or from heavy rain falling on loose pyroclastic debris. Sudden; move at tens of km/hr.	Primarily follow rivers from volcano flanks downstream. Impacts may extend many tens of km from source area. Can occur years after an eruption, if loose ash persists. Can dam side streams and create secondary floods.
Floods	Origin similar to lahars (debris flows), which may transition downstream into floods. Commonly move at < 20 km/hr. Jokulhlaup (glacier burst) floods release water from breakage of ice-dammed lakes, which may be subglacial and therefore unknown from surface observation.	Confined primarily to river valleys. May extend for hundreds of km. Can also back up waters in side streams and create secondary floods.
Ash (tephra)	Produced by vertical columns of fragments of magma propelled in part by expanding hot gas. Fine-grained ash can be carried great distances by wind. Eruption is sudden; explosive dispersal at tens of km/hr.	Can blanket all areas near and downwind from volcano; impacts may extend hundreds of km, depending on eruption volume, height, wind speed, and direction; climate change may occur globally. Many large tephra eruptions occur at rhyolitic or dacitic volcanoes.
Gases	Produced by eruptions or magma outgassing. Gases may be hot or cold and commonly contain sulphur, carbon dioxide, and other harmful compounds. Can pool locally if denser than air, or can move away from volcano through eruption or atmospheric dispersion. Carried at speeds of up to tens of km/hr.	Distribution controlled by wind speed and direction. Greatest impacts on air quality near volcano; odor, haze, and mild impacts may extend a few hundred km. If large volumes of gases are released during an eruption, climate impacts may occur globally, e.g., sulfate aerosols may cause cooling and CO_2 and other greenhouse gases (GHG) can cause warming of the atmosphere.
Phreatic and hydrothermal explosions	Usually driven by sudden burst of hot, high-pressure water changing to steam; can occur without direct eruption of magma. Rock fragments blasted out in explosion.	Typically smaller, locally confined eruptions on or near volcano; water must be at boiling temperature underground. If water comes in contact with magma, locally violent eruptions can be created.

1. Topographic control means that areas impacted by a particular product are usually limited to the bottoms and lower sides of valleys, although they may spread out over areas of low topography such as broad slopes or valley bottoms.

QUESTIONS 4, INTRODUCTION

1. Sketch representative topographic cross sections of volcanoes with high-viscosity, intermediate-viscosity, and low-viscosity lavas. Which type of magma is most explosive? Least explosive? Which type of magma has the highest silica content? The lowest?

2. Refer to Figure 4.1 and Table 4.2, and describe briefly the volcanic products and hazards for both people and property that are likely to exist in the following locations.

    a. If people or property are on a ridge top, close to a composite volcano?

    b. If people or property are on a valley bottom close to a composite volcano?

    c. If people or property are on a valley bottom 25 km from a composite volcano?

    d. If people or property are on a ridge top 10 km from a rhyolitic volcano?

3. Do basaltic lava flows present a greater hazard to people or to property? Why?

4. Refer to Figure 4.2. If you live to be 100 years old:

    a. how many eruptions the size of Kilauea or Unzen are likely in your lifetime?

    b. how many eruptions the size of Etna are likely to occur in your lifetime?

    c. how many eruptions the size of Mount St. Helens are likely to occur in your lifetime?

    d. how many eruptions the size of Pinatubo or Katmai are likely to occur in your lifetime?

5. Use the graph below, and build a histogram to illustrate the data you calculated in answering question 4 above. Determine your own vertical scale.

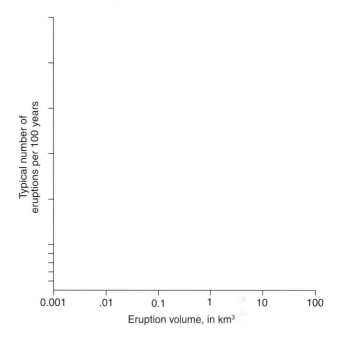

6. From the data on your graph (question number 5), describe the relationship between eruption size and eruption frequency (the typical number of eruptions per 100 years).

PART A. RHYOLITIC VOLCANOES

Rhyolitic volcanoes present many different kinds of volcanic hazards (Table 4.1). Since no major eruptions have occurred recently in the United States, the interpretation of possible hazards from rhyolitic volcanoes in the United States must be based on other data, including eruptions from elsewhere around the world and U.S. eruptions in the geologic past.

    Two of the major rhyolitic volcanic centers in the United States are the Long Valley Caldera in California and the Yellowstone Caldera in Wyoming. These sites are shown in Figure 4.3, which is a map that also shows the distribution of volcanic ash from these volcanic centers.

    Rhyolitic volcanoes may have different scales of eruptions. They may erupt small plugs and domes of rhyolite, or they may erupt as "supervolcanoes" that can have global impacts. Both Long Valley and Yellowstone have had these different scales of eruptions. Geophysical evidence suggests that there is active magma in the subsurface at both volcanoes.

A1. Supervolcano Eruptions

Table 4.3 presents data from Long Valley and Yellowstone eruptions. We focus on one aspect of these eruptions, their tephra distribution. It is important to note,

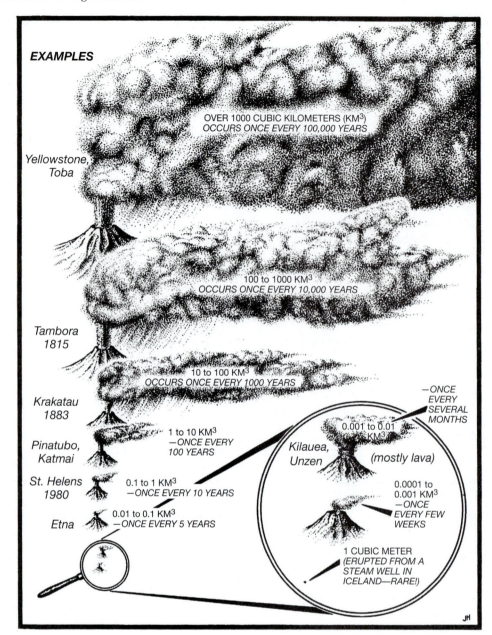

EXAMPLES

Yellowstone, Toba

OVER 1000 CUBIC KILOMETERS (KM³)
OCCURS ONCE EVERY 100,000 YEARS

100 to 1000 KM³
OCCURS ONCE EVERY 10,000 YEARS

Tambora
1815

10 to 100 KM³
OCCURS ONCE EVERY 1000 YEARS

Krakatau
1883

—ONCE
EVERY
SEVERAL
MONTHS

1 to 10 KM³
—ONCE EVERY
100 YEARS

0.001 to 0.01
KM³

Pinatubo,
Katmai

Kilauea,
Unzen   (mostly lava)

St. Helens
1980

0.1 to 1 KM³
—ONCE EVERY 10 YEARS

0.0001 to
0.001 KM³
—ONCE
EVERY FEW
WEEKS

Etna

0.01 to 0.1 KM³
—ONCE EVERY 5 YEARS

1 CUBIC METER
(ERUPTED FROM A
STEAM WELL IN
ICELAND—RARE!)

FIGURE 4.2 Magma volumes from several different volcanic eruptions (from Fischer, Heiken, and Hulen, 1997).

however, that each of these eruptions produced major pyroclastic flows that traveled far from their respective volcanoes. All eruptions created major calderas.

QUESTIONS 4, PART A1

1. How do the sizes of the larger Yellowstone and Long Valley eruptions (Table 4.3) compare with eruptions shown in Figure 4.2?

2. Use the data on Figure 4.2 to help determine the approximate "Long-Term Recurrence Interval" and record your result in Table 4.3.

3. Refer to Figure 4.3. Do you live in an area that has been impacted in the past by tephra (ash) from either volcano?

4. From the ash distribution data, were the wind directions the same for each eruption?

5. Figure 4.3 shows where tephra from Long Valley and Yellowstone are found now. Do you think that the current distribution of tephra is an accurate map of the original distribution? Explain.

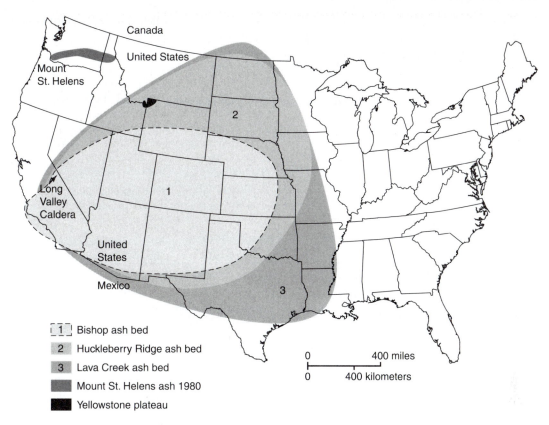

FIGURE 4.3   The distribution of the Huckleberry Ridge and Lava Creek ash beds from Yellowstone, the Bishop ash bed from Long Valley caldera, and Mount St.Helens ash of 1980 (modified from Lowenstern and others, 2005).

TABLE 4.3 Supervolcano Eruptions at Yellowstone and Long Valley

Supervolcano	Age of Eruption	Approximate Size of Eruption	Long-Term Recurrence Interval
Yellowstone	2.1 million years ago	2450 cubic km	
Yellowstone	640,000 years ago	1000 cubic km	
Long Valley	760,000 years ago	625 cubic km	

6. What geologic processes might have changed the distribution of ash since the time of eruption?

7. Given their recurrence interval, do you think it is appropriate to worry about the eruption of a supervolcano? Why or why not?

8. Figure 4.2 shows that the eruption of Tambora in 1815 produced between 100 and 1000 km^3 new material (it has been estimated at about 150 km^3 of new material). This eruption led to "the year without a summer" in some parts of the United States.

   a. The eruption of Yellowstone about 640,000 years ago was how many times larger than the eruption of Tambora?

   b. If Tambora had moderate global climatic impacts, what impacts would a Yellowstone-size eruption have today?

A2. Smaller-Scale Rhyolitic Eruptions and Ongoing Geologic Hazards: Long Valley and Yellowstone

After the cataclysmic eruptions described above, both Long Valley and Yellowstone have had rhyolitic eruptions that have filled the calderas created by the big eruptions. At Long Valley, the youngest eruptions may have been only about 250 years ago. There were larger eruptions about 600 years ago. Volcanism at Yellowstone is much older, with the youngest eruptions having occurred about 70,000 years ago. (Christiansen and others, 2007).

Several data sets, when combined, suggest the interpretation that there is active magma beneath both calderas. Long Valley and Yellowstone both have broad doming occurring that could be related to magmatic movement. Both volcanoes also have seismic patterns that suggest magma. Both also have hot springs (but only Yellowstone has numerous geysers; it also is important to note that many hot springs around the world are not located in areas with active volcanoes).

QUESTIONS 4, PART A2

Long Valley

Figure 4.4 shows the location and timing of recent volcanic activity in the Long Valley area. The named sites on this map have all been active in the last 5,000 years. The population of Lee Vining is approximately 500, the population of June Lake is somewhat lower, and the population of Mammoth Lakes is approximately 7,500. There are seasonal variations in population, however, due to tourism and skiing.

1. How many volcanic eruptions (not including steam blasts) occurred in the past 5,000 years?

2. What is the average recurrence interval between eruptions for the past 5,000 years?

3. How many volcanic eruptions occurred in the past 1,000 years?

4. What is the average recurrence interval between eruptions for the past 1,000 years?

5. According to the U.S. Geological Survey (Hill and others, 1998), it has been about 250 years since there was an eruption at Paoha Island in Mono Lake. Is this area due for an eruption or not? Explain your reasoning.

6. How many steam blasts occurred in the past 5,000 years?

7. a. What is the geographic relation of steam blasts to the 760,000-year-old caldera?

b. What does this relationship imply about the origin of steam blasts?

8. What geologic factors might contribute to there not being any steam blasts identified that are more than 1,000 years old?

9. Is there a trend in the spatial distribution (northern, southern, central region, etc.) of volcanic eruptions over the past

5,000 years that can help predict where the next eruption might take place? Explain.

10. Approximately 600 years ago eruptions from the South Deadman Creek dome included both pyroclastic flows that traveled about 5 km from the vent, and ash that traveled about 15 km from the vent. If this dome erupts again, are any populated areas at risk? If so, which one(s)?

11. If other vents were to have similar eruptions, could they place any of the populated areas at risk? Explain.

Yellowstone

Although Yellowstone has not had a magmatic eruption for the past 70,000 years, it is still geologically very active. With plentiful hot water and substantial heat, Yellowstone is subject to hydrothermal explosions. With active magma at depth, Yellowstone also has the potential for renewed, small-scale volcanic activity. Figure 4.5 is a simplified geologic map of the park.

12. Post-caldera viscous rhyolite flows have flowed up to approximately 10 miles from their vent areas. Refer to Figure 4.6. If a rhyolitic lava eruption were to occur from a vent area at the bottom of the canyon, would it likely present a hazard to developed areas on the canyon rim? Explain.

13. How would the hazards be different at Canyon if the eruption were a pyroclastic flow? Basaltic eruption? Explain.

14. Mammoth (near the north edge of the map) is located in an area of precaldera rocks. There is a series of mapped volcanic vents south of Mammoth. Based on the topography of the area (Figure 4.5), is Mammoth likely to be at risk if one of these vent areas erupts a rhyolite flow? A pyroclastic flow? A basalt flow? Explain.

15. What if there were new volcanic eruptions in the Old Faithful area? What kind of eruption(s) could threaten other developed areas?

16. If a small hydrothermal explosion like Porkchop (Figure 4.7) were to occur, is the damage likely to be widespread?

17. What impacts could a 2-mile-diameter hydrothermal explosion have if it occurred in a populated area?

18. Yellowstone Lake is the large lake on the eastern part of the caldera (Figure 4.5). What impacts might be expected if a 2-mile-diameter hydrothermal explosion occurred under the lake?

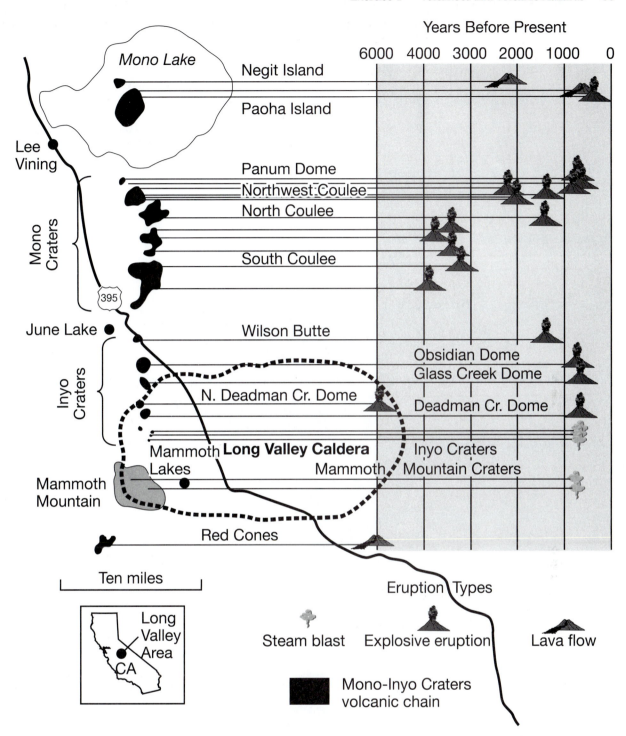

FIGURE 4.4 This map shows the location, time, and type of volcanic eruptions in the Long Valley area over the past 5,000 years. Steam blasts are also known as phreatic (pronounced "free-attic") eruptions, and are created when hot waters explode to steam. They do not involve the eruption of new magma. Explosive eruptions include both relatively small pyroclastic eruptions that extend about 5 miles from their vents, and tephra eruptions. (US Geological Survey, Long Valley Observatory)

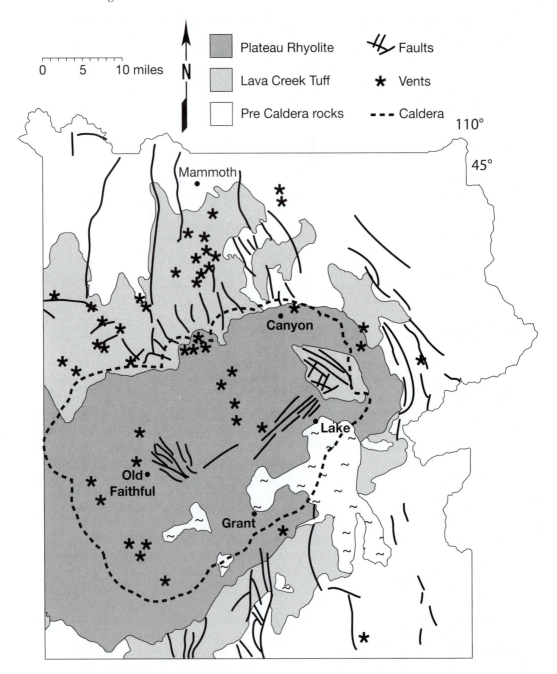

FIGURE 4.5   Simplified geologic map of Yellowstone National Park. The rock units are combined into three major categories. Pre-caldera rocks are those that were deposited before the major eruption 640,000 years ago, and include sedimentary, igneous, and metamorphic rock units. The Lava Creek Tuff is the pyroclastic flow unit that was deposited when the caldera erupted 640,000 years ago. Postcaldera rhyolite flows now fill and extend beyond the hole made by the caldera collapse. Star symbols indicate volcanic vent areas.

(Modified from Christiansen, 2001, Kciffer, 1971, and Lowenstern and others, 2005)

FIGURE 4.6   Aerial photograph of the Canyon of the Yellowstone River. For scale, the waterfall is slightly more than 300 feet (90 m) high. (Photograph © 2008 Duncan Foley, reprinted with permission)

FIGURE 4.7   Porkchop Geyser erupted in a small hydrothermal explosion in 1989. The blocks that surround the pool were thrown out of the ground during the explosion. This is one of the smaller such explosions that have occurred. Other hydrothermal explosions have been documented that are up to 2 miles in diameter. (Photograph © 2008 Duncan Foley, reprinted with permission)

PART B. COMPOSITE VOLCANOES

Mount Rainier, Washington

Mount Rainier (Figure 4.8) is located in western Washington, near the Puget Sound lowlands that include the cities of Tacoma and Seattle (Figure 4.9, in the color map section at the back of the book, is a shaded topographic map of Mount Rainier). The proximity of this active volcano to major population areas has led the U.S. Geological Survey to designate Mount Rainier as potentially one of the most hazardous volcanoes in the world (U.S. Geodynamics Committee, 1994).

Potential geologic hazards from Mount Rainier are particularly great, due to the combination of the height of the volcano (4393 m; 14,410 feet), its active nature, steep upper slopes, and heavy cover of the mountain by glaciers. Hazards at Mount Rainier exist both during and between magma eruptions (Hoblitt and others, 1995; U.S. Geological Survey, 1996).

People who live near a volcano need to know the types of volcanic events that create risks to people and/or property, the probable size of those events, how far the events are likely to extend beyond the volcano and, ideally, when the events will occur. If it cannot be determined when an event is going to occur (i.e., a forecast or prediction cannot be made), it is helpful to know when the last episode of a particular event occurred and how often similar events have occurred in the geologic past. In this section of this exercise we look at past geologic events at Mt. Rainier that have created hazards, and evaluate the risks that exist for people living in nearby towns.

QUESTIONS 4, PART B

1. Table 4.4 lists some hazards (adapted from U. S. Geodynamics Committee, 1994) that have been identified as existing at Mount Rainier. Fill in the blanks in the table, using "H" for high risk, "M" for moderate risk, and "L" for low risk. Refer to Tables 4.1 and 4.2 and Figure 4.1 to help you fill in the blanks in the table. You may also wish to refer to your textbook.

2. A major concern for people living below the slopes of Mount Rainier is the possibility of large landslides and lahars. These could occur without warning during dormant periods of the volcano, if the rocks become too weakened by heat and fluids

FIGURE 4.8   Oblique aerial photograph of the upper cone of Mount Rainier from the northwest, showing the Puyallup Glacier and the steep upper slopes. Note also the crater at the top of the volcano, which has been filled in with younger eruptive products. Photo courtesy of U.S. Geological Survey Austin Post photograph collection.

that alter them, or if the rocks are shaken during a regional (not volcano-related) earthquake. Use the topographic maps in this exercise and at the back of the book to analyze areas that may be at risk from landslides and lahars. Sketch on the topographic map (Figure 4.9 in the back of the book) areas that may be at risk from landslides and lahars from Mount Rainier. Explain your choice of areas, and your decision of how far to extend the areas from the peak of the mountain.

3. What volcanic hazards do the citizens of Orting need to be aware of?

4. How are the hazards in Orting likely to be different from hazards in Seattle or Tacoma? Refer to Figure 4.9, the color plate in the back of the book.

5. Review the aerial photograph (Figure 4.10) and topography of Orting as shown on Figure 4.9.

   a. If you were in charge of planning an evacuation because of an imminent volcanic hazard, what geological, economic, and social factors should you include in developing your plan?

TABLE 4.4 Volcanic Hazards and Their Relative Risks at Mt. Rainier, During Eruptions and in Dormant Periods

Hazard	Risk Close to Mt. Rainier	Risk Away from Mt. Rainier	Risk During an Eruption	Risk in Dormant Periods
Lava flow				
Phreatic (steam-or gas-driven) eruptions				
Volcanic bombs				
Tephra				
Pyroclastic flows				
Lahars				
Jokulhlaups (glacier-burst floods)				
Collapse of part of upper cone				
Debris avalanche				
Volcanic earthquakes				
Gases				

b. Assume that there are about 30 minutes from the time a lahar starts on the upper slopes of Mt. Rainier until it reaches Orting. Where do you suggest citizens of Orting go? Explain your choice.

c. In 1990, the population of Orting was about 2,100 people. In 2005 it was approximately 4,500 people. What additional risks are created by population growth in Orting?

PART C. SHIELD VOLCANOES

Shield volcanoes are characterized by the eruption of basaltic magmas, which have low viscosity and therefore cannot develop steep sides to their edifices. Basaltic eruptions, despite their drama, are the least explosive types of eruptions. In Hawaii, shield volcanoes typically erupt either from central vent areas or from rift zones that extend laterally from the central vents for many kilometers.

Hawaii

Figure 4.11a is a map of the Island of Hawaii, and shows general zones of volcanic hazards. It also identifies the five major volcanoes that compose the island. Table 4.5 describes how the U.S. Geological Survey has identified nine different levels of hazards on the island. Figure 4.11b is a more detailed map of lava flow hazards from Kilauea volcano.

QUESTIONS 4, PART C

1. Use the data in Table 4.5 and Figures 4.11a and b. Is the hazard from lava flows greater in the northern or southern part of the island of Hawaii?

2. Is anywhere on the island completely safe from volcanic hazards? Explain.

3. Use topographic data on Figure 4.11a and sketch on lined or graph paper a topographic profile from the City of Hilo through the summit of Mauna Loa to the ocean on the west. (See Exercise 2 for information about drawing topographic profiles.)

FIGURE 4.10 An aerial photograph of Orting. The river along the north side of town is the Carbon River, and the river along the south side of town is the Puyallup River. Both these rivers have their origins high on the slopes of Mt. Rainier. The rivers flow toward the top of the image.

a. From its profile, is Hawaii most like a basaltic shield volcano, an andesitic composite cone, or a rhyolitic volcano?

b. What is the vertical exaggeration of your profile? What would the profile look like with no exaggeration?

c. Does the amount of vertical exaggeration that you drew influence your interpretation of the kind of magma in Hawaii? _____ Explain.

4. Look at Figure 4.11b. Analyze it to answer the following questions.

a. Where are the zones of highest hazard from Kilauea and Mauna Loa?

b. Where are the zones of lowest hazard?

c. What geologic factors might be different between zones of higher hazard and zones of lower hazard?

d. Using the information on these two maps, which town(s) would likely be subject to hazards from the eruption of Pu'u O'o on the east rift zone of Kilauea? (Note that lava flows downhill.)

5. Review hazards listed on Tables 4.1 and 4.2. Lava is not the only hazard on Hawaii. What other hazards could be expected in Hilo from an eruption of:

TABLE 4.5 Hazard zones from lava flows on the Island of Hawaii are based chiefly on the location and frequency of historic and prehistoric eruptions and the topography of the volcanoes. Scientists have prepared a map that divides the five volcanoes of the Island of Hawaii into zones that are ranked from 1 through 9 based on the relative likelihood of coverage by lava flows. Caption and table from U.S. Geological Survey (http://hvo.wr.usgs.gov/hazards/LavaZonesTable.html, downloaded August 4, 2007)

	Hazard Zones for Lava Flows on the Island of Hawaii		
Zone	Percentage of area covered by lava since 1800	Percentage of area covered by lava in last 750 years	Explanation
1	Greater than 25	Greater than 65	Includes the summits and rift zones of Kilauea and Mauna Loa where vents have been repeatedly active in historic time.
2	15–25	25–75	Areas adjacent to and downslope of active rift zones.
3	1–5	15–75	Areas gradationally less hazardous than Zone 2 because of greater distance from recently active vents and/or because the topography makes it less likely that flows will cover these areas.
4	About 5	Less than 15	Includes all of Hualalai, where the frequency of eruptions is lower than on Kilauea and Mauna Loa. Flows typically cover large areas.
5	None	About 50	Areas currently protected from lava flows by the topography of the volcano.
6	None	Very little	Same as Zone 5.
7	None	None	20 percent of this area covered by lava in the last 10,000 yrs.
8	None	None	Only a few percent of this area covered in the past 10,000 yrs.
9	None	None	No eruption in this area for the past 60,000 yrs.

Source: Wright and others, 1992

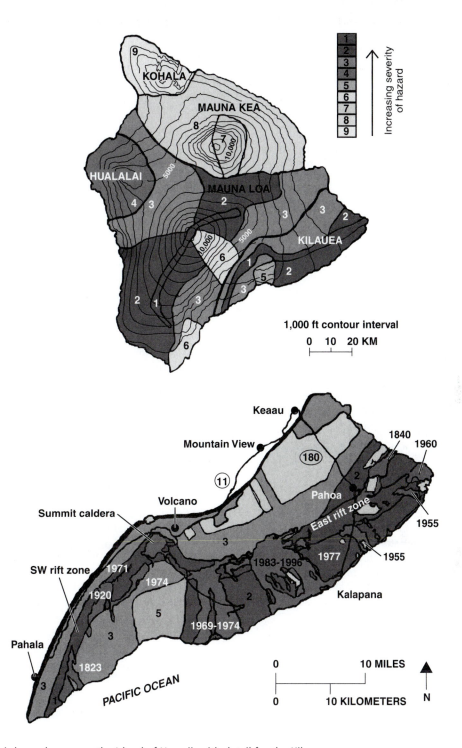

FIGURE 4.11   Volcanic hazard zones on the Island of Hawaii, with detail for the Kilauea area.

a. Mauna Loa?

b. Kilauea?

6. If Mauna Loa is showing early signs of an eruption, such as increased seismic activity and swelling of the volcano from rising magma, what actions should the residents of Hilo take?

7. What data are needed to determine the areas on the island that are at risk from tephra? Gases?

8. Are these areas likely to be large or small? What is your evidence?

9. If you plan to vacation at or move to the Island of Hawaii, what are some data that would be helpful to know before your trip?

Bibliography

Christiansen, R. L., 2001, The Quaternary and Pliocene Yellowstone Plateau volcanic field of Wyoming, Idaho, and Montana, U.S. Geological Survey Professional Paper: 729-G, 145 p., 3 plates, scale 1:125,000

Christiansen, R. L., Lowenstern, J. B., Smith, R. B., Heasler, H., Morgan, L. A., Nathenson, M., Mastin, L. G., Muffler, L. J. P., and Robinson, J. E., 2007, Preliminary assessment of volcanic and hydrothermal hazards in Yellowstone National Park and Vicinity: U.S. Geological Survey Open-File Report 2007–1071; 94 p.

Fisher, R. V., Herlan, G., and Holen, J. B., 1997, Volcanoes, crucibles of change: Princeton University Press, 317.

Hoblitt, R. P., Walder, J. S., Driedger, C. L., Scott, K. M., Pringle, P. T., and Vallance, J. W., 1995, *Volcano hazards from Mount Rainier, Washington:* U.S. Geological Survey Open-File Report 95–273, 12 p.

Keefer, W. R., 1971, The geology story of Yellowstone National Park: U.S. Geological Survey Bulletin 1347, 92 p.

Lowenstern, J. B., Christiansen, R. L., Smith, R. B., Morgan, L. A., and Heasler, H., August 25, 2008, Steam explosions, earthquakes, and volcanic eruptions—What's in Yellowstone's future? U.S. Geological Survey Fact Sheet 2005–3024, http://pubs.usgs.gov/fs/2005/3024/, downloaded August 5, 2007.

Miller, C. D., 1989, *Potential hazards from future volcanic eruptions in California:* U.S. Geological Survey Bulletin 1847, 17 p.

Tilling, R. I., Heliker, C., and Wright, T. L., 1987, *Eruptions of Hawaiian volcanoes: Past, present, and future:* U.S. Geological Survey General Interest Publication, 54 p.

U.S. Geodynamics Committee, 1994, *Mount Rainier, active cascade volcano:* Washington, D.C., National Research Council, National Academy Press, 114 p.

U.S. Geological Survey, 1996, *Perilous beauty: The hidden dangers of Mount Rainier* (video): Seattle, WA. Pacific Northwest Interpretive Association.

Wright, T. L., and Pierson, T. C., 1991, *Living with volcanoes:* U.S. Geological Survey Circular 1073, 57 p.

Wright, T.L., Chu, J.Y., Esposo, J., Heliker, C., Hodge, J., Lockwood, J.P., and Vogt, S.M., 1992, Map showing lava-flow hazard zones, island of Hawaii: U.S. Geological Survey Miscellaneous Field Studies Map MF-2193, scale 1:250,000.

Hazards of Mount St. Helens

INTRODUCTION

The goals of this exercise are, first, to look at the well-documented events associated with the May 18, 1980, and subsequent eruptions of Mount St. Helens; and second, to investigate the impact that these volcanic events had on the surrounding area. Mount St. Helens is located in southwestern Washington. Figure 5.1 illustrates the locations of and different eruption histories that characterize Cascade volcanoes in the United States. Mount St. Helens, as can be seen from its eruption history over the past 4,000 years, is very active. Events that occurred on and hazards generated by Mount St. Helens are typical of those that could be expected if other Cascade volcanoes were to erupt. Mount Rainier is close to the Seattle–Tacoma area of Washington, and Mount Hood is close to the Portland area of Oregon. Eruptions of either of these volcanoes might have major impacts on nearby metropolitan areas.

PART A. ERUPTION OF 1980

The eruption of May 18, 1980, was preceded by increasing seismic activity at the volcano, small phreatic eruptions (steam-driven eruptions created as the mountain warmed up and heated groundwater within the volcano), tilting of the volcano's slopes, and noticeable bulging on the north side of the volcano.

There were several distinct events on the morning of May 18, 1980. These chronologically were (1) an earthquake of Richter magnitude 5.1, (2) a landslide on the north side of the mountain, (3) a lateral blast to the north, (4) an eruption column that soon became vertical, with pyroclastic flows to the north, (5) lahars down many river valleys that lead from the volcano, and (6) extensive tephra deposits, carried by the wind to eastern Washington and beyond. In this exercise we investigate a few of these events and processes. Refer to your text or to reference materials suggested by your instructor if you need additional information.

QUESTIONS 5, PART A

Mount St. Helens: Topography Before and After

1. Figure 5.3a is a topographic map of Mount St. Helens prior to the eruption, and Figure 5.3b is a map of the mountain after the eruption. Compare these maps with both the oblique photographs (Figure 5.2) and with the stereo photographs (Figure 5.4). Use a Stereoscope, if available, to study Figure 5.4.

  Draw topographic profiles of the mountain from point A to point B on Figure 5.3a and from point C to point D on Figure 5.3b. Draw both profiles on the same sheet of your own graph paper; you need use only the heavy contours in your profiles. Each profile begins at 4,800 ft on the west end and ends at 4,400 ft on the east end.

2. Describe the main differences in these profiles. How are these differences reflected in the stereo photographs? Why does it help (or not) to have the oblique photos for reference in drawing these profiles?

3. What were the peak elevations of Mt. St. Helens before and after the eruption? How much elevation was lost from the old peak to the new crater floor? (Hint: Use the average elevation between the two contours that define the 1980 crater floor.)

4. From the shape of the mountain prior to eruption, is Mount St. Helens closer in profile to Hawaiian volcanoes or to composite volcanoes? (See Introduction, Exercise 4.)

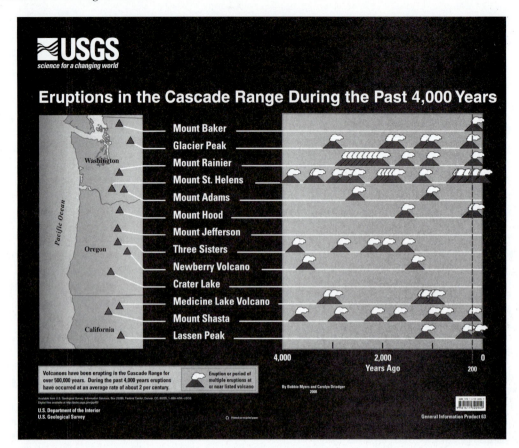

FIGURE 5.1 Regional location map of Cascade volcanoes, eruption histories of the volcanoes over the past 4,000 years, and nearby metropolitan areas. Courtesy of Bobbie Myers and Carolyn Driedger, U.S. Geological Survey Cascade Volcano Observatory. http://Pubs .usgs.gov/gip/63/, downloaded August 2008.

5. What does the shape of Mount St. Helens suggest about the composition of rock that you would expect to find here?

6. After the eruption, much of the volcano was missing. Where did the material go?

a. Compare the elevation of Spirit Lake on Figures 5.3a and 5.3b. What is the elevation before the eruption? _____ What is the elevation after the eruption? _____ Explain how the changes occurred.

b. Follow the channel of the North Fork of the Toutle River west from Spirit Lake. On Figure 5.3a, in section 18, just east of where Studebaker Creek enters the Toutle River, what is the elevation of the bench mark?

c. After the eruption (Figure 5.3b), how much elevation change occurred in the elevation of the northern third of section 18?

d. What geologic deposit occurs in this area southwest of Johnston Ridge (refer to Figure 5.5 in the colored maps section)? What is the origin of this deposit? What other eruption impacts occurred in this area?

7. Examine Figures 5.3 and 5.5 (and photos 5.2 and 5.4) to determine the distribution of glaciers near the top of Mt. St. Helens both before and after the eruption.

a. What changes took place?

b. Where did the ice go?

c. What river channel(s) was (were) heavily impacted by waters from the melting ice?

Volcanic Deposits Near the Mountain

8. Figure 5.5 (in maps section at back of book) is a map of proximal deposits and features of 1980 eruptions of Mount St. Helens published by the U.S. Forest Service. What are the major types of volcanic deposits and eruption-related features depicted on this map?

9. What are the relative geographic and topographic positions of these deposits and features in relation to the main cone of the mountain (i.e., which are nearby and which are farther away)? Which are confined to valleys and which are found on hills?

FIGURE 5.2 Oblique aerial photographs of Mount St. Helens before (top) and after (bottom) the May 18, 1980 eruption. View looking south. Compare these photographs with the topographic maps in Figures 5.3a and 5.3b.

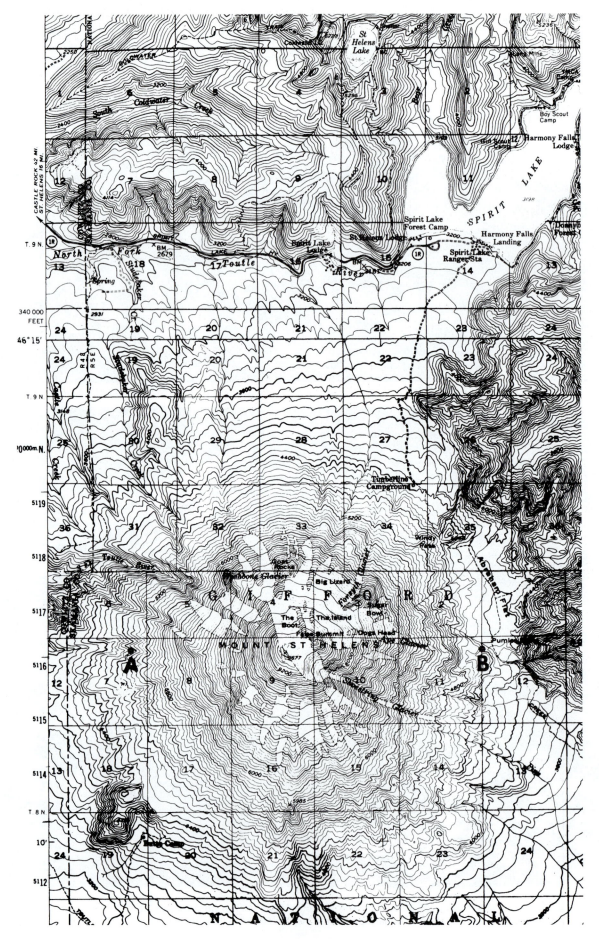

FIGURE 5.3(A)   Topographic map of Mount St. Helens prior to May 18, 1980 eruption (USGS). Sections for Scale.

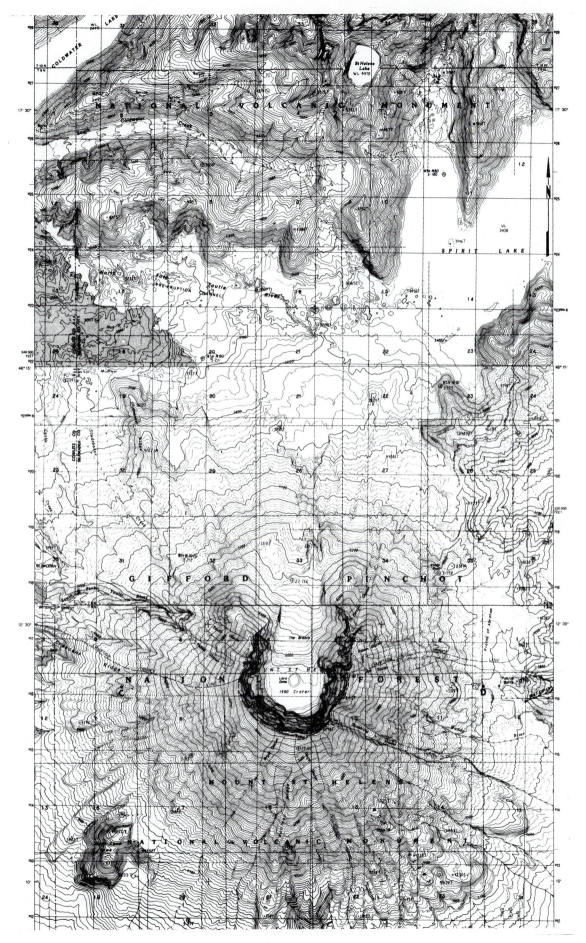

FIGURE 5.3(B)   Topographic map of Mount St. Helens after the May 18, 1980, eruption (USGS). Sections for Scale.

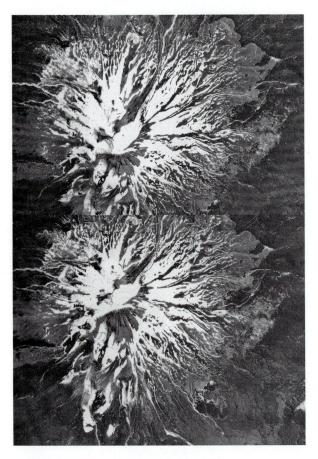

FIGURE 5.4(A)   Vertical aerial stereographic photos of Mount St. Helens prior to May 18, 1980 (USGS).

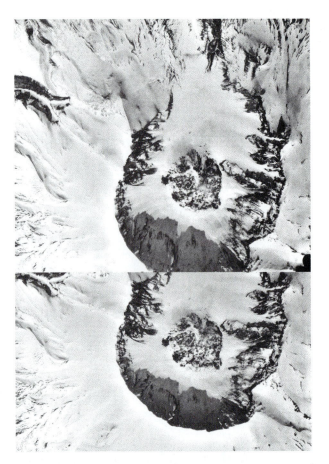

FIGURE 5.4(B)   Vertical aerial stereographic photos of Mount St. Helens after the May 18, 1980, eruption (USGS).

Eruption-Related Problems Away From the Volcano

10. Figure 5.6 shows the general area impacted by lahars (mudflows). Since the major direction of the first volcanic burst was to the north, what geologic processes caused lahars to form on all sides of the volcano?

11. The lahars transported tremendous amounts of sediment far from the volcano. Figure 5.7 shows the configuration of the Columbia River bed before and after the eruption. What differences in configuration are evident?

12. How thick did sediment deposited in 1980 in the Columbia River get?

Tephra

Figure 5.8 is a wind rose, which shows the percent of time that the wind blows in a particular direction. Imagine that you are standing in the center of the circle. The letters on the outside of the circle represent compass directions, and the numbers inside the circle represent the percent of time that the wind is blowing from you toward that particular direction. It can be seen, for instance, that the wind blows 16% of the time toward the east (E), and only 1% of the time toward the west (W).

Note that this convention is opposite what you might expect, in that a westerly wind (one coming out of the west) is plotted toward the east, in the direction the wind is going.

13. Based on the wind rose data, what percent of time would winds blow toward the ENE and E? _____ Are Olympia (NNW) or Portland (SSW) at much risk of tephra from Mount St. Helens, based on the data in the wind rose? _____ Explain.

Figures 5.9 and 5.10 are maps of tephra deposits from two Mount St. Helens eruptions. Note the little circles in the corner of each figure. These are wind directions, toward which winds are blowing at the times of eruption, given for different elevations in the atmosphere. Remember that the posteruption elevation of the top of Mount St. Helens is about 2,550 meters.
14. Compare the distributions of tephra, the wind directions at the times of the eruptions, and the wind rose. Did the eruptions occur at times of common or uncommon wind directions? Explain.

Assessing the Likely Hazard

Figure 5.11 shows potential hazard areas of Mount St. Helens; the map was published about 2 years before the major eruption. Compare it with Figure 5.5 in the map section at the back of the book, which depicts deposits from the 1980 eruptions.

15. List the predicted events for each flowage–hazard zone shown in Figure 5.11.

16. a. Was a preferred direction predicted for events in flowage zone 1? _____ Zone 2? _____ Zone 3? _____

b. What eruption processes and products that occurred, if any, were not included in the predictions?

17. a. How closely do the predictions in Figure 5.11 match the locations of actual mudflows or lahars in the southern half of zone 1 at end of the following valleys (see Figure 5.5 and 5.6).

Muddy River _____

Pine Creek _____

Swift Creek _____

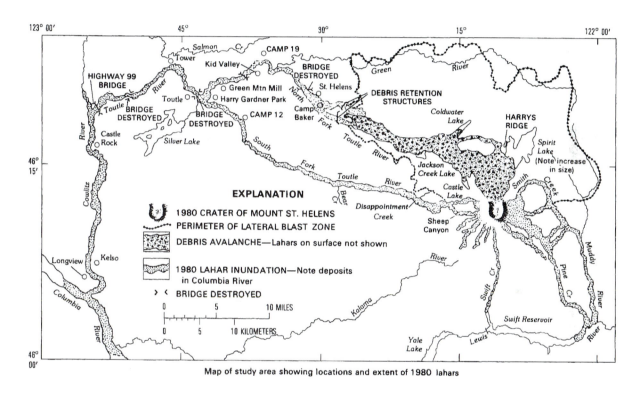

Map of study area showing locations and extent of 1980 lahars

FIGURE 5.6 The general area impacted by lahars (mudflows) from Mount St. Helens. (Scott, 1988)

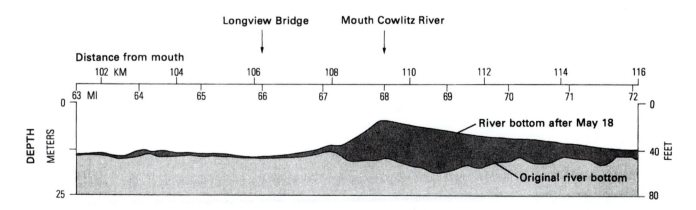

FIGURE 5.7 Longitudinal profile (along the river channel) of the Columbia River bed before and after the 1980 eruption of Mount St. Helens. Dark shaded area represents the thickness of sediment deposited as a result of the May 18, 1980, eruption (Schuster, 1981).

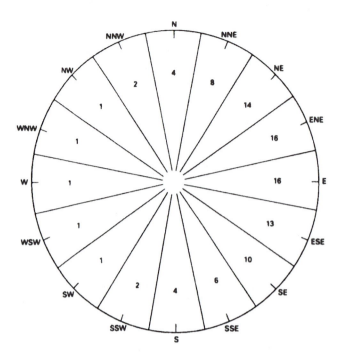

FIGURE 5.8 Approximate percentage of time, annually, that the wind blows toward various sectors in western Washington. Frequencies determined between 3,000m and 16,000m at Salem, OR, and Quillayute, WA (Crandell and Mullineaux, 1978).

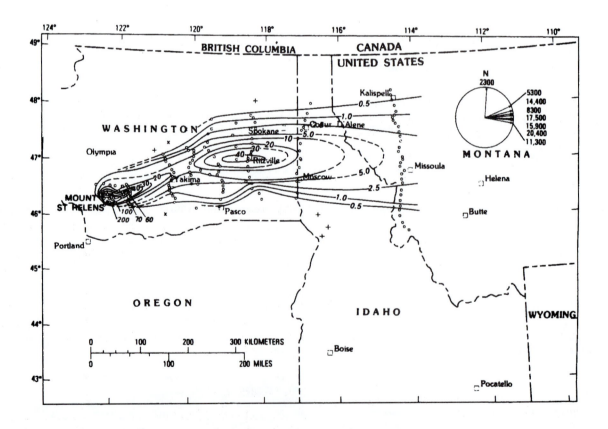

FIGURE 5.9 Map of tephra thickness in mm from May 18, 1980, eruption. + = light dusting of ash, x = no ash observed, and o = observation sites. Wind circle shows directions toward which wind is blowing, which were measured at selected altitudes (in meters) at 10:20 PDT on May 18 at Spokane, WA. Data from U.S. National Meteorological Service (Sarna-Wojcicki, and others, 1981).

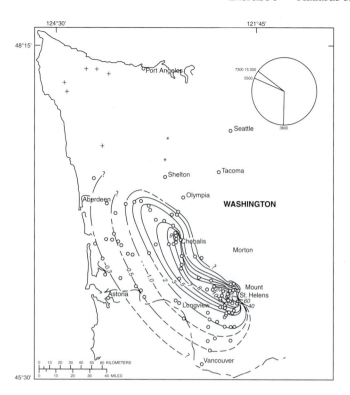

FIGURE 5.10 Distribution of tephra from the May 25, 1980, eruption. Ash thickness in mm. Circle represents measureable thickness, t = trace, and x = no ash. Wind circle shows directions toward which wind is blowing which were measured at selected altitudes (in meters). (Sarna-Wojcicki et al., 1981).

b. How closely do the predictions in Figure 5.11 match the locations of pyroclastic flows in the northern half of zone 1? (see Figure 5.5.)

18. a. How closely do the predictions for lahars in Figure 5.11 match the actual distribution of lahars as shown in Figure 5.6 for each of the following valleys?

     Toutle River (North) _____

     Toutle River (South) _____

     Cowlitz River _____

     Kalama River _____

     Lewis River _____

b. Do you think that future lahars will follow the patterns of May 18, 1980? Why or why not?

PART B. THE VOLCANO WAKES UP

After 18 years of no new magma, Mount St. Helens awoke again on September 23, 2004. A series of small earthquakes heralded the start of a new eruption sequence, which so far has largely consisted of growth of a new dome. This new dome, however, will have one of two fates. Either it will become, like the previous 1986 dome, a building block for constructing a new mountain, or it will be blown to bits in a new violent eruption.

Before we examine the rejuvenated eruption more closely, let's put the eruptions of Mount St. Helens into their context in the overall volcanic activity of the Cascade Range. Over its relatively short geological history (less than 50,000 years) Mount St. Helens has had several stages of eruptive activity, which have lasted several thousand years, separated by periods of inactivity (Mullineaux, 1996). The last 4,000 years have been a time of increased eruptive activity (Pringle, 1993).

QUESTIONS 5, PART B

1. Figure 5.1 shows larger eruptions of Cascade volcanoes, including Mount St. Helens, for the past 4,000 years.

     a. How many eruptions are shown for Mount St. Helens?

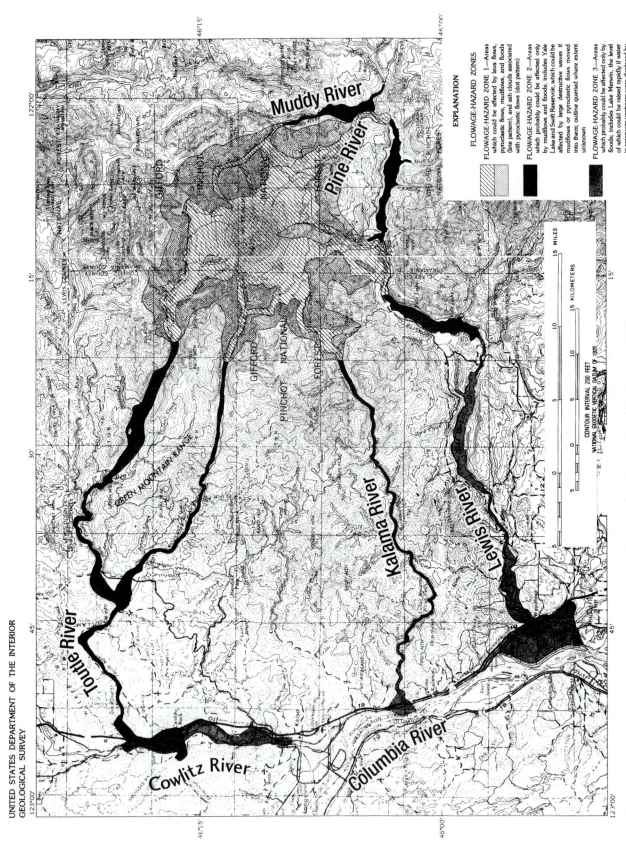

FIGURE 5.11 Predicted hazard areas from Mount St. Helens eruptions (Crandell and Mullineaux, 1978).

b. Which is the second most active volcano? _____
How many eruptions does it have? _____

c. What is the average interval between eruptions of Mount St. Helens for the past 4,000 years?

d. Prior to the 1980 eruption (shown at the present on Figure 5.1), approximately how many years before was the previous eruption?

e. Does the interval between the previous eruption and 1980 fit the pattern for average intervals during the past 4,000 years?

2. The dome-building eruption that began in 2004 is continuing in 2008. The eruption is producing about 0.6 cubic yards of new magma every second. Scientists with the U.S. Geological Survey estimate that about 3.5 billion cubic yards of the mountain was lost in the original eruption in 1980. Since 1980, dome-building eruptions have produced almost 200 million cubic yards of new rock. At the rate of 0.6 cubic yards per second, how long will it take the mountain to rebuild itself to its pre-1980 eruption shape?

3. A record of the growth of the 2004 dome is seen in a series of images from a LIDAR-sensing mission flown by NASA in conjunction with the U.S. Geological Survey (Figure 5.12a-f). The dates of each image are indicated in the upper left-hand corner of each image. The LIDAR image has a horizontal resolution of 2 m and a vertical resolution of 20–30 cm.
  a. Examine Figures 5.12a (September 3) and 5.12b (September 24). Mark on Figure 5.12b any changes in topography that you see.

  b. Describe what seems to have happened in the time between these two images.

4. On each of Figures 5.12c, d, e, and f, trace the outline of the dome that is building.

5. For each of Figures 5.12c, d, e, and f, describe the change from the previous image. Consider the area or size of the dome, the appearance of the surface of the dome, and the position of the dome in the crater. The circles in Figures 5.12a, b, c, and d provide a reference for the change.

Image c (changes from b)

Image d (changes from c)

Image e (changes from d)

Image f (changes from e)

6. As a journalist or a scientist, you have been asked to describe, on camera near the site, the changes that have taken place between September 24 and November 20. You think that percentage increase in the area of the new dome would be one interesting and important fact to report for viewers. You set out to determine the change in area. You recall that areas can be estimated from a formula, either for a rectangle $[A = L \times W]$ or for a circle $[A = \pi r^2]$, where $\pi = 3.14$ and r is the radius. Begin by calculating the areas of the dome in Figures 5.12c and 5.12f. Then calculate the percentage increase between the area of the dome in c and in f. Show your work.
   (Note: The diameter of the circle is 300 m. You can use this as a scale to determine the sizes of the dome in Figures 5.12c and f.)

Optional online access required for the next question.
7. Compare your answer with new images that may have been posted on Mount St. Helens web sites http://vulcan.wr.usgs.gov/Volcanoes/MSH or http://www.fs.fed.us/gpnf/mshnvm/. Describe below the current status of the volcano. What possible events could occur today? What do the alert level and aviation color code mean?

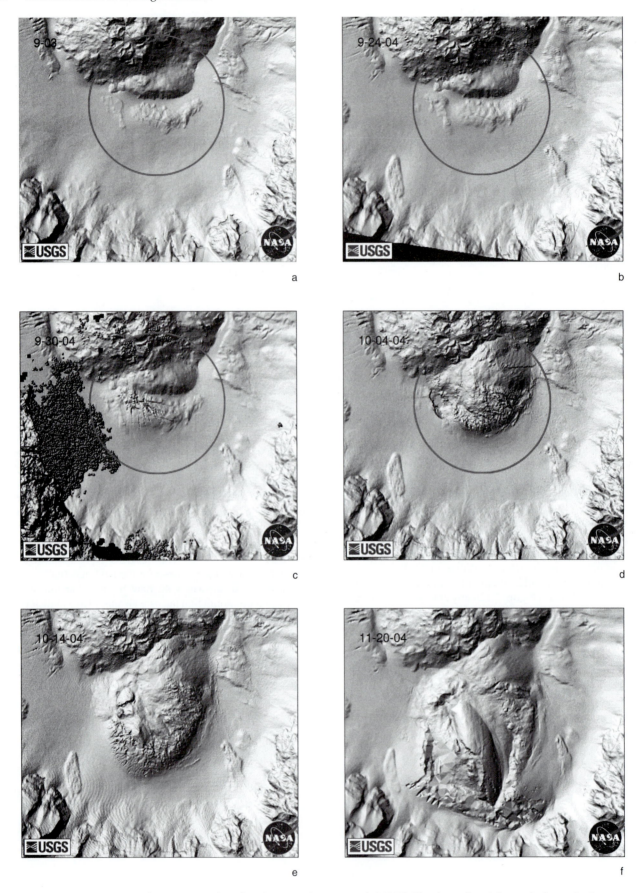

FIGURE 5.12   LIDAR images of Mount St. Helens showing 2004 dome growth (USGS). The date of each image is shown in the upper left corner.

Bibliography

Crandell, D. R., and Mullineaux, D. R., 1978, *Potential hazards from future eruptions of Mount St. Helens volcano, Washington:* U.S. Geological Survey Bulletin 1383-C, 26 p.

Haugerud, R., Harding, D., Queija, V., and Mark, L., 2004, Elevation change at Mt. St. Helens September 2003 to October 4–5, 2004. Accessed on 3/12/05 at http://vulcan .wr.usgs.gov/Volcanoes/MSH/Eruption04/LIDAR/Maps/ LIDAR_prelim_msh_elevation_change_map_2003-04.pdf

Janda, R. J., Scott, K. M., Nolan, K. M., and Martinson, H. A., 1981, Lahar movement, effects and deposits. In *The 1980 eruptions of Mount St. Helens, Washington:* eds. P. W. Lippman and D. R. Mullineaux. U.S. Geological Survey Professional Paper 1250, p. 461–478.

Lipman, P. W. and Mullineaux, D. R., (eds.), 1981, *The 1980 eruptions of Mount St. Helens,* Washington: U.S. Geological Survey Professional Paper 1250, 844 p.

Mullineaux, D.R., 1996, *Pre: 1980 tephra fall deposits erupted from Mount St. Helens, Washington:* U.S. Geological Survey Professional Paper 1563, 99p.

Pallister, J. S., Hoblett, R. P., Crandell, D. R., and Mullineaux, D. R., 1992, Mount St. Helens a decade after the 1980 eruptions: Magmatic models, chemical cycles, and a revised hazards assessment: *Bulletin of Volcanology,* v. 54, no. 2, p. 126–146.

Pringle, P. T., 1993, *Roadside geology of Mount St. Helens National Volcano Monument and vicinity:* Washington Division of Geology and Earth Resources, Information Circular 88, 120 p.

Sarna-Wojcicki, A. M., Shipley, S., Waitt, R. B., Jr., Dzurisin, D. and Wood, S. H., 1981, Areal distribution, thickness, mass, volume, and grain size of air fall ash from the six major eruptions of 1980. In *The 1980 eruptions of Mount St. Helens, Washington:* eds. P. W. Lipman, and D. R. Mullineaux. U.S. Geological Survey Professional Paper 1250, p. 577–600.

Schuster, R. L., 1981, Effects of the eruptions on civil works and operations in the Pacific Northwest. In *The 1980 eruptions of Mount St. Helens, Washington:* eds. P.W. Lipman and D. R. Mullineaux. U.S. Geological Survey Professional Paper 1250, p. 701–718.

Scott, K. M., 1988, *Origins, behavior, and sedimentology of the lahars and lahar-runout flows in the Tootle-Cowlitz River System:* U.S. Geological Survey Professional Paper 1447-A, 74 p.

U.S. Geological Survey, 1980, *Preliminary aerial photographic interpretative map showing features related to the May 18, 1980 eruption of Mount St. Helens, Washington:* U.S. Geological Survey Map MF-1254.

Wolfe, E. W., and Pierson, T. C., *Volcanic-hazard zonation for Mount. St. Helens, Washington, 1995:* U.S. Geological Survey Open-File Report 95–497, 12 p.

Earthquake Epicenters, Intensities, Risks, Faults, Nonstructural Hazards and Preparation

INTRODUCTION

Zachary Grey, writing in 1750, said "An earthquake is a vehement shake or agitation of some considerable place or part of the Earth, from natural causes, attended with a huge noise, like thunder; and frequently with an eruption of water, fire, smoke or wind. They are looked upon to be the greatest and most formidable phenomena of nature." Although our present understanding of earthquakes is much more refined, they are still considered to be formidable phenomena. An earthquake is the ground shaking caused by elastic waves propagating in the Earth generated by a sudden release of stored strain energy. The sudden release of stored strain energy is the result of an abrupt slip of rock masses along a break in the Earth called a *fault*. Most fault slip occurs below the Earth's surface without leaving any surface evidence. The place where this slippage occurs is known as the *hypocenter* or *focus* of the earthquake, and the point on the surface vertically above the focus is the *epicenter*.

In this exercise we review earthquake wave types, locate an earthquake epicenter, determine earthquake intensities, assess seismic risk, examine fault types, and study fault zone characteristics.

Earthquake Waves

The energy released at the focus of an earthquake sets up several types of vibrations or waves that are transmitted through the Earth in all directions. Some waves travel through the Earth to the surface and are known as body waves. Others travel along the Earth's surface and are known as surface waves (Figure 6.1).

One type of body wave is a compressional wave in which the particles of rock vibrate back and forth in the direction of wave travel; the motion is similar to that of sound waves that alternately compress and dilate the medium—solid, liquid, or gas—through which they travel. Compressional waves are also called longitudinal or *primary* waves (P waves); the latter name is given because these waves appear first on seismograms (Figure 6.2) that record earthquake waves. Another type of wave is the shear or transverse wave, in which the particles vibrate at right angles to the direction of wave progress, in the same manner as a wave moving along a stretched string that is plucked. Because these waves are the second waves to appear on the seismogram, they are called *secondary* waves (S waves).

After the body waves, another class of seismic waves, the surface waves, arrive. They have frequencies of less than 1 cycle per second and often approximate the natural frequency of vibration in tall buildings. Surface waves in general decrease in amplitude more slowly than body waves. The surface waves consist of *Love* waves (horizontal lateral vibrations perpendicular to direction of transmission; they travel forwards but shake sideways) and *Rayleigh* waves (rotational displacement of particles to produce a wavy or undulating surface; they travel up and down in small circles).

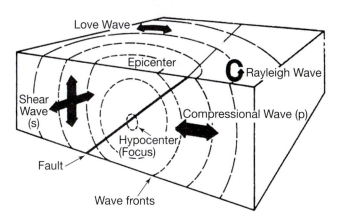

FIGURE 6.1 Diagram of directions of vibrations of body (P and S) and surface (Love and Rayleigh) waves (Hays, 1981).

PART A. EPICENTER, INTENSITY, AND SEISMIC RISK

Epicenter (6, Part A1)

After an earthquake, seismologists are faced with the task of finding when and where the shaking began. They do this by examining the *seismograms* from several seismograph stations. Because the P and S waves travel at different rates, the difference in arrival times varies from station to station depending on the distance from the source.

The average travel times of P and S waves compiled from many earthquake records are used to make travel-time graphs and tables showing the time required for waves to travel various distances from a hypocenter. These records show that P waves travel more rapidly than S waves. Therefore, travel-time curves will show P and S waves as separate curves. Surface waves travel at about 90 percent of the velocity of S waves because the surface waves are traveling through lower velocity materials located at the Earth's surface.

If arrival times are available from several seismograph stations, the distances given by the travel-time curves may be used to determine the earthquake's location. The distance provides the radius of a circle about the seismograph station. The *epicenter* is located somewhere on that circle. With at least three stations, the location of the epicenter may be determined as the point where the three circles intersect.

We can also arrive at the distance to the epicenter by using simple subtraction and a proportional relationship. Because of their different velocities, there is a time lag between arrival of the first P and first S wave at a seismograph station. The time lag (time of S minus time of P) can be determined from seismograms. This time lag can be used to compute the distance to the epicenter, provided the average velocity of each wave type is known. In the first part of the exercise, we will use seismograms from four different stations to locate the epicenter and time of an earthquake.

QUESTIONS (6, PART A1)

Epicenter

1. In Figure 6.2, use the time scale to determine the lag in arrival time between the P and S waves at four stations: St. Louis, Missouri (SLM); Bloomington, Indiana (BLO); Minneapolis, Minnesota (MNM); and Bowling Green, Ohio (BGO). The first major impulse on the left in the seismogram indicates the arrival of the first P wave at the station, the second impulse, the arrival of the first S wave. The lag time, T, is given by the difference between S and P times. Enter the lag time value for each station below:

SLM: _____ sec	BLO: _____ sec
MNM: _____ sec	BGO: _____ sec

2. To determine the distance from the earthquake to each seismograph station we must first determine the time lag between P and S wave arrivals at a given distance from an earthquake, say 100 km, knowing the average velocities of the P and S waves. If the average velocity of the P wave is 6.1 km/sec and the average velocity of the S wave is 4.1 km/sec, what is the time required for each wave to travel 100 km? (It may help to think of this problem like a very fast driving trip: if you want to go 100 km, and you drive at a rate of 6.1 km/sec, how long, in seconds, will it take you to get to your destination?)

P waves (6.1 km/sec) travel 100 km in _____ seconds.

S waves (4.1 km/sec) travel 100 km in _____ seconds.

Thus the time lag between the arrival of P and S waves at a distance 100 km from the hypocenter (T_{100}) is _____ seconds.

3. Remembering that for longer distances there is a proportionally longer lag time, we can construct a simple equation to calculate the unknown distance x to each station:

$$\frac{x}{T_X} = \frac{100\,\text{km}}{T_{100}}$$

where x = unknown distance in km; Tx = lag time for distance x; T_{100} = lag time at 100 km

Since values for Tx are known from Question 1 and the value of T_{100} is known from Question 2, the equation can be solved for x for each station. More than one station is needed to determine the epicenter since the information from one station can only give the distance to the earthquake and not the direction. The minimum number of stations needed to locate an epicenter is three.

Using the data from Figure 6.2 and the equation above, determine the distance to the earthquake epicenter from each station and enter below.

SLM: _____ km
MNM: _____ km
BLO: _____ km
BGO: _____ km

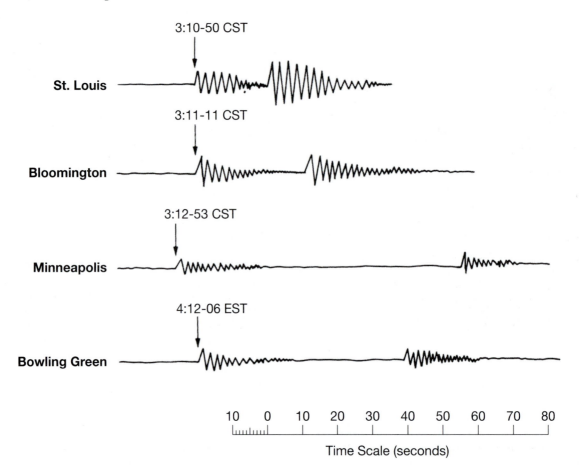

FIGURE 6.2   Partial seismograms for an earthquake. The P wave arrived at the St. Louis seismograph at 10 minutes and 50 seconds after 3:00 P.M. CST. The second disturbance on the seismogram represents the arrival of the S waves.

4. a. The epicenter of the earthquake can be pinpointed by drawing compass arcs from three of the stations with radii corresponding to the distances calculated in Question 3. The intersection of these radii marks the epicenter. Do this in Figure 6.3.

b. Where is the epicenter? (Give location within a state.)

c. Label it on the map (Figure 6.3).

d. At what time did the earthquake occur? (Refer to Figure 6.2.)

Intensity (6, Part A2)

The *intensity* of an earthquake at a site is based on the observations of individuals during and after the earthquake. It represents the severity of the shaking, as perceived by those who experienced it. It is also based on observations of damage to structures, movement of furniture, and changes in the Earth's surface as a result of geologic processes during the earthquake. The Modified

Mercalli Intensity Scale is commonly used to quantify intensity descriptions. It ranges from I to XII (Table 6.1).

An *isoseismal* map shows the distribution of seismic intensities associated with an earthquake. The greatest impact of an earthquake is usually in the epicentral region, with lower intensities occurring in nearly concentric zones outward from this region. The quality of construction and variation of geologic conditions affect the distribution of intensity.

Seismic risk maps have been based on the distribution and intensities of past earthquakes or on the probability of future earthquake occurrences (of a given ground motion in a given time period). In this exercise the first type of map is adequate for our examination of seismic risk in middle North America; however, maps based on the probabilistic approach may be needed in other investigations. The latter maps do not express intensity. Rather, they show probability of occurrence of ground shaking that has a 10 percent probability of being exceeded in 50 years.

Note that we also use the term *magnitude* to describe an earthquake. The magnitude of an earthquake is a measure of the amplitude of an earthquake wave on a seismograph (Bolt, 1988). The Richter magnitude scale is a commonly used standardized system of amplitude measurement, and allows for comparison of different

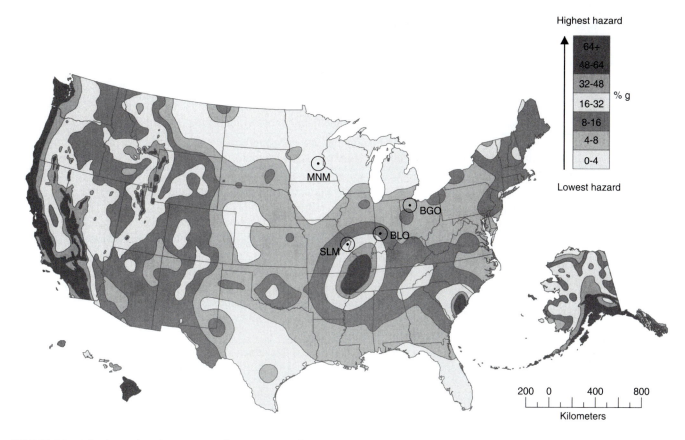

FIGURE 6.3 Seismic acceleration, expressed as a percent of gravity, that can be expected during a 50 year period. Higher numbers indicate greater potential for shaking (From Peterson and others, 2008).

earthquakes around the world. The Richter scale is a logarithmic scale, which means that each increase in number, for example from M5 to M6, represents a 10-fold increase in amplitude (and about a 30-fold increase in actual energy released by the earthquake).

QUESTIONS (6, PART A2)

1. Following are some historical descriptions of earthquakes (a–d). Such statements, made to scientists or reporters or recorded in diaries or on survey forms distributed by government agencies, allow scientists to determine the intensity of an earthquake. Using the Modified Mercalli Intensity Scale (Table 6.1), assign each of the quakes an intensity number. Pick the lowest number exhibiting the characteristics given. The first quotation describes the observations of an eyewitness to a California earthquake around 1913. The second, third, and part of the fourth descriptions are from data gathered by the U.S. Coast and Geodetic Survey after the Daly City, California, earthquake of 1957 (Richter magnitude 5.3).

   a. "There was a keen frost, and when we reached the water-hole a thin film of ice was seen upon the water. I dismounted and led my horse by the bridle, and walked to the edge of the water. Just as I reached it, the ground seemed to be violently swayed from east to west. The water splashed up to my knees; the trees whipped about and limbs fell on and all around me. I was affected by a fearful nausea, my horse snorted and in terror struggled violently to get away from me, but I hung to him, having as great a fear as he had himself. The lake commenced to roar like the ocean in a

storm, and, staggering and bewildered, I vaulted into the saddle and my terrified horse started, as eager as I was to get out of the vicinity." (Eisman, 1972)

Intensity:

   b. "The shock seemed to be a sort of gentle swaying back and forth, causing hanging fixtures to swing, but doing no damage." (Iacopi, 1971)

Intensity:

   c. "The earthquake was very intense . . . a heavy oak china cabinet and massive table moved 2 to 3 inches away from original positions; kitchen stove moved 2 inches; furnace in basement moved two inches off base and water heater tilted off base." (Iacopi, 1971)

Intensity:

TABLE 6.1   Modified Mercalli Intensity Scale of 1931.

Intensity	Description of Effects
I	Not felt by people, except under especially favorable circumstances. Sometimes birds and animals are disturbed. Trees, structures, liquids, and bodies of water may sway gently, and doors may swing slowly.
II	Felt indoors by a few people, especially on upper floors of multistory buildings. Birds and animals are disturbed, and trees, structures, liquids, and bodies of water may sway. Hanging objects may swing.
III	Felt indoors, usually as a rapid vibration that may not be recognized as an earthquake at first, similar to that of a light truck passing nearby. Movements may be appreciable on upper levels of tall structures.
IV	Felt indoors by many, outdoors by few. Awakens a few individuals. Characterized by vibration like that due to passing of heavy or heavily loaded trucks, a heavy body striking building, or the falling of heavy objects inside. Dishes, windows, and doors rattle. Walls and house frames creak. Hanging objects often swing. Liquids in open vessels are disturbed slightly. Stationary automobiles rock noticeably.
V	Felt indoors by practically everyone, outdoors by most people. Awakens many or most sleepers. Frightens a few people; some persons run outdoors. Buildings tremble throughout. Dishes and glassware break to some extent. Windows crack in some cases, but not generally. Vases and small or unstable objects overturn in many instances. Hanging objects and doors swing generally. Pictures knock against walls, or swing out of place. Pendulum clocks stop, or run fast or slow. Doors and shutters open or close abruptly. Small objects move, and furnishings may shift to a slight extent. Small amounts of liquids spill from well-filled containers.
VI	Felt by everyone, indoors and outdoors. Awakens all sleepers. Frightens many people; there is general excitement, and some persons run outdoors. Persons move unsteadily. Trees and bushes shake slightly to moderately. Liquids are set in strong motion. Plaster cracks or falls in small amounts. Many dishes and glasses, and a few windows, break. Books and pictures fall. Furniture may overturn or heavy furnishings move.
VII	Frightens everyone. There is general alarm, and everyone runs outdoors. People find it difficult to stand. Persons driving cars notice shaking. Trees and bushes shake moderately to strongly. Waves form on ponds, lakes, and streams. Suspended objects quiver. Damage is negligible in buildings of good design and construction; slight to moderate in well-built ordinary buildings; considerable in poorly built or badly designed buildings. Plaster and some stucco fall. Many windows and some furniture break. Loosened brickwork and tiles shake down. Weak chimneys break at the roofline. Cornices fall from towers and high buildings. Bricks and stones are dislodged. Heavy furniture overturns.
VIII	There is general fright, and alarm approaches panic. Persons driving cars are disturbed. Trees shake strongly, and branches and trunks break off. Sand and mud erupt in small amounts. Flow of springs and wells is changed. Damage slight in brick structures built especially to withstand earthquakes; considerable in ordinary substantial buildings, with some partial collapse; heavy in some wooden houses, with some tumbling down. Walls fall. Solid stone walls crack and break seriously. Chimneys twist and fall. Very heavy furniture moves conspicuously or overturns.
IX	There is general panic. Ground cracks conspicuously. Damage is considerable in masonry structures built especially to withstand earthquakes; great in other masonry buildings, with some collapsing in large part. Some wood frame houses built especially to withstand earthquakes are thrown out of plumb, others are shifted wholly off foundations. Reservoirs are seriously damaged, and underground pipes sometimes break.
X	Most masonry and frame structures and their foundations are destroyed. Ground, especially where loose and wet, cracks up to widths of several inches. Landsliding is considerable from riverbanks and steep coasts. Sand and mud shift horizontally on beaches and flat land. Water level changes in wells. Water is thrown on banks of canals, lakes, rivers, etc. Dams, dikes, and embankments are seriously damaged. Well-built wooden structures and bridges are severely damaged, and some collapse. Railroad rails bend slightly. Pipelines tear apart or are crushed endwise. Open cracks in cement pavements and asphalt road surfaces.
XI	Few if any masonry structures remain standing. Broad fissures, earth slumps, and land slips develop in soft wet ground. Water charged with sand and mud is ejected in large amounts. Sea waves of significant magnitude may develop. Damage is severe to wood frame structures, especially near shock centers, great to dams, dikes, and embankments, even at long distances. Supporting piers or pillars of large, well-built bridges are wrecked. Railroad rails bend greatly and some thrust endwise. Pipelines are put out of service.

TABLE 6.1 Modified Mercalli Intensity Scale of 1931. (Continued)

Intensity	Description of Effects
XII	Damage is nearly total. Practically all works of construction are damaged greatly or destroyed. Disturbances in the ground are great and varied, and numerous shearing cracks develop. Landslides, rockfalls, and slumps in riverbanks are numerous and extensive. Large rock masses are wrenched loose and torn off. Fault slips develop in firm rock, and horizontal and vertical offset displacements are notable. Water channels, both surface and underground, are disturbed and modified greatly. Lakes are dammed, new waterfalls are produced, rivers are deflected, etc. Surface waves are seen on ground surfaces. Lines of sight and level are distorted. Objects are thrown upward into the air.

(Modified from Cluff and Bolt, 1969, p. 9)

d. "It was as if giant hands took the house and shook it . . . the pea soup jumped out of the pot and the grandfather clock was silenced." (modified from Iacopi, 1971)

Intensity:

2. Not all earthquakes occur in areas where high levels of risk have been identified. On July 27, 1980, an earthquake of Richter magnitude 5.1 shook Kentucky, Ohio, and adjacent states. The earthquake epicenter was determined to be at latitude 38.2° N, longitude 83.9° W, near Sharpsburg, Kentucky (approximately 30 miles southwest of the Ohio River town of Maysville, Kentucky). It had a focal depth of 13 km. Damage to structures along the Ohio River in Maysville, Kentucky, and in the Ohio communities of Aberdeen, Manchester, Ripley, and West Union, consisted of chimneys being knocked down, cracks in plaster and concrete blocks, and merchandise being toppled from store shelves. In Cincinnati a cornice reportedly fell from city hall.

a. Based on the reported damage, what was the intensity of this earthquake along the Ohio River?

b. Locate the earthquake epicenter with an X on Figure 6.3.

Isoseismal Maps (6, Part AB)

Large earthquakes have the potential for significant damage. This damage varies with the geologic nature of the earthquake and the rocks between the focus and the site, types and properties of the materials at a site, and the nature of the buildings. In this part of the exercise we use data from 1949 and 1965 earthquakes in western Washington to construct isoseismal maps.

QUESTIONS (6, PART A3)

1. The intensity of an earthquake is a measure of the impact of seismic shaking on the ground, structures, and people. It is described on a scale of I to XII (in Roman numerals), where I is only rarely felt and XII is total destruction. Use the Modified Mercalli Intensity Scale (Table 6.1) and the descriptions of site damage for the April 13, 1949, earthquake (Table 6.2) to determine the intensity at each site. Record the intensity and the primary evidence used in determining the intensity for each site, beside the names of the sites in Table 6.3. Several intensities, with evidence, are given.

2. Place the intensity values from Table 6.3 on the map of Washington (Figure 6.4). Then draw boundaries between these intensities to produce an isoseismal map.

3. What was the maximum intensity from the 1949 earthquake?

4. Where does the epicenter for the 1949 earthquake appear to have been?

5. What observation in Table 6.2 was the most interesting or surprising to you? Why?

6. Using intensity numbers from the April 29, 1965, western Washington earthquake shown in Table 6.4, enter the intensity values on the map of Washington (Figure 6.5).

7. Draw the approximate boundaries of the intensity zones as determined by the values you entered for each locality. Part of one boundary is given for you in Figure 6.5.

TABLE 6.2   Impact of the 1949 Earthquake in Western Washington at Various Sites.

Aberdeen	One death. Scores of chimneys tumbled at roof level. Broken dishes and windows.
Bellingham	Hanging objects swung. Swaying of buildings. Pendulum clocks stopped or ran fast or slow.
Bremerton	One death. Considerable falls of plaster. Elevator counterweights pulled out of guides. Swaying of buildings. Trees shaken moderately to strongly.
Buckley	Part of high school building fell. Most chimneys in town toppled at roofline. Cracked plaster and ground.
Centralia	One death; 10 persons hospitalized. Very heavy damage. Collapse of building walls and many chimneys. Water mains broken; Water and sand spouted from ground. Violent swaying of buildings and trees. Many objects moved, including pianos. Objects fell from shelves. Pendulums swinging east–west stopped. Many persons panic-stricken. Four miles southwest of town, water spouted 18 in. high in middle of field, leaving a very fine sand formation around each hole (1–3 in. in diameter). Gas or air boiling up through river.
Cle Elum	Pendulum clocks stopped. Small objects and furnishings shifted. Trees and bushes shaken moderately.
Eatonville	Chimneys toppled. Plaster fell in large pieces in schoolhouse. People had difficulty in maintaining balance.
Hyak	Few windows broke. Trees and bushes shaken moderately. Furnishings shifted.
Longview	Two minor injuries. Gable of community church fell. Water main broke, beams cracked in school. Extensive but scattered damage to business buildings, industrial properties, and residences. Considerable damage to irrigation ditches. Landslides on cuts along highway. Objects fell in all directions. Some heavy furniture overturned. Glass figurine on mantle thrown 12 ft.
Olympia	Two deaths; many persons injured. Conspicuous cracks in ground and damage in masonry structures. Capitol buildings damaged. Nearly all large buildings had cracked or fallen walls and plaster. Two large smokestacks and many chimneys fell. Streets damaged extensively; many water and gas mains broken. Portion of a sandy spit in Puget Sound disappeared during the earthquake.
Port Townsend	Pendulum clocks facing northeast stopped. Hanging objects swung. Slight damage in poorly built buildings. Subterranean sounds heard. Bells rang in a small church.
Puyallup	Many injured. High school stage collapsed. Nearly every house chimney toppled at roof line. Several houses were jarred off foundations. Minor landslides blocked roads. Water mains broke. Multiple-story brick buildings most severely damaged. Some basement floors raised several feet, driving supports through floor above. Plaster badly damaged. Water spouted in fields, bringing up sand.
Randle	Twisting and falling of chimneys; about one-fourth of all chimneys fell. Damage considerable. Water spilled from containers and tanks. Plaster and walls fell; dishes and windows broke. Lights went out.
Satsop	Cracked ground. Pendulum clocks stopped. Trees and bushes shaken strongly. Furnishings overturned.
Seattle	One death; many seriously injured with scores reporting shock, bruises, and cuts. Many houses on filled ground demolished; many old buildings on soft ground damaged considerably. Collapse of top of one radio tower and one wooden water tank with damage to many tanks on weak buildings. Many chimneys toppled. Heavy damage to docks (fractures in decayed pilings). Several bridges damaged; many water mains in soft ground broken. Telephone and power service interrupted. Large cracks in filled ground; some cracking of pavement. Water spouted 6 ft or more from ground cracks. At the federal office building, bookcases thrown face down. Very heavy furniture overturned. Plaster badly cracked and broken with pieces 1–3 ft square thrown from walls. Pictures on north–south walls canted; those on east–west walls— little cant. Some doors did not fit after shock. Many old brick buildings partially destroyed.
Snoqualmie	Most damage confined to brick chimneys, windows, and plaster. Overturned vases and floor lamps. Coffee shaken out of cups. Rockslides on Mt. Si. Trees and bushes shaken strongly.
Tacoma	One death. Many buildings damaged and parts fell. Many chimneys toppled. Several houses slid into Puget Sound. One smokestack fell. One 23-ton cable saddle was thrown from the top of tower at Tacoma Narrows Bridge, causing considerable loss. Railroad bridges thrown out of line. Tremendous rockslide, a half-mile section of a 300-ft cliff, into Puget Sound. Considerable damage to brick; plaster, windows, walls, and ground cracked.

(Modified from Murphy and Ulrich, 1951)

TABLE 6.3 Intensities from the April 13, 1949, Earthquake

Location (symbol)	Intensity	Primary Evidence for 1949 Earthquake
Aberdeen (Ab)		
Bellingham (Be)		
Bremerton (Br)		
Buckley (Bu)	VIII	Walls and chimneys fall; cracked ground
Centralia (Ce)	VIII–IX	Sand and mud eruption: pipes break; damage considerable
Cle Elum (Cl)		
Eatonville (Ea)	VII	Plaster falls in large pieces; difficulty maintaining balance
Hyak (Hy)	VI	Trees/bushes shake moderately; few windows break
Longview (Lo)	IX	
Olympia (Ol)		
Port Townsend (Po)		
Puyallup (Pu)		
Randle (Ra)	VIII	Twisted and fallen chimneys; walls fall
Satsop (Sa)		
Seattle (Se)		
Snoqualmie (Sn)	VII	Damaged chimneys/windows; trees/bushes shaken strongly
Tacoma (Ta)		

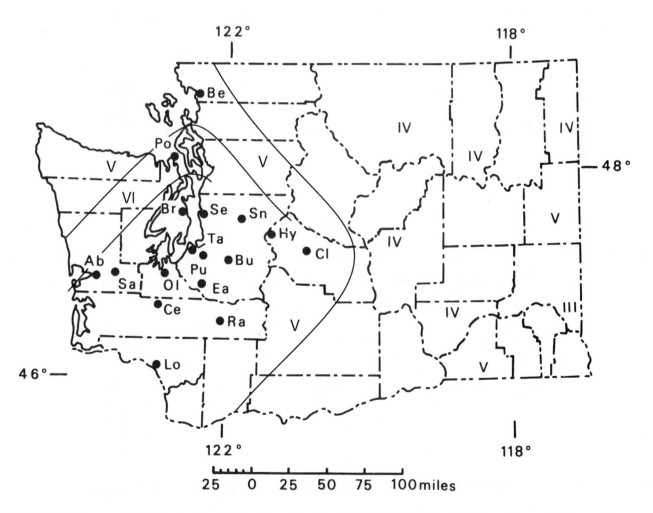

FIGURE 6.4   Index map of Washington showing locations of the sites listed in Table 6.3. Modified Mercalli Intensities for April 13, 1949, earthquake are given for eastern Washington. Some boundaries of intensity zones shown. Completed map is an isoseismal map.

TABLE 6.4 Locations and Intensity Data for the April 29, 1965, Western Washington Earthquake

Location	Intensity	Location	Intensity
Aberdeen (Ab)	V	Longview (Lo)	V
Arlington (Ar)	VI	Olympia (Ol)	VI
Bellingham (Be)	V	Port Angeles (Pa)	V
Bremerton (Br)	VI	Port Townsend (Po)	V
Buckley (Bu)	VI	Puyallup (Pu)	VII
Centralia (Ce)	VI	Randle (Ra)	V
Cle Elum (Cl)	V	Satsop (Sa)	VI
Concrete (Co)	VI	Seattle (Se)	VII
Eatonville (Ea)	VI	Snoqualmie (Sn)	VII
Forks (Fo)	IV	Tacoma (Ta)	VII
Hyak (Hy)	VI	Vancouver (Va)	V

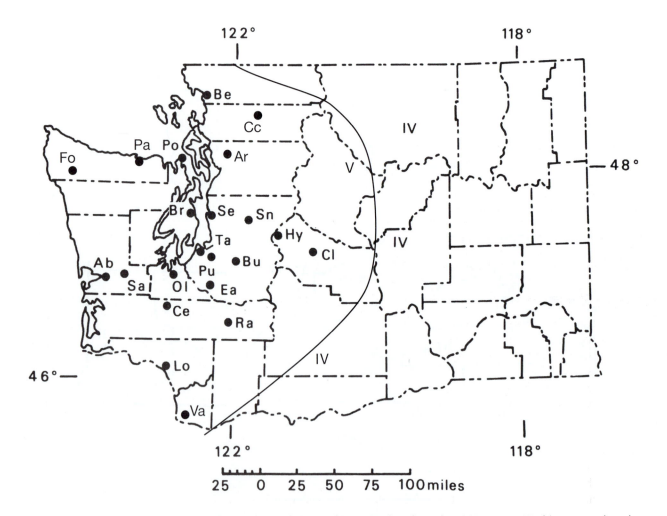

FIGURE 6.5   Index map of State of Washington showing the sites of intensity data from the 1965 western Washington earthquake.

8. a. What was the maximum intensity for the 1965 earthquake?

b. Where does the epicenter for the 1965 earthquake appear to have been?

9. On the Web you will find additional information on these two earthquakes (and others) at http://earthquake.usgs.gov/eqcenter/dyfi.php. Complete the following blanks for these two western Washington earthquakes.

a.   **1949 Earthquake**   **1965 Earthquake**

Date

Maximum Intensity

Name (Earthquake)

Magnitude

Number of Reports

b. You might still be curious about earthquake intensities. Below write a question for your TA or instructor about some aspect of the exercise that you don't understand.

Note: If you experience an earthquake you can report what you saw/felt during the earthquake and what damage you noted. It is useful to write your account of the event as soon as you are safe and can make notes. Then use those notes when completing the form at http://pasadena.wr.usgs.gov/shake/pnw/html/unknown_form.html.

Earthquake Shaking Hazard Maps (6, Part A4)

National maps of earthquake shaking hazards provide information that helps to save lives and property by providing data for building codes. Buildings designed to withstand severe shaking are less likely to injure occupants. These hazard maps are also used by insurance companies, FEMA (for support of earthquake preparedness), EPA (for landfill design), and engineers (for landslide potential).

The map shows the hazard by zones (or in some maps contour peak values) of the levels of horizontal shaking. The higher the number the stronger is the shaking. The number is % g or percent of acceleration due to gravity (in this case as horizontal acceleration). Acceleration is chosen, because building codes prescribe how much horizontal force a building should be able to withstand during an earthquake. 10% g is the approximate threshold for damage to older (pre-1965) structures. Additional information on these maps is available from Frankel et al. (1997) and from the USGS (Fact Sheet 183-96).

Figure 6.3 is a ground-shaking hazard map that shows a 10 percent probability of exceeding a given value in a 50-year period (Peterson and others, 2008). That is, over the next 50 years there is a 1 in 10 chance that the acceleration given for any area will be exceeded. Use information in Figure 6.3 to help answer the following questions.

QUESTIONS (6, PART A4)

1. Which areas of the country have the lowest hazard from earthquake shaking (where 4% g, or less, peak acceleration is expected)?

2. If damage to older (pre-1965) structures can be expected with horizontal accelerations of 10% g or more, which areas of your home state are:
   a. at some risk?

   b. at greatest risk?

   c. what is your home state?

3. What three or four regions of the country have the highest accelerations?

4. What geologic processes, other than shaking and fault displacement, could produce a hazard in an earthquake? List two.

5. The geologic material on which a building rests plays a role in the type of shaking that occurs during an earthquake. Weak materials amplify the shaking. Which of the following foundation materials would most likely result in less shaking and a safer building? (circle one)

artificial fill, poorly consolidated sediments, marine clays, unweathered bedrock

6. If the Internet is available, now or after class, determine and list (places and magnitudes) where the largest two earthquakes have occurred in the last 2 weeks. Also list what processes, other than shaking, contributed to the loss of structures and life. A possible source to begin the search is: http://earthquake.usgs.gov/recenteqs/

7. The Mississippi Valley is indicated as a high-risk area because of earthquake activity that is associated with stress within the continental lithospheric plate. Consider the types of plate margins in the plate tectonics model to answer the following questions. (See a geology text for basic details on plate margins.)

a. What is the tectonic explanation for the major shaking hazard in southern California?

b. What is the tectonic explanation for the major shaking hazard around Seattle?

PART B. FAULTS AND FAULT DETECTION

Earthquakes are related to movements along fault zones. Diagrams of several types of faults are shown in Figure 6.6. If necessary, the review questions in this portion of the exercise may be answered with the aid of a standard introductory geology textbook.

Fault zones can be recognized on aerial photographs, satellite and LIDAR images, geologic maps, and topographic maps, as well as by field observations. Features that might indicate a fault zone area are (1) scarps (cliff or break in slope) formed by horizontal or vertical movement; (2) steep mountain fronts; (3) offset streams and ridges; (4) sag ponds and lakes; (5) lineaments of vegetation; (6) valleys in fault zones; (7) changes in rock type, structure, moisture, and vegetation and (8) faceted spurs. These features are shown in Figure 6.7, which depicts a strike-slip fault (largely horizontal displacement). Normal faults also can have distinct features, which are illustrated in Figure 6.8.

QUESTIONS (6, PART B)

Fault Diagrams

1. The freshly exposed cliff of bedrock or regolith along a fault line is known as a fault _____.

2. Following an earthquake, the horizontal distance between two utility poles on opposite sides of a fault trace (not a strike-slip fault) had increased. Are the regolith (soils) and bedrock in this area in a region of compression (squeezing together) or tension (pulling apart)? Explain your reasoning.

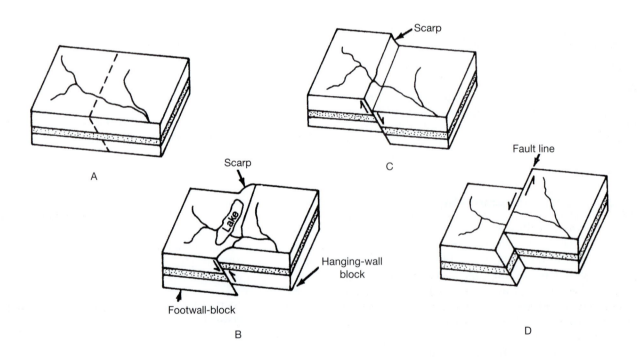

FIGURE 6.6 Types of fault movement: (a) block before movement; (b) reverse fault, or thrust fault, in which the hanging-wall block has moved up relative to the footwall block; (c) normal fault, in which the hanging-wall block has moved down relative to the footwall block; (d) strike-slip fault, in which the blocks on either side of the fault have moved sideways past each other. Arrows indicate relative motion of the blocks.

(Modified in part from McKenzie, Pettyjohn, and Utgard, 1975)

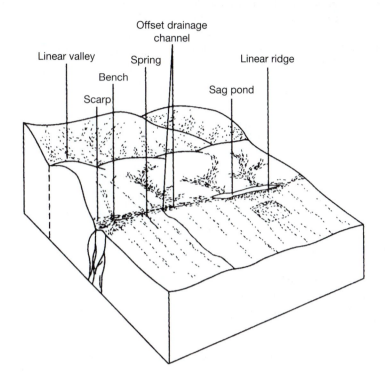

FIGURE 6.7 Distinctive landforms and drainage patterns aligned along a strike-slip fault are visible evidence that fault movement is recent enough to have interrupted the more gradual processes of erosion and deposition.

(Brown and Kockelman, 1983)

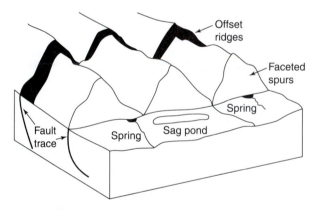

FIGURE 6.8 Typical geologic features found associated with a normal fault in mountainous terrain.

3. Sketch two utility poles on Figure 6.6a, placing one on each side of the fault. Use fault diagrams Figure 6.6b and c, and determine if the relative motion of the fault blocks in Question 2 indicates that it is a normal fault or reverse fault. Explain your reasoning.

San Andreas Fault

Study Figure 6.9, which is a photo of the San Andreas Fault, and then answer the questions below. It will help to review Figure 6.7.

4. What geological features can be used to identify the location of the fault? Outline the fault zone in Figure 6.9.

5. Does the fault zone consist of a single fracture or several parallel fractures? What is your evidence?

6. In addition to faults, what other natural or human-made features can create straight lines in topography? Are any of these features present in this photograph?

7. Indicate the direction of movement along the fault by drawing arrows on either side of the fault in Figure 6.9.

FIGURE 6.9   Aerial photograph of the San Andreas Fault in the Carrizo Plain at Wallace Creek, California. The fault runs horizontally across the middle of the photograph; the horizontal white line in the lower part of the photograph is a road. Agricultural features on the photo include fences and crop harvest patterns (Wallace, 1990).

Figure 2.17, on the back cover of this book, is a satellite image of the San Francisco Bay area. The yellow dots on this image are earthquake epicenters. Use this figure to answer the following questions.

8. Which features of strike-slip faults shown of Figure 6.7 can be seen easily on the satellite image? Why are some features easier than other features to see on the image?

9. How do the locations of earthquake epicenters help you determine geographic features that are related to faults?

10. Mark on the back cover (Figure 2.17) or on a tracing such as Figure 2.18, the traces of several major faults in this area.

Study the radar image of southern California in Figure 6.10. This image was made from an airplane that bounced radar waves off the earth. The radar is able to penetrate vegetation and clouds, so the image is very clear. The surface of the land is shown as if the sun lighted it, with bright slopes facing the sun and dark areas of shadow. Of course, in a radar image it is not sunlight but is the location of the airplane sending out the radar that creates the bright and shadow areas.

Different geologic and land use patterns are distinctive on the image. Areas with large shadows indicate high topography (mountains). Linear alignments of valleys, rivers, lakes, or mountains may indicate a fault zone. Flat, dark areas may be lakes, reservoirs, or the ocean. Mottled gray patterns, with very small rectangles, are urban areas, broken into blocks and dissected by rivers and freeways. Larger rectangular patterns can represent agricultural fields. Rivers often have winding patterns through mountain, agricultural, and urban areas.

11. What evidence of faults do you see on the radar image?

12. Draw on the image all the fault traces you can find.

13. Use the Web, or material provided by the instructor, to label faults including the Elsinore and San Andreas on Figure 6.10.

14. Find and label an example of the following: agriculture pattern (A), urban region (U), major highway (H), lake (L), and river (R).

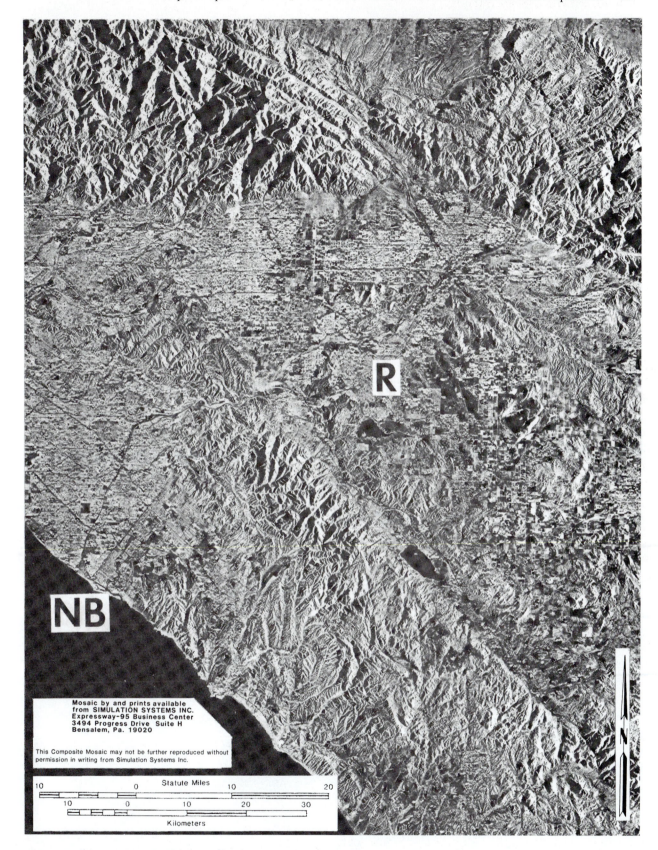

FIGURE 6.10   Radar image of southern California. R = Riverside, NB = Newport Beach.

(Courtesy of and copyrighted by Simulation Systems, Inc.)

Wasatch Fault

15. Study the map of the Draper, Utah, quadrangle (Figure 6.11 in the colored plates section of this book) and mark on the map the location of the Wasatch Fault along the mountain front. (Hint: Refer to Figure 6.8 for features found along normal faults, and begin your identification at Beaver Ponds Springs.)

Figure 2.16, also in the colored plates section at the back of the book, is a color stereo photograph of this area. Locate Beaver Ponds Springs on the photograph, and extend the fault based on topographic features.

Faults can also be seen on color oblique photographs. Use Figure 6.12 and identify on it the fault scarp and extend trace of the fault both north and south.

16. Using information on normal faults from Figures 6.6, 6.7, and 6.8, review the photographs of the site (Figures 2.16 and 6.12) and the map (Figure 6.11), and describe the location and appearance of at least one of each of the following features of a normal fault:

spring

sag pond

fault scarp

faceted spur

FIGURE 6.12 Salt Lake City oblique photo, Little Cottonwood Canyon. North to left. (U.S. Department of Agriculture, Soil Conservation Service, Salt Lake City, UT).

17. In the San Andreas Fault example (questions 4 and 5 above), lateral stream offset was an important clue to movement along the fault. Follow the traces of streams in Figure 6.11 as they flow west from the mountains to the valley. Is there any offset of streams where they cross the fault(s)? Why or why not?

18. Can you tell from the map (Figure 6.11) which side of the fault has moved up and which side has moved down? Describe your evidence. What additional kinds of information might be helpful in determining the movement along the fault?

Faults in Forests (Bainbridge Island, Washington)

Urban areas and thick forests are two environments in which it can be difficult to locate active or potentially active faults. In urban areas, faults may only be exposed in excavations for construction, or detectable if there is offset of roads or structures. In forested areas, virtual deforestation provided by LIDAR can allow the identification of faults. The example below is from such a forested area.

19. Look at the aerial photograph in Figure 6.13. Identify on this figure any linear zones you see that might be faults, being careful to avoid roads and property lines that are marked by cut forests.

20. Figure 6.14 is a LIDAR image of the same area, processed to remove the forest from the image. Mark on this figure any traces of faults that you see. Compare the trace of the fault with the drawings in Figures 6.7 and 6.8. What are the surface features that you see along the fault? What kind of fault movement and offset is most likely here? (Circle One) Normal, Reverse, Strike slip?

PART C. RECOGNITION OF NONSTRUCTURAL HAZARDS

This exercise looks at the recognition of nonstructural earthquake hazards and approaches the question, how safe are we in our daily lives? Recognition of nonstructural hazards, however, is something that everyone should know to be able to lead safer lives in earthquake country.

The structure of a building consists of those parts that help it stand up and withstand the forces of weight, wind, and earthquakes that may impact it. Everything else in a building is nonstructural. Structural failure can cause partial or total collapse of a building and injury or death to occupants and those

near a building. Nonstructural hazards can be caused by things the occupants of a building do, such as hanging plants or positioning bookshelves, or they may arise from the failure of the integral components of buildings such as water pipes, ventilation systems, and electrical systems. Nonstructural hazards may also be external, such as decorative trim that can fall off a building (Figure 6.15) or glass that can break out of windows. In summary, nonstructural hazards include building furniture, utility systems, and internal and external trim and decoration. The behavior of these items in an earthquake can cause damage, destruction, and injury, even if the building remains standing during and after an earthquake.

According to the U.S. Geological Survey, falling objects and toppling materials present the greatest hazards in earthquakes. Falling objects account for about two-thirds of the casualties from earthquakes. Also, replacement of these materials and loss of building use can be very expensive.

In this exercise you will draw a careful sketch of a room (or other indoor site) and complete a checklist to identify and comment on the hazards you find.

QUESTIONS (6, PART C)

1. Select a site that you frequently use. Make a sketch of this room on the graph paper provided in Figure 6.16. The choice of site is up to you. It could be a dorm room, a bedroom at home, a place where you work, a place where you study, or some other room or facility. You may do a map view or an elevation, but there is no need to do both. Identify and label the nonstructural hazards that you find. Include a scale on your drawing.

2. The list in Table 6.5 identifies some of the nonstructural hazards that you may encounter in your search. This list is not intended to be comprehensive; there are undoubtedly some missing hazards. Space is left at the bottom of the list for you to add other hazards that you discover.

Identify the room and building in the table title. Then identify all hazards in the room that you select. In the space provided, make brief comments about the specific nature of each hazard and your vulnerability (such as "Textbooks in environmental geology may fall off shelf and hit me"). Remember, in analyzing earthquake hazards you need to imagine what would happen to the objects around you if they were suddenly launched horizontally.

PART D. EARTHQUAKE PREPARATION AND HAZARD REDUCTION

This exercise explores specific coping techniques to reduce earthquake hazards and improve safety during and after an earthquake. Consider situations in which electric power and water supplies are not available; ATMs, credit cards and other sources of money are not

FIGURE 6.13 Aerial photograph of a portion of Bainbridge Island, WA. (USGS)

functioning; and the only transportation is by foot or bicycle. By imagining the conditions in a disaster, seeking information from various sources on preparation and coping, and preparing for the disaster, you and your family, friends, and coworkers or students can minimize losses. And remember, many of the preparations for an earthquake can be transferred to other types of natural and human-induced disasters. Even if you do not live in an earthquake hazard area, an earthquake could occur when you are traveling. It makes sense to understand the nature of disasters and to prepare for them.

Use a separate sheet of paper for answers to Questions 1 and 2.

1. Assume that you have three different amounts to spend on earthquake preparation for you or your family: $25, $100, and $200. Develop specific earthquake safety strategies for each amount. Explain what you would buy, how you would store

it, and how you would cope with safety and survival issues in and after an earthquake. Are there simple lifestyle changes before the earthquake that would also help you? You may find it helpful to use your textbook or other sources from a library to do research on earthquake hazard reduction.

2. Imagine that an earthquake occurs while you are in a class. What would you want from your educational institution immediately? What would you want over the next few days? How soon would you want to be back in class? Develop a separate list of items that you believe your educational institution should be ready to provide or situations that it should be prepared to deal with after an earthquake. Divide your list into two sections, with the first section itemizing needs during the first 72 hours, and the second section listing longer-term considerations. Who do you contact at your school to determine which preparations on your list have been made and which are still needed?

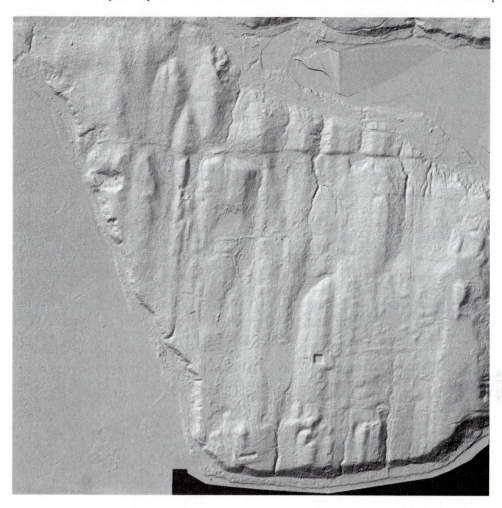

FIGURE 6.14   LIDAR image of part of Bainbridge Island, WA. (Image courtesy of Puget Sound Lidar Consortium)

FIGURE 6.15   The statue of Louis Aggasiz at Stanford University after the 1906 earthquake illustrates what can happen when large unsecured objects fall.
(W.C. Mendenhall, USGS).

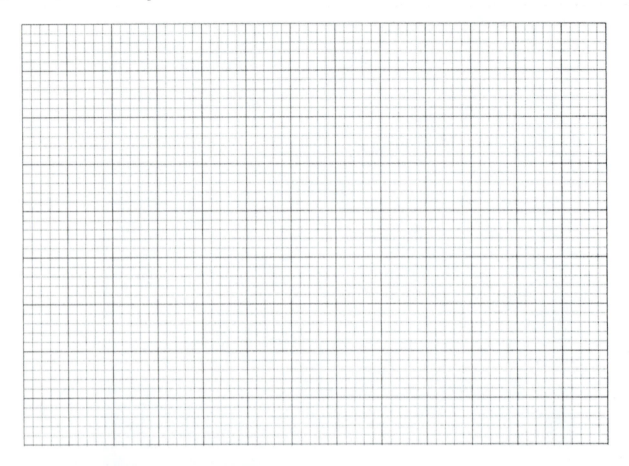

FIGURE 6.16   Sketch of _____ (your caption)

TABLE 6.5 List of Possible Hazards in _____.

Type of Hazard	Nature and Effect of Hazard	Possible Methods to Reduce Hazard
Windows: may implode in large pieces; consider proximity to furniture, especially beds		
Bookcases: may topple if not secured		
Books: may fall out of bookshelves		
Furniture: may slide across room; if tall and narrow, such as files, may tip over		
Cabinets: doors may open and contents may fall out if unsecured		
Storage items on desks, shelves, or other surfaces: may fly or fall if unsecured; special hazards if toxic or flammable		
Fixtures: heavy objects may fall if unsecured or poorly secured		
Hanging objects: if unsecured, plants, pictures, mirrors may fall		
Lights: can swing, shatter, and fall		
Exits: consider conditions that would block room exits, such as toppled furniture		
Lifelines: problems in room if adjacent to lifelines, such as water pipes, break		
Other items:		

Bibliography

Algermissen, S. T., 1969, *Seismic risk studies in the United States*. Fourth World Conference on Earthquake Engineering: Santiago, Chile, v. 1, no. 26.

Algermissen, S. T., and Perkins, D. M., 1976, *A probabilistic estimate of maximum acceleration in rock in the contiguous United States*: U.S.G.S. Open File Report 76–416, 45 p.

Algermissen, S. T., Perkins, D. M., Thenhaus, P. C., Hanson, S. L., and Bender, B. L., 1982, *Probabilistic estimates of maximum acceleration and velocity in rock in the contiguous United States*, U.S. Geological Survey Open-File Report 82-1033, 99 p.

Berthke, P. R., and Beatley, T., 1992, *Planning for earthquakes, risk, politics and policy*: Baltimore, MD, Johns Hopkins University Press, 210 p.

Bolt, B. A., 2003, *Earthquakes*: New York, W. H. Freeman and Company, 320 p.

Brown, R. D., and Kockelman, W. J., 1983, *Geologic principles for prudent land use—A decision maker's guide for the San Francisco Bay region*: U.S.G.S. Professional Paper 946, 97 p.

Coburn, A., and Spence, R., 1992, *Earthquake protection*: New York, Wiley, 355 p.

Cluff, L. S., and Bolt, B. A., 1969, Earthquakes and their effects on planning and development. In *Urban environmental geology in the San Francisco Bay region*: ed. E. A. Danehy. Sacramento, CA, Special Publication of the San Francisco Section of the Association of Engineering Geologists, p. 25–64.

Earthquake Engineering Research Institute, 1996, *Scenario for a magnitude 7.0 earthquake on the Hayward Fault*: Pub. HF-96, Oakland, CA, 109 p.

Eisman, D. B., 1972, Dust from antelope: *California Geology*, v. 25, no. 8, p. 171–73. Originally printed in *Historic fact and fancies*, (1913?), California Federated Women's Clubs.

Executive Office of the President, Office of Science and Technology, 1970, *Report of the task force on earthquake hazard reduction*: Washington, D.C., U.S. Government Printing Office, 54 p.

Frankel, Arthur, Mueller, Charles, Barnhard, Theodore, Perkins, David, Leyendecker, E.V., Dickman, Nancy, Hanson, Stanley, and Hopper, Margaret, 1997, Seismic-hazard maps for the conterminous United States, U.S. Geological Survey Open-File Report 97-131-F.

Grey, Z., 1750, *A chronological and historical account of the most memorable earthquakes that have happened in the world, from the beginning of the Christian period to the present year 1750*: Cambridge, UK, J. Bentham.

Hanks, T. C., 1985, *National earthquake hazard reduction program - Scientific status*: U.S.G.S. Bulletin 1659, 40 p.

Hays, W. W., 1980, *Procedures for estimating earthquake ground motions*: U.S. Geological Survey Professional Paper 1114, 77 p.

Hays, W. W., 1981, Hazards from earthquakes. In *Facing geologic and hydrologic hazards*, ed. W. W. Hays. U.S.G.S. Professional Paper 1240-B, 108 p.

Hays, W. W., and Gori, P. L., eds., 1986, *Earthquake hazards in the Puget Sound, Washington, area*. Workshop presented at Conference XXXIII: U.S. Geological Survey Open-File Report 86-253, 238 p. plus appendices.

Hopper, M. G., Langer, C. J., Spence, W. J., Rogers, A. M., and Algermissen, S. T., 1975, *A study of earthquake losses in the Puget Sound, Washington, area*: U.S. Geological Survey Open-File Report 75-375, 298 p.

Iacopi, R., 1971, *Earthquake country* (3rd ed.): Menlo Park, CA, Lane Books, 160 p.

Keller, E. A., 2000, *Environmental geology* (8th ed.): Upper Saddle River, NJ, Prentice Hall, 562 p.

Keller, E. A., and Pinter, N., 1996, *Active tectonics*: Upper Saddle River, NJ, Prentice-Hall, 338 p.

McKenzie, G. D., Pettyjohn, W. A., and Utgard, R. O., 1975, *Investigations in environmental geoscience*: Minneapolis, Burgess, 180 p.

Murphy, L. M. and Ulrich, F. P., 1951, *United States earthquakes 1949*, U.S. Department of Commerce, Coast and Geodetic Survey Serial 748, p. 20–28.
Reprinted in Thorsen, G. W., comp., 1986, *The Puget lowland earthquakes of 1949 and 1965*: Washington Division of Geology and Earth Resources Information Circular 81, p. 25–34.

Nichols, D. R. and Buchanan-Banks, J. M., 1974, *Seismic hazards and land-use planning*: U.S.G.S. Circular 690, 33 p.

Peterson, M. D,. Frankel, A. D., Harmsen, S. C., Mueller, C. S., Haller, K. M., Wheeler, R. L., Wesson, R. L., Seng, Y., Boyd, O. S., Perkins, D. M., Luco, N., Field, E. H., Wills, C. J., and Rukstales, K. S., 2008, Documentation for the 2008 Update of the United States National Seismic Hazard Maps: U.S. Geological Survey Open-File Report 2008–1128, 120 p.

Reiter, L., 1991, *Earthquake hazard analysis*: New York, Columbia University Press, 254 p.

Reitherman, R., 1985, *Reducing the risks of nonstructural earthquake damage: A practical guide*: Federal Emergency Management Agency, Earthquake Hazards Reduction Series 1, 87 p.

Rogers, A. M., Walsh, T. J., Kockleman, W. J., and Priest, G. K., 1998, Assessing earthquake hazards and reducing risk in the Pacific Northwest, Volume 2: USGS Professional Paper 1560. Available on line http//pubs.usgs.gov/prof/p1560/p1560po.pdf

Solomon, B. J., Storey, N., Wong, I., Silva, W., Gregor, N., Wright, D., and McDonald, G., 2004, Earthquake-hazards scenario for a M7 earthquake on the Salt Lake City segment of the Wasatch fault zone, Utah, CD-ROM, ISBN 1-55791-704-3.

The Federal Emergency Management Agency offers a series of brochures on earthquake preparation. Two that may be particularly helpful are FEMA 113, Family earthquake safety home hazard hunt and drill, and FEMA 46, Earthquake safety checklist.

Wallace, R. E., 1974, *Goals, strategy, and tasks of the earthquake hazard reduction program*: U.S.G.S. Circular 701, 26 p.

Wallace, R. E., ed., 1990, *The San Andreas fault system*: U.S.G.S. Professional Paper 1515, 283 p.

Weaver, C. S., and Smith, S. W., 1983, Regional tectonic and earthquake hazard implications of a crustal fault zone in southwestern Washington: *Journal of Geophysical Research*, v. 88, no. B12, p. 10371–83.

Wong, I., Silva, W., Olig, S., Thomas, P., Wright, D., Ashland, F., Gregor, N., Pechmann, J., Dober, M., Christenson, G., and Gerth, R., Earthquake scenario and probabilistic ground shaking maps for the Salt Lake City, Utah, metropolitan area, CD-ROM, ISBN 1-55791-666-7.

Yanev, P., 1991, *Peace of mind in earthquake country*: San Francisco, Chronicle Books, 218 p.

Yanev, P. I., 1991, *Peace of mind in earthquake country*: San Francisco, Chronicle Books, 218 p.

Yeats, R. S., 2004, *Living with earthquakes in the Pacific Northwest*: Corvallis, Oregon State University Press, 390 p.

The Loma Prieta Earthquake of 1989

INTRODUCTION

As many people prepared to watch World Series baseball on October 17, 1989, the earth in the San Francisco Bay region shook from a major earthquake. The U.S. Geological Survey has named this earthquake after the small community of Loma Prieta near the epicenter. Although San Francisco and Oakland received the most media coverage for their damage, it is important to note that these large communities were about 100 km away from the epicenter. Closer communities, such as Watsonville and Santa Cruz, also suffered great damage and fatalities in the earthquake. This exercise looks at the regional geology of the San Andreas Fault and the geology of areas in San Francisco and Oakland that were damaged.

Damage that results from an earthquake is a result of many factors. Among these are

1. The geologic nature of the earthquake and the rocks between the focus and a site

   a. the type of fault movement: different earthquakes generate different frequencies

   b. the size of the earthquake: larger earthquakes will release more energy, and therefore cause more acceleration and have longer duration

   c. the distance from the source of the earthquake to a site: in general, sites closer to the focus fare worse

   d. characteristics of the rocks through which the seismic waves travel from the source to the site: some rocks transmit seismic energy better than others

2. The types and properties of the materials at the site determine the ground motions at the site

   a. soil thickness: shallow soils may not perform as well as deep, well-consolidated soils

   b. soil saturation: saturated soils will typically perform less well than dry soils

   c. soil grain size and sorting: well-sorted, fine-grained sands and coarse silts that are geologically young have the highest potential to liquefy (flow) if saturated; soils that liquify lose their ability to hold up structures such as buildings and bridges

   d. types and properties of the bedrock at the site: unweathered igneous and other massive rocks typically perform better than weak and fractured or jointed rocks

3. The nature of the building

   a. architectural simplicity of the building: seismic response of simple buildings is easier to predict and control than the response of complex buildings

   b. size and use of the building: a single family home will behave differently than a large commercial or industrial building

   c. type of construction: small wood structures will typically flex more and therefore perform better in earthquakes than brick or concrete block buildings

   d. type of seismic design considerations: is the building new or old, is it built to appropriate seismic standards or not?

The emphasis in this exercise is on the geology of the earthquake and the geologic reasons for building failures in different areas. The actual damage that occurred to specific buildings and other structures is not covered.

This exercise is divided into several parts. We first explore the nature of the San Andreas Fault, general seismic activity in the area, the specific sequence of Loma Prieta earthquakes, and seismic attenuation in this area. We will investigate the geology and hazards in the San Francisco Bay region in general, and focus in particular on the Marina district.

PART A. THE GEOLOGY AND SEISMIC ATTENUATION OF THE LOMA PRIETA EARTHQUAKE

Geology (7, Part A1)

Figure 7.1 shows and names many of the major faults in this area. Figure 7.2 shows the distribution of earthquakes along the San Andreas Fault before and immediately after the Loma Prieta earthquake. Figure 7.3 shows the distribution of the aftershocks of the Loma Prieta earthquake in map view and cross section. Use these figures to answer the following questions.

QUESTIONS (7, PART A1)

1. a. Describe the geographic orientation (compass direction) of the faults shown in Figure 7.1.

b. What is the large-scale geological process that is occurring in this part of California that accounts for fault orientation and motion?

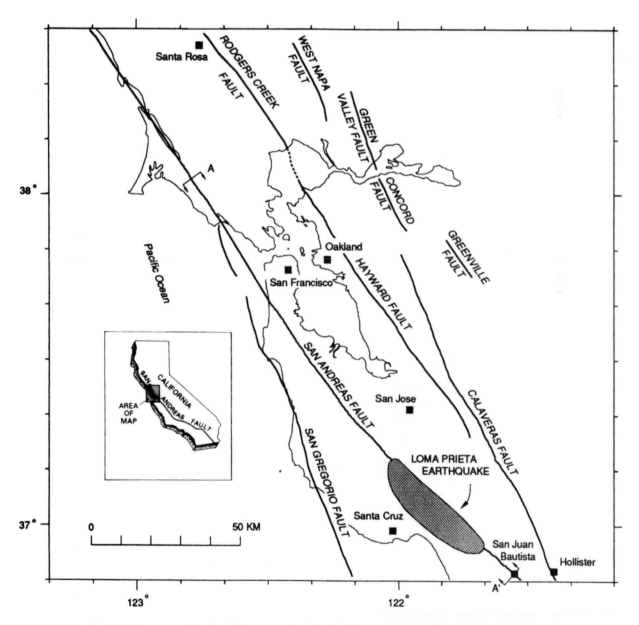

FIGURE 7.1   Major faults in the area of the Loma Prieta earthquake (Working Group, 1990).

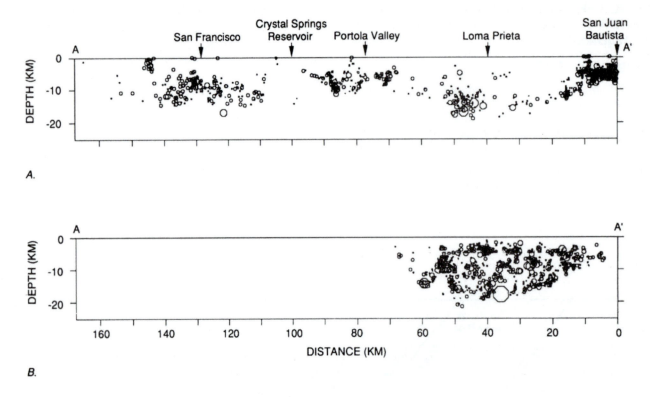

FIGURE 7.2 (a) Seismicity (1969–89) before the Loma Prieta earthquake along the San Andreas Fault. See Figure 7.1 for A-A'; size of symbols increases with increasing magnitude. (b) The Loma Prieta earthquake and its aftershocks (Working Group, 1990).

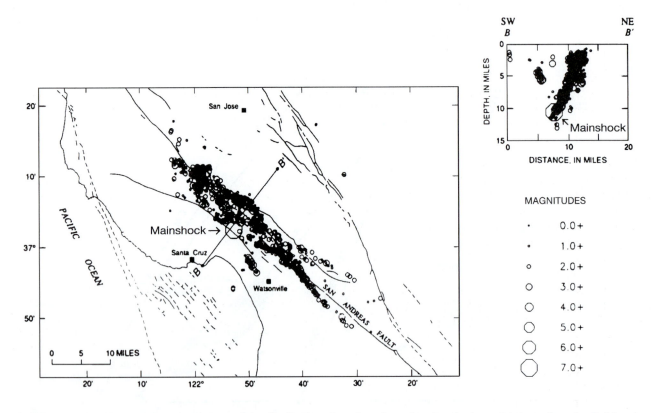

FIGURE 7.3 Map and cross-section (B-B') views of the distribution of earthquakes related to the Loma Prieta earthquake. Solid and dashed lines represent faults (except coast) (Pflaker and Galloway, 1989).

2. One theory for the prediction of earthquakes suggests that new events will fill in gaps where earthquakes have not been recently recorded. The gaps are presumed to be zones where stress is accumulating, which will be released in a future earthquake. In areas along faults with frequent earthquakes, stress is released in each event.

a. Describe the differences in locations of epicenters for earthquakes that occurred before the Loma Prieta earthquake (Figure 7.2a) and the main shock and aftershocks (Figure 7.2b and shaded zone in Figure 7.1).

b. Do the Loma Prieta earthquake and its aftershocks fit the pattern of new events filling gaps? Explain.

3. Figure 7.3 shows a map and an insert with a cross-section view along line B-B'; of the Loma Prieta earthquake and its aftershocks.

a. Looking at the map view, would you infer that all of the earthquakes occurred along the San Andreas Fault?

b. Compare the map view with the cross-section view. What would you infer from the cross-section view about the number of faults?

c. What is the evidence for your inferences?

Seismic Attenuation and Acceleration (7, Part A2)

Attenuation is the decrease in the amplitude of seismic waves as distance from the source increases. In general, there is a geometric decrease in amplitude with distance. Selected conditions, however, tend to cause local variations in amplitude such that sites distant from the epicenter will show an amplification of ground motion.

Seismologists use sophisticated instruments to measure many different components of the earthquake waves that travel within and along the surface of the Earth. These components include acceleration, velocity, and total displacement at a site. For example, in automobiles acceleration is how fast the vehicle starts from a stop (the faster it starts, the more you are pushed back in your seat), velocity is how fast it is going (has it reached 25 mph?), and displacement is how far it has gone (has it moved a mile?). These are practical measurements for an earthquake because the damage a building sustains can be related to how quickly it is hit by the shock waves, how fast it moves, or how far it moves.

Accelerographs, which measure acceleration of the ground, record data that is typically presented as a percentage of the force of gravity (g). Buildings are designed to withstand the vertical force of gravity (1 g = 980 cm/sec/sec); that is, they are designed to stand up under gravitational forces. They are not, however, necessarily designed to be moved sideways, and earthquakes generate strong sideways forces against a building.

Figure 7.4 is a map showing the horizontal acceleration from the Loma Prieta earthquake as recorded at many sites in western California.

QUESTIONS (7, PART A2)

1. Describe the general relationship between the recorded acceleration at different sites and the epicentral area, which is indicated by shading.

2. a. Draw a contour line around the sites with 20% g in Figure 7.4. 20% g is a value for moderate damage.

b. What is the general shape of the contour?

c. How does its shape compare with the orientation of the faults that are shown in Figure 7.1?

3. What geological factors contribute to the difficulties in drawing the irregular 20% g isoseismal line?

4. From the data presented in Figure 7.4, summarize the role of distance from the epicenter in attenuating earthquake waves.

5. Compare the data shown in Figure 7.5, which is an isoseismal map showing areas of similar damage to structures as measured by the Modified Mercalli Intensity scale, with the acceleration numbers shown in Figure 7.4. Describe the correlation between acceleration and seismic impacts as recorded on the Mercalli scale.

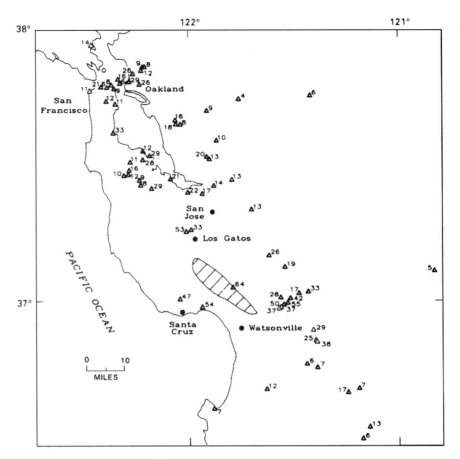

FIGURE 7.4 Horizontal acceleration as a percent of gravity (980 cm/sec/sec) at measured sites for the Loma Prieta earthquake. The lined region is the epicentral area (Pflaker and Galloway, 1989).

PART B. GEOLOGY AND SEISMIC IMPACTS: SAN FRANCISCO BAY AND MARINA DISTRICT

San Francisco Bay Region (7, Part B1)

Figures 7.6 (geologic map), 7.7 (maximum predicted intensity), and 7.8 (1906 apparent intensity) show different factors of the geology and the expected and actual impacts of earthquakes in the San Francisco Bay region. The regional setting for these maps is shown on Figure 2.17, the satellite image on the back cover of this book.

QUESTIONS (7, PART B1)

Read and compare the maps in Figures 7.6, 7.7, and 7.8 to answer the following questions.
1. a. What two geologic units in San Francisco and Oakland have the maximum predicted intensities for earthquakes?

b. What are the geologic conditions that have lead these areas to have high predicted maximum intensities?

2. Compare Figure 7.6 with Figure 7.8.
a. What geologic conditions (materials and setting) led to the maximum intensities during the 1906 earthquake?

b. What geologic conditions led to the minimum intensities during the 1906 earthquake?

3. a. Using data from all three maps, determine typical geologic conditions that are likely to create areas with maximum seismic impacts. Explain these conditions below.

b. Using data from all three maps, determine typical geologic conditions that are likely to create areas with minimum seismic impacts. Explain these conditions below.

The Marina District (7, Part B2)

Impressive photographs and videos were made of damage from the Loma Prieta earthquake in the Marina District in northern San Francisco. This was an area of extensive collapsed and damaged buildings

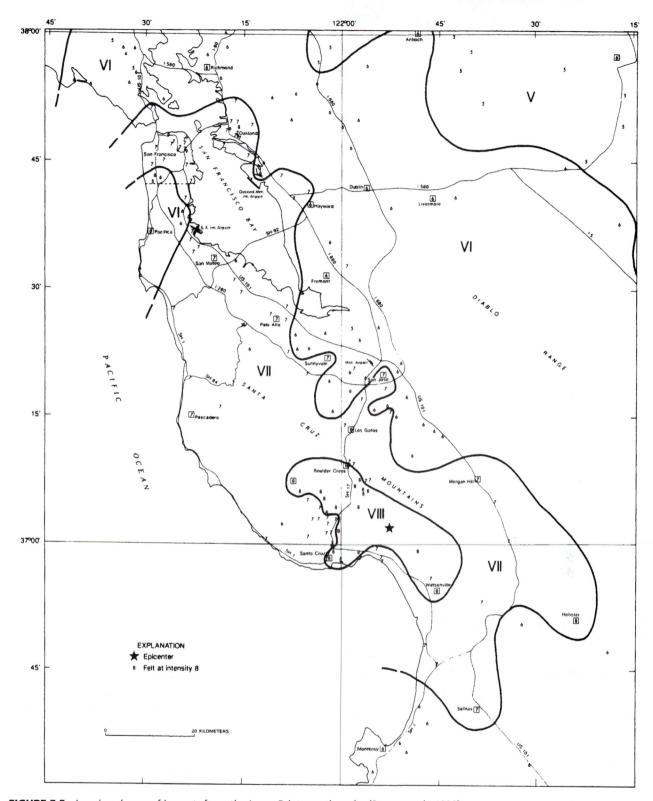

FIGURE 7.5   Isoseismal map of impacts from the Loma Prieta earthquake (Stover et al., 1990).

and dramatic fires. Although much of the problem with building collapse in this area was due to the structural style of the buildings, local geology also played an important role.

Figure 7.9 is a map of the Marina District, showing in detail the dates when artificial fill was placed in San Francisco Bay (this area is shown only as Qm on the larger geologic map and is located just west of Fort Mason; see map labels). Use this figure, and Figures 7.10 (types of fills) and 7.11 (location of sand boils) to answer the following questions. Figure 7.12 shows changes in seismic response at three temporary seismograph stations. You may also wish to refer to earlier figures in this exercise.

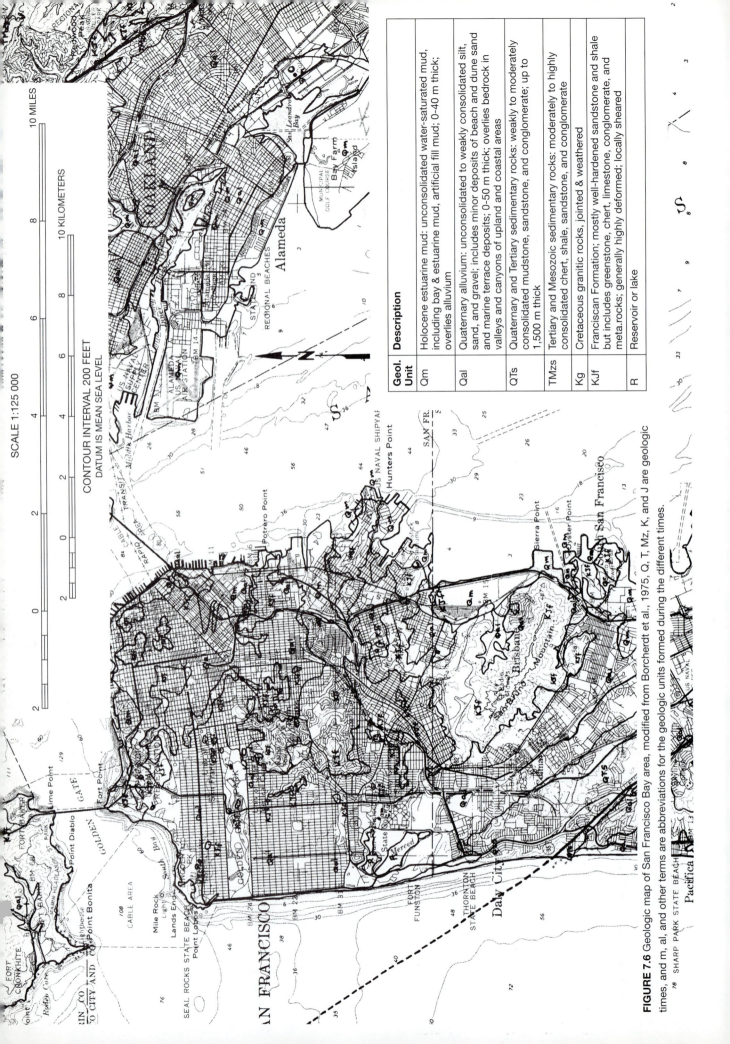

FIGURE 7.6 Geologic map of San Francisco Bay area, modified from Borcherdt et al., 1975. Q, T, Mz, K, and J are geologic times, and m, al, and other terms are abbreviations for the geologic units formed during the different times.

SCALE 1:125 000

CONTOUR INTERVAL 200 FEET
DATUM IS MEAN SEA LEVEL

Geol. Unit	Description
Qm	Holocene estuarine mud: unconsolidated water-saturated mud, including bay & estuarine mud, artificial fill mud; 0-40 m thick; overlies alluvium
Qal	Quaternary alluvium: unconsolidated to weakly consolidated silt, sand, and gravel; includes minor deposits of beach and dune sand and marine terrace deposits; 0-50 m thick; overlies bedrock in valleys and canyons of upland and coastal areas
QTs	Quaternary and Tertiary sedimentary rocks: weakly to moderately consolidated mudstone, sandstone, and conglomerate; up to 1,500 m thick
TMzs	Tertiary and Mesozoic sedimentary rocks: moderately to highly consolidated chert, shale, sandstone, and conglomerate
Kg	Cretaceous granitic rocks, jointed & weathered
KJf	Franciscan Formation; mostly well-hardened sandstone and shale but includes greenstone, chert, limestone, conglomerate, and meta.rocks; generally highly deformed; locally sheared
R	Reservoir or lake

SCALE 1:125 000

CONTOUR INTERVAL 200 FEET
DATUM IS MEAN SEA LEVEL

This map delineates potentially hazardous areas during large
earthquakes on the San Andreas and the Hayward fault.
Below, the earthquake zones are given with predicted
intensity and approximate Mercalli Scale intensity.

Zone	Predicted Intensity	Mercalli Intensity
A	Very violent	X–XII
B	Violent	VIII–XI
C	Very strong	VIII
D	Strong	VI–VII
E	Weak	VI

FIGURE 7.7 Predicted maximum earthquake intensity, San Francisco Bay area (Borcherdt et al., 1975).
This map delineates potentially hazardous areas during large earthquakes on the San Andreas and the Hayward Faults.
The earthquake zones are given on the right with predicted intensity and approximate Mercalli scale intensity.

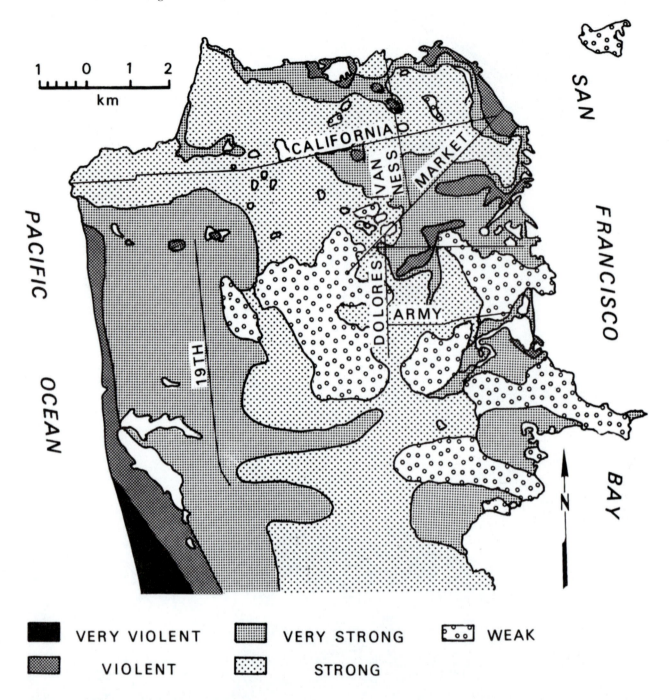

FIGURE 7.8   Apparent intensities of the 1906 San Francisco earthquake.

(After Wood, 1908; in Borcherdt et al., 1975)

QUESTIONS (7, PART B2)

1. What are the ages of fills in the Marina District (Figure 7.9)? Was any fill emplaced in the last 50 years? How well engineered for seismic response do you think the fill is? Explain your answer.

2. a. What is the relationship between the age and kinds of fills?

b. How do you interpret the term *hydraulic*?

c. What is the natural coastal landform that existed at "A" (Figure 7.9) prior to 1912?

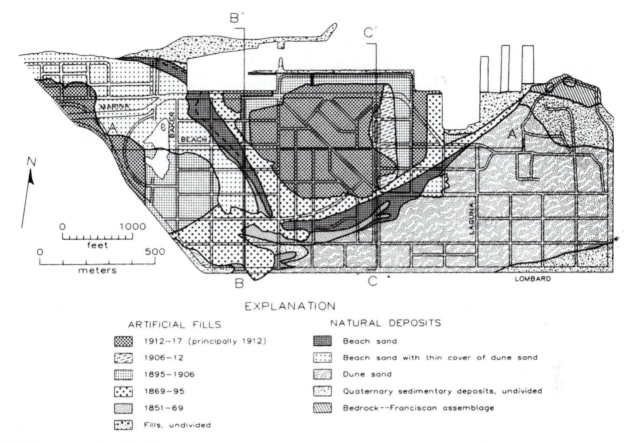

FIGURE 7.9   Dates of artificial fills and types of natural deposits, Marina District, San Francisco (Bonilla, 1992).

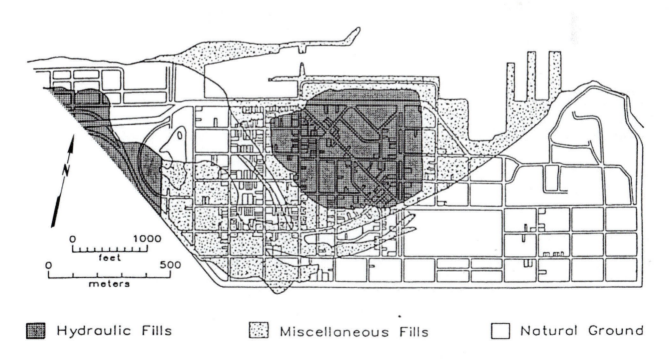

FIGURE 7.10   Types of artificial fills, Marina District, San Francisco (Seekings et al., 1990).

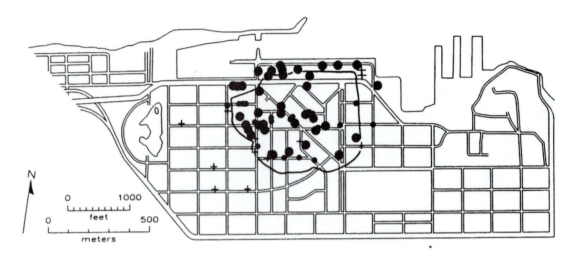

EXPLANATION

+ Brown sand boils, definite

+ Brown sand boils, probably
   from pipe breakage

● Gray sand boils, definite

• Gray sand boils, probably
   from pipe breakage

FIGURE 7.11 Locations of sand boils, Marina District, San Francisco (Bennett, 1990).

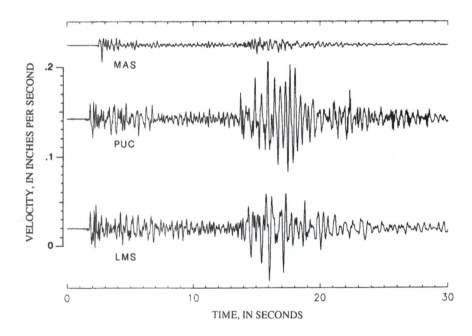

FIGURE 7.12 Vertical velocities at three seismic stations in and near the Marina district, from a magnitude 4.6 aftershock to the Loma Prieta earthquake. MAS is a site on bedrock. PUC is located on natural sand deposits and LMS is located on fill (Pflaker and Galloway, 1989).

3. Sand boils are sites of liquefaction (where the ground changes from behaving as a solid to behaving as a fluid, like quicksand, when shaken). Use Figure 7.11 and explain the relationship of sand boils to areas of fill. Why does this relationship exist?

types of geologic deposit? If you wanted to build near the Marina district, but yet be relatively safe from earthquake damage (assuming that you were building an appropriate, seismic-resistant structure), where would be a safer place to build? Mark the site on Figure 7.9, and explain your choice.

4. Figure 7.12 presents three seismic responses. The larger squiggles on the figure represent greater velocities when the sites were shaken. What is the relationship of velocity to

Bibliography

Baldwin, J. E., and Sitar, N., 1991, *Loma Prieta earthquake: Engineering geologic perspectives:* Association of Engineering Geologists, Special Publication No. 1, 170 p.

Bennett, M. J., 1990, Ground deformation and liquefaction of soil in the Marina District, in *Effects of the Loma Prieta Earthquake on the Marina District, San Francisco, California:* U.S. Geological Survey Open-File Report 90–253, p. 39–43.

Bonilla, M. G., 1992, Geological and historical factors affecting earthquake damage, in O'Rourke, T. D., ed., 1992, *The Loma Prieta California, Earthquake of October 17, 1989—Marina District:* U. S. Geological Survey Professional Paper 1550-F, p. 7–28.

Borcherdt, R. D., 1975, *Studies for seismic zonation of the San Francisco Bay region:* U.S. Geological Survey Professional Paper 941A, 102 p.

Borcherdt, R. D., Gibbs, J. F., and Lajoie, K. R., 1975, *Maps showing maximum earthquake intensity predicted in the Southern San Francisco Bay region, California, for large earthquakes on the San Andreas and Hayward Faults:* U.S. Geological Survey Miscellaneous Field Studies Map MF-709.

Johnston, M. J. S., ed., 1993, *The Loma Prieta, California, Earthquake of October 17, 1989—Preseismic observations:* U. S. Geological Survey Professional Paper 1550-C, 85 p.

Kisslinger, C., 1992, Sizing up the threat: *Nature,* v. 355, no. 6355, p. 18–19.

National Research Council Geotechnical Board, 1994, *Practical lessons from the Loma Prieta earthquake:* Washington, DC, National Academy Press, National Academy of Sciences, 274 p.

Pflaker, G., and Galloway, J. P., eds., 1989, *Lessons learned from the Loma Prieta, California, earthquake of October 17, 1989:* U.S. Geological Survey Circular 1045, 48 p.

Seekings, L., Lew, F., and Kornfield, L., 1990, Areal distribution of damage to surface structures, In *Effects of the Loma Prieta Earthquake on the Marina District, San Francisco, California:* U.S. Geological Survey Open-File Report 90–253, p. 39–43.

Sleeter, B. M., Calzia, J. P., Walter, S. R., Worg, F. L., and Saucedo, G. J., 2004, *Earthquakes and faults in the San Francisco Bay area (1970–2003):* U.S. Geological Survey Scientific Investigations Map 2848.

Stover, C. W., Reagor, B. G., Baldwin, F. W., and Brewer, L. R., 1990, *Preliminary isoseismal map for the Santa Cruz (Loma Prieta), California, earthquake of October 18, 1989, UTC:* U.S. Geological Survey Open-File Report 90–18, 8 p.

U.S. Geological Survey, 1990, The *next big earthquake in the Bay area may come sooner than you think:* (insert in local newspaper): U.S. Geological Survey, Menlo Park, California, 24 p.

Wallace, R. E., ed., 1990, *The San Andreas Fault System, California:* U. S. Geological Survey Professional Paper 1515, 283 p.

Working Group on California Earthquake Probabilities, 1990, *Probabilities of large earthquakes in the San Francisco Bay region, California:* U.S. Geological Survey Circular 1053, 51 p.

Landslides and Avalanches

INTRODUCTION

Downslope movement, or mass movement, of earth materials is taking place continuously at rates that vary from the very slow creep of soil and rock to extremely rapid rockfalls and avalanches. The movement may involve only a few cubic yards or as much as 5 cubic miles of widely differing consolidated and unconsolidated materials. The primary purpose of this exercise is to understand the nature of mass wasting and some of the ways in which it impacts the human colony. We will investigate the types and causes (including human activity) of mass wasting and options for damage reduction with examples of landslides from California, Montana, Ohio, and Washington and snow avalanches from Colorado and Utah. Snow avalanches are a major winter hazard in mountainous terrain throughout the world. More people than just skiers, snowmobilers, and owners of mountain cabins or homes are at risk. Because many avalanche paths cross highways or railroads, snowslides may impact travelers as well.

Avalanche risks can be high throughout the year in mountains with glaciers or permanent snowfields.

Geologists and engineers have classified slope movements (e.g., Sharpe, 1938; Varnes, 1978). Although numerous classification schemes have been devised, the one that is probably used most today is the scheme developed by the National Research Council (Table 8.1).

The most common types of slope movement are falls; slides, which are classified as rotational (slumps) and translational; and flows (Table 8.1 and Figures 8.1 and 8.2). The term landslide, which will be used in this exercise, remains a widely used nontechnical term for most perceptible forms of downslope movement even though some by definition involve little or no sliding. Landslides are a geologic hazard that annually causes millions of dollars of damage to structures and substantial loss of life. The most destructive landslides in recorded history occurred in Kansu Province, China. There more than

TABLE 8.1 Classification of Slope Movements

Type of Movement			Type of Material		
			Bedrock	Engineering Soils	
				Predominantly Coarse	Predominantly Fine
Falls			Rock fall	Debris fall	Earth fall
Topples			Rock topple	Debris topple	Earth topple
Slides	Rotational	Few units	Rock slump	Debris slump	Earth slump
	Translational	Many units	Rock slide	Debris slide	Earth slide
Lateral Spreads			Rock spread	Debris spread	Earth spread
Flows			Rock flow	Debris flow	Earth flow
Complex	Combination of two or more principal types of movement				
(modified from Varnes, 1978)					

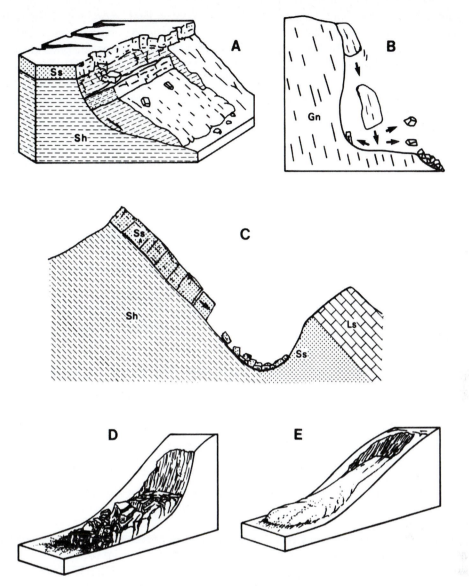

FIGURE 8.1   Common types of landslides. A, rock slump: rotational slide of coherent or intact masses that move downslope along a curved surface; B, rock fall: rock masses that move by falling through air; C, rock slide: translational movement downhill along a more or less planar surface; D, debris slide: broken masses of rock and other debris that move downslope by sliding on a surface under the deposit; E, earth flow: soil and other colluvial materials that move downslope in a manner similar to a viscous fluid. Ss = sandstone; Sh = shale, Gn = granite, Ls = limestone.

(Modified from Nilsen and Brabb, 1972; Varnes, 1978)

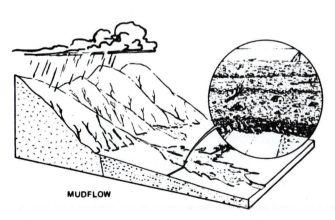

FIGURE 8.2   Mudflow deposits.

(From Van Horn, 1972)

100,000 people were killed in their homes dug into loess (weak wind-blown deposits) during an earthquake in 1920. The identification of former landslides and the assessment of the potential for future landslides, particularly under new land uses, are very important in land-use planning.

Landslides occur when the pull of gravity on earth materials overcomes their frictional resistance to downslope movement. Slope stability is affected by

1. *Type of earth materials present.* Unconsolidated deposits will move downslope more easily than bedrock.

2. *Structural properties of earth materials.* The orientation of layering of some rocks and sediments relative to slope directions, as well as the extent of fracturing of the materials, will affect landslide potential.

3. *Steepness of slopes.* Landslides are more common on steeper slopes.

4. *Water.* Landslides are generally more frequent in areas of seasonally high rainfall because the addition of water to earth materials commonly decreases their resistance to sliding by decreasing internal friction between particles (such as soil or sand grains) and by decreasing the cohesive forces that bind clay minerals together. Water also lubricates surfaces along which failure may occur; adds weight to the material; reacts with some clay minerals, causing volume changes in the material; and mixes with fine-grained unconsolidated materials to produce wet, unstable slurries.

5. *Ground shaking.* Strong shaking, for example during earthquakes, can jar and loosen earth materials making them less stable.

6. *Type of vegetation present.* Trees with deep, penetrating roots tend to hold bedrock and surficial deposits together, thereby increasing ground stability. Removing such trees can increase the likelihood of landslides. Trees add weight, however, and roots can break up rock.

7. *Proximity to areas undergoing active erosion.* Rapid undercutting and downcutting along stream courses and shorelines makes the resulting slopes particularly susceptible to landsliding. (Nilsen & Brabb, 1972)

The parts of a landslide are shown in Figure 8.3. The key nomenclature follows (from Nilsen and Brabb, 1972):

PART A. RECOGNIZING LANDSLIDES ON TOPOGRAPHIC MAPS, AERIAL PHOTOGRAPHS, AND LIDAR IMAGES

According to Nilson and Brabb (1972), landslide deposits are commonly characterized by one or more of the following features:

1. Small, isolated ponds, lakes, or other closed depressions
2. Abundant natural springs

Main scarp	The steep surface between the slide and the undisturbed ground. The projection of the scarp surface under the slide material is the rupture surface.
Minor scarp	A steep surface on the displaced material produced by differential movements within the slide mass.
Head	The upper part of the slide material along the contact between the displaced material and the main scarp.
Crown	The material that is still in place, undisturbed, adjacent to the highest parts of the main scarp.
Toe	The margin of displaced material most distant from the main scarp.

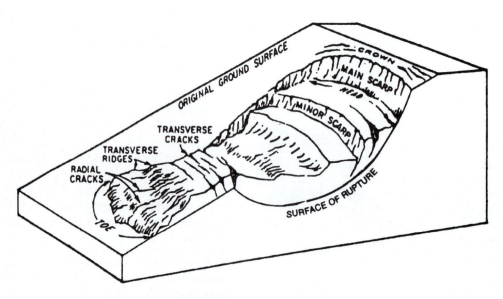

FIGURE 8.3 Parts of a complex (slump-earth flow) landslide (Nilsen and Brabb, 1972).

FIGURE 8.4 On an oblique photograph, contours on the upper, scarp, area of a landslide bend into the hill, and contours on the lower, hummocky, area of the landslide bend out from the hill. See Figure 8.3 for the parts of this landslide. Photograph courtesy of http://www .geog.ucsb.edu/~ jeff/projects/la_conchita/1995/95_slideshots.html

3. Abrupt and irregular changes in slope and drainage pattern

4. Hummocky irregular surfaces

5. Smaller landslide deposits that are commonly younger and form within older and larger landslide deposits (in other words, if an area has slid in the past, it is subject to both small and large renewed slides)

6. Steep, arcuate scarps at the upper edge

7. Irregular soil and vegetation patterns

8. Disturbed vegetation

9. Abundant flat areas

QUESTIONS (8, PART A)

La Conchita, California

Figure 8.4 shows a landslide that occurred in 1995 above the small town of La Conchita, California. This landslide reactivated in the wet winter of 2005, with disastrous consequences. Note the shape of contours as they cross the landslide. Figure 8.5 shows a vertical aerial photograph of the same site, and note how contours appear on this photograph.

1. a. Can you recognize the shape of contours that may indicate a landslide? Check with your Teaching instructor before going on to the next question.

b. In Figure 8.4 and 8.5, were any homes damaged? Explain.

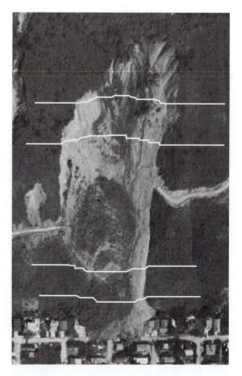

FIGURE 8.5 Contours that cross landslides, as shown on vertical photos, bend to follow concave part of slope in the upper scarp area of the slide. Contours in the lower part of a landslide will be convex and irregular over hummocky terrain. La Conchita, California, 1995; Same site as in Figure 8.4. Photograph courtesy of Pacific Aerial Surveys, http://www.geog.ucsb.edu/~ jeff/projects/la_conchita/1995/95_slideshots.html

Gardiner, Montana

Figure 8.6 is a photograph of the lower part of a landslide complex south of Gardiner, Montana. Direction of view in the photograph is to the south from the townsite. This is an old landslide, but it well illustrates hummocky terrain related to landslide deposition. Refer to Figure 8.7, the aerial photograph, and Figure 8.8, the topographic map.

2. Mark on the aerial photograph (Figure 8.7) the area of the landslide(s). Look for irregular terrain, typified by hummocky topography.

3. On the topographic map (Figure 8.8), mark areas of the landslide(s). Particularly look for areas of highly crenulated (folded, angular) contours rather than smooth contours. The 5,400-foot contour south of the town of Gardiner is an example of a crenulated contour. An example of an area with smoother contours is the area northwest of the town of Gardiner.

4. What other landslide features identified by Nielson and Brabb (listed above) are present in this landslide complex?

Green River Gorge, Washington

This gorge, southeast of Seattle, has many landslides along its slopes. They can be hard to identify on aerial photographs and topographic maps due to forest cover. A relatively new technology, LIDAR, is expected to improve identification.

5. Figure 8.9 is a vertical aerial photograph of the Green River Gorge. Mark on this photograph areas that you think, based on topography, may have landslides. What landslide characteristics helped you make your identification?

6. Figure 8.10 is a topographic map of the Green River area. Use this map and see if you can find any landslide areas. Are there any areas that have similar topographic contours to those shown in Figures 8.4, 8.5, and 8.8? If so, mark them.

7. LIDAR images can show landslides more clearly than either aerial photographs or topographic maps. LIDAR allows "virtual deforestation," that is computer processing that removes trees and lets the ground surface be mapped. Figure 8.11 is a LIDAR image of the Green River Gorge. Note that Figure 8.11 covers a larger area than Figure 8.9. Mark on this image all the landslides you can find. Do you agree or not that LIDAR allows identification of more landslides than maps or photographs?

8. Use a colored pencil, and transfer these landslides back to the topographic map (Figure 8.10) and the aerial photograph (Figure 8.9). How many more landslides could you identify on LIDAR than on the map? The photograph?

FIGURE 8.6 Gardiner, MT, landslide(s). View to SSW. (Photograph by © Duncan Foley)

FIGURE 8.7 Landslides southwest of Gardiner, MT. See Figure 8.8. (USGS Digital Ortho Quarter Quad downloaded August 11, 2008, from http://nris.state.mt.us/nsdi/doq.asp?Srch24=4511016)

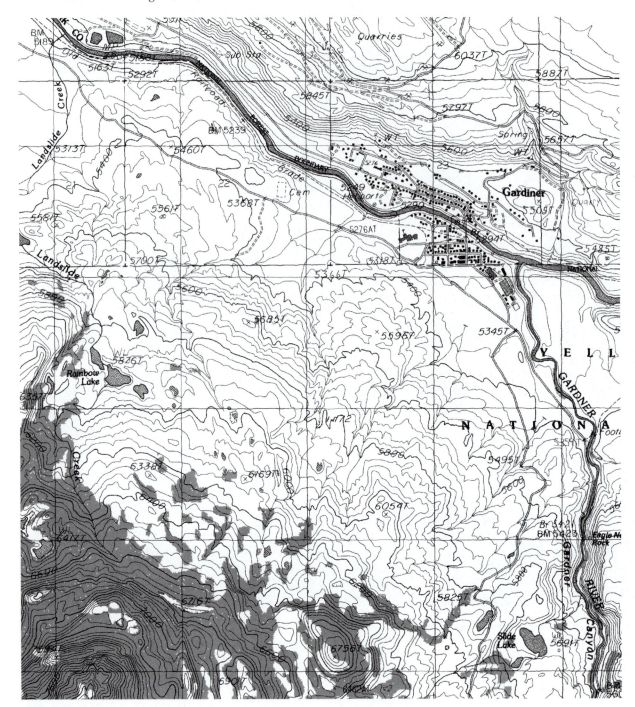

FIGURE 8.8 Gardiner, MT. Topographic map. Contour Interval = 40 ft. See Figure 8.7.

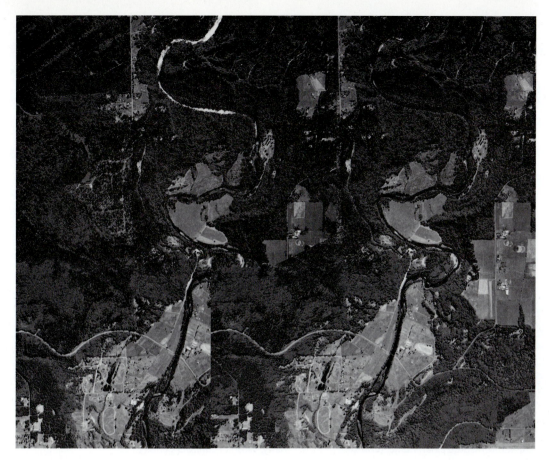

FIGURE 8.9 Stereo aerial photos of Green River Gorge, Washington. Photos courtesy of Washington Department of Transportation.

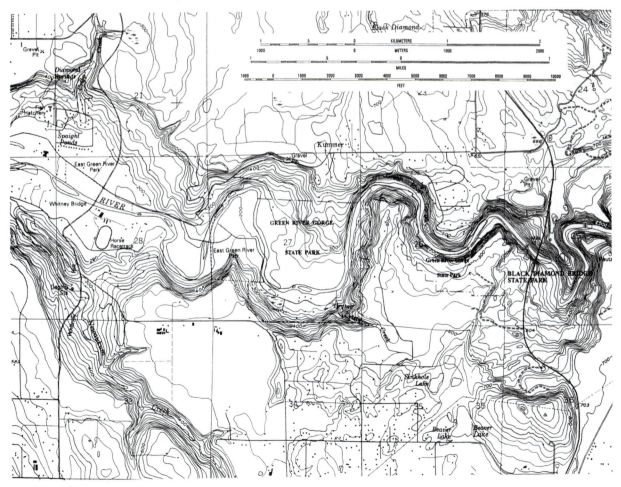

FIGURE 8.10 Topographic map, Green River Gorge. Contour Interval = 25 ft.

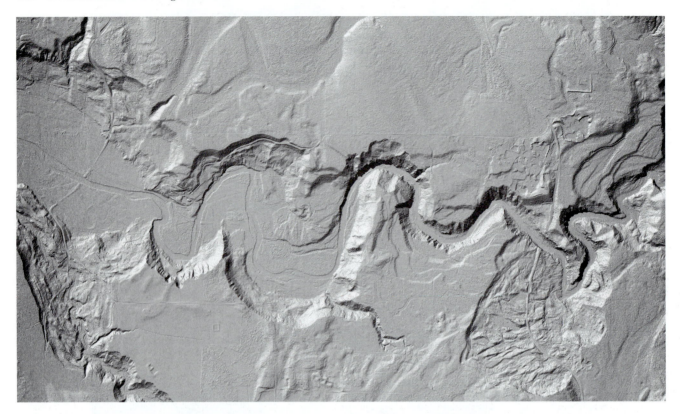

FIGURE 8.11   Green River Gorge Lidar Image. (Courtesy of King County, Washington, GIS Center)

PART B. LANDSLIDES IN SAN JOSE, CALIFORNIA

Landslide deposits in the northeastern part of the city of San Jose were first shown on a geologic map done by Crittenden (1951). A subsequent map, based on inter-pretation of aerial photographs and field investigations, included many additional landslide deposits that are shown as lighter areas with arrows (Figure 8.12, Nilsen and Brabb, 1972). Damage to urban structures within part of the landslide-prone area is shown in Figure 8.14. The landslides, which may be continuously or intermit-tently moving, or not moving at all, are primarily the result of natural processes. Human activities may alter these processes and even render some areas unstable. In order to determine the stability of a particular site, a landslide deposit map, such as Figure 8.12, would be used in conjunction with other information concerning soils, vegetation, hydrology, and other geologic factors.

In general, fewer of these characteristics will be noted in smaller deposits. Detailed site studies, of course, are required for predicting the behavior of landslide deposits under changing conditions.

QUESTIONS (8, PART B)

1. Review Part A, carefully examine Figure 8.12, the explana-tion for this figure, and the topographic map for this area (Figure 8.13 in the map section at the back of the book). Note that two different contour intervals are used on each map.
   a. What are these contour intervals?

b. Explain why two different contour intervals were nec-essary for each map.

2. Place an X in the area of the largest landslide deposit within the enclosed area in the center of Figure 8.12. What is the approximate area of this slide, in square miles?

3. What materials make up the landslide deposit?

4. The density of contour lines changes from the northeast part of the map to the southwest part, indicating a change from steep to gentle slopes and suggesting an increase in potential for landslides. What is the difference in elevation between A and B?

5. What is the distance from A to B in miles?

6. What is the average gradient (slope), in feet per mile, from A to B?

7. The gradient from X to Y is _____, and from K to L it is _____.

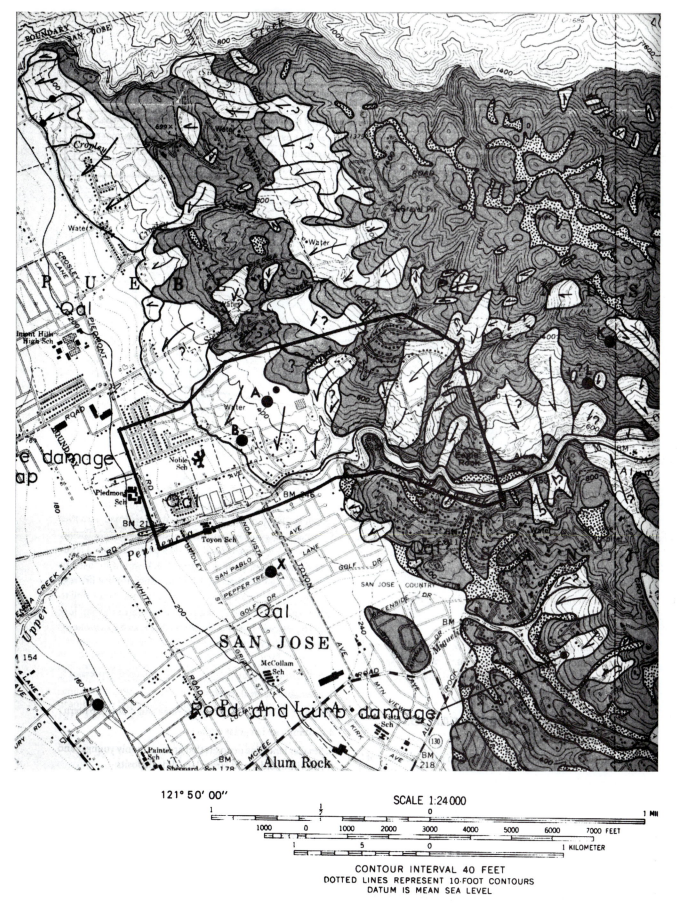

FIGURE 8.12 Landslide and other surficial deposits in northeastern San Jose, California.

(Nilsen and Brabb, 1972).

Explanation for Figure 8.12

Alluvial deposits

Irregularly stratified, poorly consolidated deposits of mud, silt, sand, and gravel deposited in stream and river beds and on adjoining flood plains. Includes older and younger alluvial fan deposits that form broad, extensive, gently sloping surfaces. Deposition is continuing on the younger parts of these fans and in the major alluvial channels that cut across the fan surfaces.

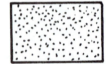

Colluvial deposits

Colluvial deposits: unstratified or poorly stratified, unconsolidated to poorly consolidated deposits composed of fresh and weathered rock fragments, organic material, sediments, or irregular mixtures of these materials that accumulate by the slow downslope movement of surficial material predominantly by the action of gravity, but assisted by running water that is not concentrated into channels. Colluvial deposits have been mapped only where they form a distinct apron near the base of slopes or where they fill and flatten canyon, ravine, and valley bottoms. Colluvial deposits are probably forming on almost every slope in the bay region, but only the thicker and extensive accumulations that are recognizable on aerial photographs have been mapped.

Landslide deposits

Arrows indicate general direction of downslope movements; queried where identification is uncertain

Debris composed of fresh and weathered rock fragments, sediment, colluvial material, and artificial fill, or any combinations thereof, that has been transported downslope by falling, sliding, rotational slumping, or flowing. Landslide deposits smaller than approximately 200 feet in longest dimension are not shown on the map. Complex landslide deposits, which result from combinations of different types of downslope movement, are perhaps the most common type of landslide deposit in the bay region. In particular, materials near the head of landslide deposits typically move in a different manner than materials at the toe. The landslide deposits shown on this map have not been classified according to either type of movement or type of material of which the deposit is composed. The deposits vary in appearance from clearly discernible, largely unweathered and uneroded topographic features to indistinct, highly weathered and eroded features recognizable only by their characteristic topographic configurations. The time of formation of the mapped landslide deposits ranges from possibly a few hundred thousand years ago to 1970. No landslide deposits that formed since 1970 are shown. The thickness of the landslide deposits may vary from about 10 feet to several hundred feet. The larger deposits are generally thickest; many small deposits may be very thin and may involve only surficial materials.

Bedrock
(queried where identification uncertain)

Igneous, metamorphic, and sedimentary rocks of various ages, physical properties, and engineering characteristics. Areas not shown on the map as covered with surficial deposits probably contain bedrock either exposed at the surface or mantled by a thin veneer of surficial deposits, most commonly colluvial material. The bedrock is commonly weathered to a considerable depth, so that there is a gradual change downward from highly weathered organic-rich soil to fresh bedrock. Thus, many of the small landslide deposits and some of the large landslide deposits that are shown on the map to lie within bedrock areas probably involve only material derived from weathered bedrock and other colluvial material.

Artificial fill

Highway, railroad, and canal fills composed of rock and soil derived from nearby cuts or quarries; only large fill areas are shown on the map.

8. According to gradient, is the potential for landslide greater at X or K? Explain.

9. What is the material labeled "Qal" on the map? (Q is an abbreviation for Quaternary, a geologic period.)

10. What are the two ways in which Qal is formed (see description)?

11. Locate the deposit Qaf. What is it?

These deposits are becoming more common as humans continue to alter the landscape. Some geologists use the term anthropogenous deposits for these materials.

12. Review the drawing of the mudflow in Figure 8.2. Observe the topography shown on Figure 8.12, and mark with "X" as two areas that might be subject to mudflows. Explain your decisions. (Hint: See Colluvial deposits on the explanation for Figure 8.12.)

   a.

   b.

13. a. How far (in feet) is Noble School from a landslide deposit or an area of damage from landslides (refer to Figures 8.12 and 8.14).

   b. Are any schools on or closer to mapped landslides?

14. Would you expect a landslide to develop under Noble School? Explain.

15. Note that a road in the subdivision had to be abandoned. What road was extended to accommodate the traffic?

16. In what subdivision are the badly damaged and abandoned houses?

17. Would you purchase a house on the north side of Boulder Drive, east of Sophist Drive (see Figures 8.12 and 8.14)? Explain.

18. Why have the utility lines been placed above ground on the west side of Boulder Drive?

19. Money and resources are still being used to make this landslide region habitable. What short-term and long-term solutions to the community's problems might be appropriate?

20. On Figure 8.15 (1963), outline three major landslides.

Hint: Use the maps (Figures 8.12 and 8.14) showing areas of movement or damage for clues.

21. Are there landslides on Figure 8.16 (1993) that did not appear on the 1963 stereopair? If so outline them.

22. Mark two areas of additional houses and one other cultural feature that appear in the 1993 photo that are not on the 1963 photo.

23. Using a colored pencil draw the contact between the base of the mountain and the alluvial fan on which San Jose is built. What is the approximate elevation of this line (compare with the map, Figure 8.13 in the back of the manual or Figure 8.12)?

24. a. What advice would you give to those seeking a building site in the mountain area?

   b. As a consulting geologist, what advice would you give to the local zoning commission in this area?

PART C. LANDSLIDES IN ATHENS, OHIO

Landslides in Athens, Ohio, are representative of those throughout much of the Appalachian Plateau Physiographic Province. They include many of the types of mass wasting listed in the Introduction to Exercise 8. The most common types in Athens and southeastern Ohio are slump-earthflow, rock falls, and rockslides. Recognition of geomorphic features of landslide deposits (see Part A) is important if we are to avoid potentially hazardous sites and understand the processes of change in the landscape. In this exercise we investigate, using a series of photographs, the changes in an apartment complex constructed in the 1960s adjacent to the Hocking River near Ohio University. Study the indicated aerial photographs and maps and Table 8.2 to answer the questions.

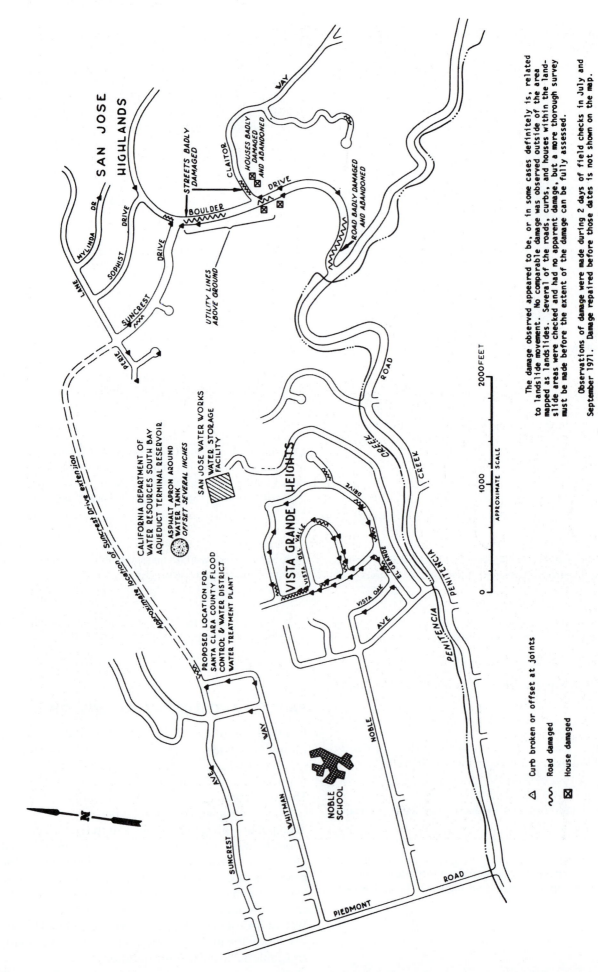

FIGURE 8.14 Map showing damage observed in July and September 1971 (Nilsen and Brabb, 1972).

FIGURE 8.15 Stereopairs for northeastern San Jose, California, in 1963.

FIGURE 8.16 Stereopairs for northeastern San Jose, CA in 1993.

TABLE 8.2 Some Representative Descriptions of Members and Formations of Systems in Ohio

Description	Formation or Member	System
Green plastic shale; sandstone	Fulton	Pennsylvanian
Coal	Meigs Creek No. 9	Pennsylvanian
Sandstone	Pomeroy	Pennsylvanian
Red plastic clay, clay shale	Pittsburgh	Pennsylvanian
Limestone	Cambridge	Pennsylvanian
Plastic clay and shaly sandstone	Sciotoville	Pennsylvanian
Conglomerate and shaly sandstone	Black Hand	Mississippian
Shale	Sunbury	Mississippian
Limestone	Berea	Mississippian
Shale	Ohio	Devonian
Limestone	Columbus	Devonian
Dolomite	Greenfield	Silurian
Dolomite	Guelph	Silurian
Limestone	Brassfield	Silurian
Limestone and calcareous shale	Richmond	Ordovician
Limestone and calcareous shale	Maysville	Ordovician

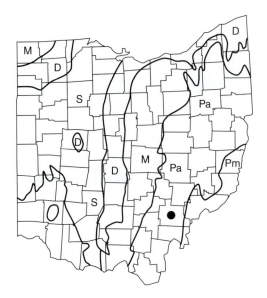

FIGURE 8.17   Geologic map, of Ohio. Athens marked by dot. Geologic systems marked by letters, Pa-Pennsylvanian, Pm=Permian.

QUESTIONS (8, PART C)

1. Locate Athens, Ohio, on Figure 8.17 the geologic map of Ohio.

a. What is the geologic system (age) of rocks in Athens?

b. From your knowledge of geologic materials (clays, shales, limestones, sandstones, etc., which are reviewed in Exercise 1) and the detailed descriptions in Table 8.2, which of the four youngest geologic systems would be the most susceptible to landsliding? Explain.

c. Which formations or members would most likely deform by flow? (Hint: Engineers in Ohio refer to them as "those d——d redbeds"; Delong 1996.)

d. Much of the Appalachian Plateau (Physiographic Province) is underlain by Pennsylvanian age and similar rocks. Would you expect similar landslide conditions in other area of the plateau such as Pittsburgh, Pennsylvania, to the east of Athens County? Explain.

2. Knowing that the bedrock in Ohio is nearly flat, list those slope stability factors (select from those in the Introduction to Exercise 8) that are important in causing landslides in Ohio.

3. List two human activities that could increase the potential for landslides.

4. Figure 8.18 shows an apartment complex under construction near the top of a hill that has been graded to produce a flat area for the buildings. On this aerial photo taken in 1968 near Athens, Ohio, identify

a. a pond on the floodplain (P)

b. the apartment buildings with dark roofs at the construction site (A) (one is given)

c. two landslides (L)

5. How many buildings are roofed or partially roofed?

FIGURE 8.18 Aerial photo (1968) of apartment site overlooking the Hocking River, south of Ohio University. North arrow in the upper middle of the photo points to lower left of photo. Building "A" is marked on this photo and on Figure 8.23 (in color plate section at the back of the book). Hocking River "HR" in lower left side of bottom edge of photo.

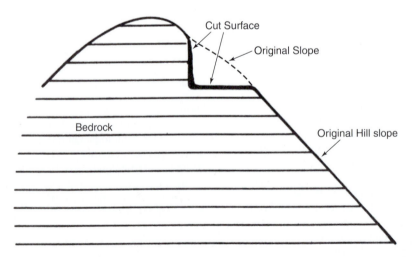

FIGURE 8.19 Sketch of cross section of apartment site adjacent to the Hocking River. Complete the profile to show placement of the material removed from the cut in the slope.

6. The rock and regolith graded from the high areas of the hillside were used as fill to provide additional building sites on the slope of the hill adjacent to the cut surface. Complete the sketch (Figure 8.19) to illustrate the site.

7. a. Where would the landslides observed in the 1968 photo (Figure 8.18) be in your sketch (Figure 8.19)?

b. What is the material of this landslide: bedrock, fill or regolith?

8. On the 1971 stereo triplet (Figure 8.20),
a. Mark the area of rockfall hazard for the apartments with (R).

b. Compared to the 1968 (Figure 8.18) photo, where is the increased area of mass movement?

c. How many apartment buildings are there in the completed complex in 1971 (Figure 8.20)?

9. On the 1975 stereopair (Figure 8.21)
a. Outline the areas of landslides (L)

b. How many different landslides can you identify?

c. Identify any site(s) where a building has been removed.

d. Mark buildings with an "X" that you think will be lost or removed because of mass wasting.

e. What is the purpose of the chain-link fence (F) behind the apartment?

f. What type of mass wasting deposit would you find adjacent to the fence? Explain.

10. From the 1976 image (Figure 8.22),
a. Is the road at the base of the hill suitable for automobiles?

b. How many apartments are left?

c. Why did they not build these structures where it is flat, such as at B?

d. What is the type of mass wasting that destroyed the road?

Consult the maps (Figure 8.23 in back of book) to answer questions 11–13.
11. a. Circle the location of the apartment complex on each map.

b. From the maps, how many buildings were on the site in 1961? ___ In 1995? ___

c. On the 1995 map, add the apartment buildings that once existed (see 1975 photo). The maximum number was ___?

d. The thin dashed line represents the city boundary. How does the river influence this boundary or does it?

e. The Hocking River underwent a major change in form and position between 1961 and 1995. Describe these changes near Ohio University.

f. On the 1995 map, mark one location where natural change in the river has occurred following channelization by humans in the early 1970s. What is the change and what is your evidence that it has occurred?

12. From what you know of the landscape in the vicinity of the apartment site between 1968 (photo) and 1995(map), mark the location where highway 33/50 passes over the toe of a landslide.

13. Briefly describe what happened to the river near the row of trees on the river bank between 1968 (Figure 8.18) and 1975 (Figure 8.21).

14. Now that you have nearly completed the exercise you should be able to see the development of the major landslide, active before all the apartments were built. Describe changes in the road and flood plain from the photos:
a. 1968–1971

b. 1971–1975

c. 1975–1976

15. What recommendation would you give to anyone preparing a building site in the vicinity of and at the same level (and presumably with the same strata) as the apartment site?

FIGURE 8.20   Stereo triplet (1971) of the apartment site.

FIGURE 8.21   Stereo pair (1975) of apartment site. The channelized Hocking River is adjacent to the flood plain at bottom of the hill.

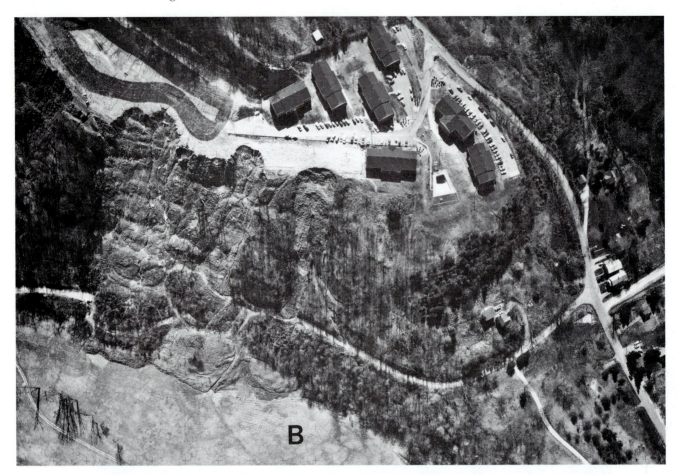

FIGURE 8.22 Aerial photo (1976) of apartment site.

PART D. THE NATURE AND OCCURRENCE OF SNOW AVALANCHES

The goals of part of the exercise are to familiarize you with some of the basic causes of avalanches and introduce you to various types of avalanche terrain.

The following text is slightly edited from the U.S. Forest Service brochure "Snow avalanche: General rules for avoiding and surviving snow avalanches."

There are two principal types of snow avalanches: loose snow avalanches and slab avalanches. These types are illustrated in Figure 8.24. Loose snow avalanches start at a point or over a small area. They grow in size and the quantity of snow involved increases as they descend. Loose snow moves as a formless mass with little internal cohesion. Slab avalanches, on the other hand, start when a large area of snow begins to slide as a mass.

Loose snow avalance

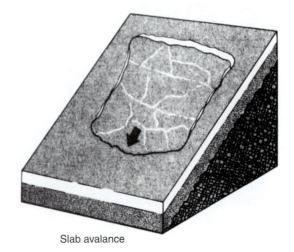

Slab avalance

FIGURE 8.24 Two modes of snow slope failure: (left), loose-snow avalanche and (right), slab avalanche (Perla and Martinelli, 1978).

A well-defined fracture line occurs where the moving snow breaks away from the stable snow. In slab avalanches snow crystals tend to stick together. Angular blocks or chunks of snow may be part of the slide. According to Perla and Martinelli (1978), the slab may range from 100 to 10,000 square meters in area and from 0.1 to about 10 m thick. Slab avalanches are the more dangerous of the two types.

Avalanches are the result of both the underlying terrain and the conditions of the snowpack. This exercise concentrates on terrain factors, including slope steepness, slope profile, slope aspect, and ground cover.

The relevant terrain factors, as described by the U.S. Forest Service, are:

Slope steepness. Avalanches most commonly release from slopes of 30 to 45 degrees (60 to 100 percent), but may begin on slopes ranging from 25 to 65 degrees (45 to 215 percent). LaChapelle (1985) notes that many avalanches begin on slopes of 35 to 40 degrees (70 to 85 percent).

Slope profile. Dangerous slab avalanches are more likely to begin on convex slopes but may also begin on concave slopes. Short slopes may be as dangerous as long slopes; 42 percent of all avalanche fatalities result from slides with a slope distance of less than 300 ft (100 m).

Slope aspect. Snow on north-facing slopes may be slower to stabilize than snow on slopes that face other directions. South-facing slopes are especially dangerous in the spring due to solar heating. Leeward (downwind, or the direction the wind is blowing toward) slopes are dangerous because they accumulate wind-deposited snows that add depth and may create unstable slabs of snow. Windward (upwind, or the direction the wind is coming from) slopes generally have less snow; the snow is more compacted and usually more stable than snow deposits on leeward slopes.

Ground cover. Large rocks, trees, and heavy brush help anchor the snow. Smooth, open slopes are more dangerous, but avalanches can start even among trees.

Old slide paths and recent avalanche activity. Generally, avalanches may repeatedly occur in the same areas. Watch for avalanche paths. Look for pushed-over small trees, trees with limbs broken off. Avoid steep, open gullies and slopes. If you see new avalanches, suspect dangerous conditions. Beware when snowballs or "cartwheels" roll down the slope.

If you are planning to travel in avalanche country, you should also be aware of the impact of different weather conditions on the occurrence of avalanches. Some of these factors, as described by the U.S. Forest Service, are:

Old snow. When the old snow depth is sufficient to cover natural anchors, such as rocks and brush, additional snow layers will slide more readily. The nature of the old snow surface is important. For example, cold snow falling on hard, refrozen snow surfaces, such as sun or rain crusts, may form a weak bond. Also a loose under-lying snow layer is more dangerous than a compacted one. Check the underlying snow layer with a ski pole, ski, or rod.

Wind. Sustained winds of 15 mi/hr and over may cause avalanche danger to increase rapidly even during clear weather, if loose surface snow is available for the wind to transport. Snow plumes from ridges and peaks indicate that snow is being moved onto leeward slopes. This can create dangerous conditions.

Storms. A high percentage of all avalanches occur shortly before, during, and shortly after storms. Be extra cautious during these periods.

Rate of snowfall. Snow falling at the rate of 1 in/hr or more increases avalanche danger rapidly.

Crystal types. Observe general snow-crystal types by letting them fall on a dark ski mitt or parka sleeve. Small crystals—needles and pellets—often result in more dangerous conditions than the classic, star-shaped crystals.

New snow. Be alert to dangerous conditions with a foot or more of new snow. Remember that new snow depth may vary considerably with slope elevation, steepness, and direction.

Temperature. Cold temperatures will maintain an unstable snowpack while warm temperatures (near freezing) allow for snow settlement and increasing stability. Storms starting with low temperatures and dry snow, followed by rising temperatures and wetter snow, are more likely to cause avalanches. The dry snow at the start forms a poor bond to the old snow surface and has insufficient strength to support the heavier snow deposited late in the storm. Rapid changes in weather conditions (wind, temperature, snowfall) cause snowpack changes. Therefore, be alert to weather changes. Snowpack changes may affect snow stability and cause an avalanche.

Temperature inversion. Higher temperatures with higher elevations can occur when warm air moves over cold air, which is trapped near the ground and in valleys. This weather situation may produce dramatic variations in local snow stability.

Wet snow. Rainstorms or spring weather with sunny days, warm winds, and cloudy nights can warm the snow cover. The resulting free and percolating water may cause wet snow avalanches. Wet snow avalanches are more likely on south slopes and slopes under exposed rock.

The U.S. Forest Service suggests that "the safest routes for travel in avalanche terrain are on ridgetops and slightly on the windward side of ridgetops, away from cornices (accumulations of drift snow on the leeward side of the crest of a ridge). Windward slopes are usually safer than leeward slopes. If you cannot travel on ridges, the next safest route is out in the valley, far from the bottom of slopes. Avoid disturbing cornices from below or above. Gain access to ridgetops by detouring around cornice areas . . . .

If you must cross dangerous slopes, stay high and near the top. If you see avalanche fracture lines in the snow, avoid them and similar snow areas. If the snow sounds hollow, particularly on a leeward slope, conditions are probably dangerous. If the snow cracks and the snow cracks run, this indicates slab avalanche danger is high. If you must ascend or descend a dangerous slope, go straight up or down along the side; do not make traverses back and forth across the slope. Take advantage of areas of dense timber, ridges, or rocky outcrops as islands of safety. Use them for lunch and rest stops. Spend as little time as possible on open slopes."

QUESTIONS (8, PART D)

1. What are the most common slopes for avalanches in degrees and in percent?
    degrees
    percent

2. Why are avalanches not as common on slopes that are less steep or that are more steep?

3. Where on a slope profile is an avalanche most likely to begin?

4. If the winds from a storm are blowing out of the west, on which side of a mountain will most of the snow accumulate? Why?

5. Describe three possibilities for a safe route through avalanche terrain.
    a.

    b.

    c.

PART E. AVALANCHE PATH IDENTIFICATION

Avalanche hazards in the Aspen, Colorado, area have been simplified for this exercise; actual hazards from avalanches depend on the local snow and terrain conditions, and may be more or less than those used for illustration in this exercise. There is still much to learn about avalanches, and caution in areas or times of increased hazard is imperative.

Avalanche paths are separated into three zones: the starting zone, the track, and the runout zone. Figures 8.25, 8.26, and 8.27 illustrate these and their captions discuss important aspects of avalanche travel and hazards. Use these figures as you investigate the maps discussed below.

QUESTIONS (8, PART E)

Aspen, Colorado

There are two zones of avalanche hazard presented on the Aspen map (Figure 8.29 in the colored maps section). Darker zones are known avalanche areas, and lighter zones are areas that may sometimes have small avalanches. Lighter areas also represent extensions of the known areas. Bryant (1972) notes, "The most obvious avalanche paths are in gullies or on steep treeless [or sparsely vegetated] slopes below treeline."

1. Calculate slopes in degrees or percent for the following paths or potential slide areas:
    a. the path marked A, just south of Tourtellotte Peak

    b. the path marked B, for its full length from the highest elevation in the shaded zone above the B to its lowest point near the word "Fork."

    c. the lighter-colored zone immediately south of Aspen, beneath the ski lift east of Pioneer Gulch.

    d. Describe the relationship between the results you calculated above and your answer to question 1, part A. Based on your calculations, are the avalanche areas depicted on the map justifiable on the basis of their slope (as the only data)?

2. Now inspect other slopes that have mapped avalanche paths depicted on Figure 8.29. Are these slopes likely to have avalanches based on the answer to question 1, part D?

3. In addition to slope, what other kinds of data might the author have used in depicting avalanche areas on this map?

4. Mears (1976) gives the following formula for calculating the runout distance (defined as the lower boundary of the track to the outer limit of impact) of an avalanche with a confined path:

$$S = 214 + 11.4A$$

where S = runout distance in meters, and A = area of the starting zone in hectares.

This formula is based on snow and terrain conditions found in Colorado, and may not be as applicable in other locations. A hectare is an area of 10,000 square meters, or 100 m on a side. It compares with an acre, which has 43,560 square feet, and has about 209 ft on a side. One hectare equals 2.47 acres and 1 meter is 39.4 in. See Appendix I for additional conversions. Use the formula given by Mears to calculate the runout distance of the avalanches from McFarlane Creek. Use Figures 8.26b and 8.27 to help identify the points of the McFarlane Creek path. Figure 8.28 is a topographic

FIGURE 8.25 Large avalanche path showing various features. The smaller runout zone, reached more frequently, is cut through the aspen forest. The larger zone is reached much less frequently, is revegetated by aspen, and must be considered in planning any development (Mears, 1976; after Salm, 1975).

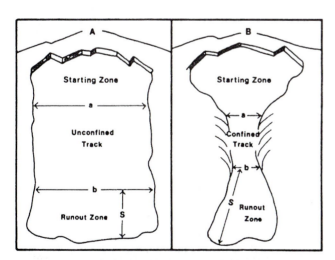

FIGURE 8.26 Comparison of an unconfined avalanche path (A) and a confined avalanche path (B). In path A, the runout distance, S, is not affected by the width of the starting zone because concentration of discharge does not occur at (a) and (b). In path B all the released snow is conveyed through the confined track at (a) and (b). Therefore, the runout distance, S, in path B depends on the size of the starting zone (Mears, 1976, after Salm, 1975).

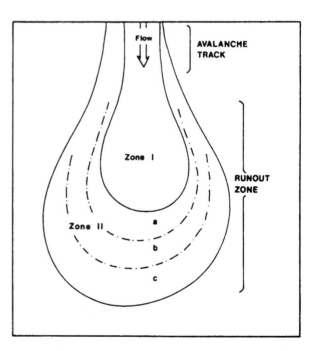

FIGURE 8.27 Runout zones of an avalanche path. Zone I is high hazard affected by avalanches with either short return periods or large impact pressures. Zone II is affected by both longer return periods and lesser impact pressures than Zone I. Hazard level decreases toward the outer margin of Zone II (Mears, 1976).

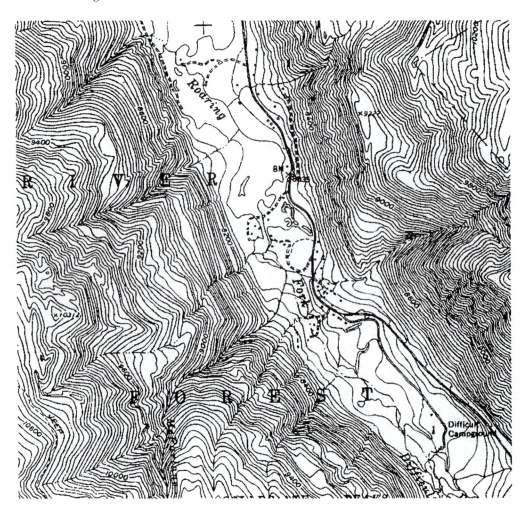

FIGURE 8.28   Topographic map of part of the Aspen, Colorado, quadrangle.

map of the area. Figure 8.29, in the maps section at the back of the book, shows avalanche paths near Aspen. You must first measure the area of the starting zone in square feet or meters and convert this measurement to hectares. Outline your starting zone on the map of Aspen. Also outline the beginning and end of the runout zone. Show your calculations.

Runout zone length _____

5. a. How does the runout distance you calculated for the McFarlane Creek runout zone compare with the length of the runout zone shown on the map?

   b. Do the results suggest any changes for the map?

6. Notice that the contours at the mouth of McFarlane Creek bend away from the range front out into the valley. They represent a debris fan or alluvial fan. Are there other debris fans along the valley of Roaring Fork that could be related to avalanche activity? If so, where are they?

7. What evidence would you look for in the field to determine whether debris fans, such as at the mouth of McFarlane Creek, is related to stream deposition or avalanches?

8. The colored map of the Aspen area was published in 1972. Did McFarlane Creek avalanches present a hazard to people at that time? Explain your answer.

9. Compare Figures 8.28 and 8.29 (the colored map). Identify changes in human use of the land at the base of McFarlane Creek that occurred between 1972 (the date of the original base map used in compiling Figure 8.29) and 1987, the date of the revised map in Figure 8.28.

10. How have changes in human use of the land at the base of McFarlane Creek increased or decreased hazards from avalanches since 1972?

Alta, Utah

11. First, let's look at a typical small-avalanche site on the Alta map (Figure 8.30). Carefully identify, in the general area

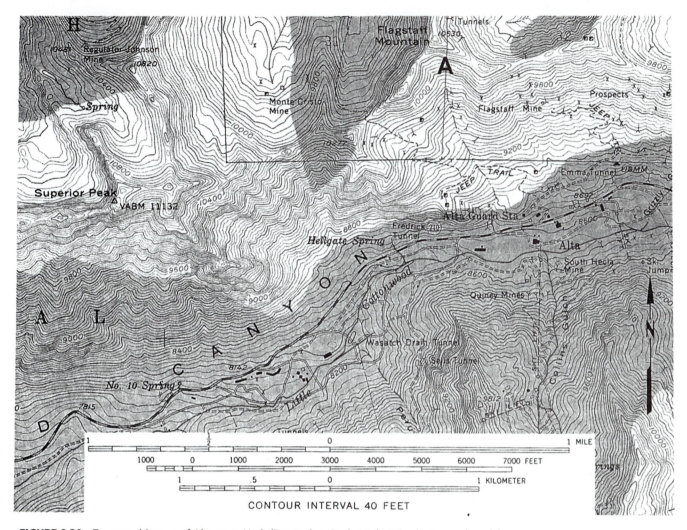

FIGURE 8.30 Topographic map of Alta area, Utah (Dromedary Peak, Utah, 7.5-minute quadrangle).

marked *A* (the gully west of the Flagstaff Mine, on the southeast side of Flagstaff Mountain), a starting zone for an avalanche that might travel down the gully. Also mark the track and identify a runout zone. Determine the slope angle for the track and calculate the length of the runout zone, based on your data and using the formula given in Question 4. (The formula may not really apply here, but it gives an approximation.) Show your work.

Slope _____

Runout zone length _____

12. The Alta Guard Station and other buildings are shown along the road. Are these buildings at risk for a small avalanche? Explain.

13. Now let's look at the risk of an unconfined avalanche on the same map. From the east side of peak 10277, which is north of Hellgate Spring and southwest of Flagstaff Mountain, draw a line along the 10,000 ft contour for 1 mile east. Refer to Figure 8.24 as a guide, and mark clearly on the map about how far you think that the runout zone from an unconfined avalanche with a mile-long starting zone might have gone. Was the town of Alta at risk for this type of an avalanche? Explain.

Bibliography

Armstrong, B. R., and Williams, K., 1992, *The Avalanche book:* Golden, Colorado, Fulcrum Publishing, 240 p.

Bovis, M. J., and Mears, A. I., 1976, Statistical prediction of snow avalanche runout from terrain variables: *Arctic and Alpine Research,* v. 8, no.1, p. 115–120.

Bryant, B., 1972, *Map showing avalanche areas in the Aspen quadrangle, Pitkin County, Colorado:* U.S. Geological Survey Map I-785-G.

Brabb, E. E., and Harrod, B. L. eds., 1989, Landslides: Extent and economic significance. *Proceedings of the 28th*

International Geological Congress Symposium on Landslides: Washington, D.C., 17 July, Rotterdam, Balkema, 399 p.

Crittenden, M. D., Jr., 1951, Geology of the San Jose-Mount Hamilton area, California. *California Division of Mines and Geology Bulletin,* no. 157, 74 p.

DeLong, R. M., 1996, "Those d——d redbeds:" *Ohio Geology,* Summer, p. 5–6

Jochim, C. L., 1986, *Debris-flow hazard in the immediate vicinity of Ouray, Colorado:* Colorado Geological Survey, Special Publication 30, 36 p.

Jochim, C. L., Rogers, W. P., Truby, J. O., Wold, R. L., Jr., Weber, G., and Brown, S. P., 1988, Colorado landslide hazard mitigation plan: *Colorado Geological Survey Bulletin 48,* 149 p.

LaChapelle, E. R., 1985, *The ABC of avalanche safety:* Seattle, Washington, Mountaineers, 112 p.

Mears, A. I., 1976, *Guidelines and methods for detailed snow avalanche hazard investigations in Colorado:* Colorado Geological Survey, Bulletin 38, 125 p.

Morton, D. M., and Streitz, R., 1967, Landslides: *Mineral Information Service,* v. 20, no. 10, p. 123–129, and no. 11, p. 135–140.

Nilsen, T. H., and Brabb, E. E., 1972, *Preliminary photointerpretation and damage maps of landslide and other surficial deposits in northeastern San Jose, Santa Clara County California:* U.S. Geological Survey Miscellaneous Field Studies Map, MF-361.

Perla, R. I., and Martinelli, M., Jr., 1978, *Avalanche handbook:* U.S. Department of Agriculture (Forest Service), Handbook 489, 254 p.

Salm, B., 1975, Principles of structural control of avalanches, in *Avalanche Protection in Switzerland:* U.S. Forest Service Gen. Tech. Rept. RM-3, 54 p.

Schultz, A. P., and Southworth, C. S., eds., 1987, *Landslides in Eastern North America:* U.S. Geological Survey Circular 1008, 43 p.

Sharpe, C. F. S., 1938, *Landslides and related phenomena:* New York, Columbia University Press, 137 p.

U.S. Department of Agriculture, Forest Service, Pacific Northwest Region, 1987, *Snow avalanche: General rules for avoiding and surviving snow avalanches,* 7 p.

Van Horn, R., 1972, *Landslide and associated deposits map of the Sugar House Quadrangle, Salt Lake County, Utah:* U.S. Geological Survey Map I-766-D

Voight, B., ed., 1990, *Snow Avalanche Hazards and Mitigation in the United States:* Panel on snow avalanches, Committee on Ground Failure Hazards Mitigation Research, National Research Council, Washington, D.C., National Academy Press, 84 p.

Varnes, D. J., 1978, Slope movement types and processes, in *Landslides: Analysis and control,* eds. R. L. Schuster and R. J. Krizek: Transportation Research Board, National Academy of Sciences, National Research Council, Special Report 176, p. 11–33.

Wold, R. L., Jr., and Jochim, C. L., 1989, *Landslide loss reduction: A guide for state and local government planning:* Colorado Geological Survey, Special Publication 33, 50 p.

Subsidence

INTRODUCTION

Subsidence, or sinking of the land surface, is a hazard commonly associated with resource extraction. Subsidence has been attributed to several factors, including decline of hydrostatic pressure in oil- and water-producing zones due to withdrawal of fluids, underground mining, solution and compaction of soil by irrigation water, oxidation of organic materials, and formation of sinkholes by collapse of bedrock in karst regions. Well-known examples of subsidence include sinkholes in Winter Park, Florida; collapse into abandoned underground coal mines in eastern Ohio, Colorado, Pennsylvania, Utah and Wyoming; and groundwater and petroleum withdrawal in California. Specific areas of subsidence due to fluid withdrawals include oil fields in Goose Creek, Texas, Long Beach, California, and Lake Maricaibo, Venezuela; gas fields in Niigata, Japan, and the Po Delta, Italy; and groundwater reservoirs near Houston-Galveston, Phoenix, Las Vegas, Mexico City, Tokyo, and the San Joaquin and Santa Clara valleys in California. The magnitude of the problem is illustrated by the following statement:

> Subsidence in the San Joaquin Valley probably represents one of the greatest single manmade alterations in the configuration of the Earth's surface in the history of man. It has caused serious and costly problems in construction and maintenance of water-transport structures, highways, and highway structures; also many millions of dollars have been spent on the repair or replacement of deep water wells. Subsidence, besides changing the gradient and course of valley creeks and streams, has caused unexpected flooding, costing farmers many hundreds of thousands of dollars in recurrent land leveling. (Ireland, et al., 1984)

With increased need for the resources extracted from the subsurface and increased human use of the surface, the problem of land subsidence can be expected to increase.

The objective of the subsidence exercise is to investigate the causes and hazards of land subsidence associated with fluid withdrawals and with mining. After an overview of the nature and occurrence of subsidence due to groundwater extraction (Part A), this exercise examines three cases of subsidence: Part B, groundwater removal in Santa Clara Valley, California; Part C, sulfur extraction from the Orchard Salt Dome, Texas; and Part D, coal mining in Illinois and Colorado.

PART A. LAND SUBSIDENCE DUE TO GROUNDWATER EXTRACTION

The causes of subsidence by water withdrawals are related to the compressibility of rocks and sediments and their contained water. When the water level is lowered by pumping wells, water not only drains from the aquifer but is also squeezed from the fine-grained confining beds. The withdrawal of water, which may be under considerable pressure, may lead to subsidence because the reduction of hydrostatic pressure results in compaction of the compressible aquifer and the confining beds. The volumetric reduction is exemplified by subsidence (Figure 9.1). General characteristics of major areas of subsidence due to groundwater extraction are given in Table 9.1.

When wells tapping a confined aquifer remove water from the sand and gravel of the aquifer, a widespread reduction in the water pressure in the aquifer and an increase in the load on the aquifer's framework commonly results (Figure 9.1). The increased load on the aquifer framework of sand and pebble grains causes the decrease in volume (compaction), but this decrease may be stopped or even slightly reversed if the fluid pressure is increased enough to cause the aquifer to return to its original shape. Thus, the aquifer portion of the unit behaves in a nearly elastic manner.

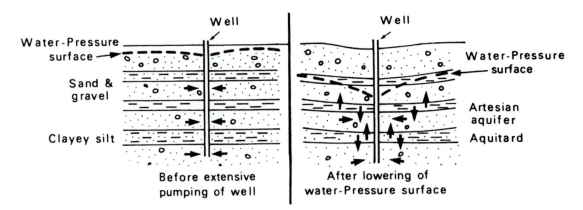

FIGURE 9.1 Change in water-pressure surface, compaction of confining beds, and associated land subsidence. (Original art and sketch. With the 1975 Burgess edition)

A large part of the compaction (and the surface subsidence) is due to the behavior of the clayey confining layers. The amount of compaction or volume reduction depends partly on the type of clay present in the confining beds. For example, the clay mineral montmorillonite (smectite) may contain more water in its structure and be more porous and compressible under a given load than clay composed mainly of the minerals illite or kaolinite. The chemistry of the clays, the texture of the sediment, and the overburden load also affect the reduction of volume and therefore the amount of subsidence.

QUESTIONS (9, PART A)

Refer to Table 9.1 to answer questions 1–5.
1. The most common types of deposits involved in ground water-induced subsidence are:

TABLE 9.1 Description of Areas of Major Land Subsidence Due to Groundwater Extraction (modified from Poland, 1970, pp. 11–21)

Location	Depositional Environment and Age	Depth Range of Compacting Beds Below Land Surface (m)	Maximum Subsidence (m)	Area of Subsidence (sq. km)	Time of Principal Occurrence
Japan: Osaka and Tokyo	Alluvial (?); Quaternary (?)	10–200 (?)	3–4	?	1928–1943, 1948–1965+
Mexico: Mexico City	Alluvial and lacustrine; late Cenozoic	Chiefly 10–50	8	25+	1938–1968+
Taiwan: Taipei Basin	Alluvial, late Cenozoic	30–200 (?)	1	100±	?–1966+
Arizona: central	Alluvial and lacustrine (?); late Cenozoic	100–300+	2.3	?	1952–1967+
California: Santa Clara Valley	Alluvial; late Cenozoic	50–300	4	600	1920–1967+
California: San Joaquin Valley (three areas)	Alluvial and lacustrine; late Cenozoic	90–900	9	5,200	1925–1977+
Nevada: Las Vegas	Alluvial; late Cenozoic	60–300 (?)	1	500	1935–1963+
Texas: Houston-Galveston area	Fluviatile and shallow marine; late Cenozoic	50–600+	1–2	10,000	1943–1964+
Louisiana: Baton Rouge	Fluviatile and shallow marine; Miocene to Holocene	40–900 (?)	0.3	500	1934–1953+

2. Would you expect these deposits to be consolidated (cemented and compacted to solid rock) or unconsolidated? Explain.

3. The geologic ages of these deposits are:

4. What is the maximum depth of deposits that have been compacted?

5. The maximum recorded subsidence is _____ meters at _____.

Clayey Silt
a.
b.
c.
d.

Sand and Gravel
a.
b.
c.
d.

6. The sediments in major areas of subsidence consist of interbedded clayey silt and sand and gravel. Place the following characteristics under the appropriate column above: aquifer, low compressibility, low permeability, high permeability, confining bed, fine-grained texture, coarse-grained texture, and high compressibility.

7. Subsidence induced by groundwater withdrawal generally occurs in areas characterized by confined aquifer systems. What is a confined aquifer?

8. Confining beds consisting of clayey silt respond differently than sand and gravel. They may contain a great deal of water (have high porosity), but it is not readily transmitted. Such rocks are said to have a low _____.

9. When the water pressure in the aquifer decreases, the water in the confining beds may move slowly/rapidly (circle one) into the aquifer. The result is a _____ in thickness of the confining bed and possibly _____ of the land surface (see Figure 9.1).

PART B. SUBSIDENCE IN THE SANTA CLARA VALLEY, CALIFORNIA

The Santa Clara Valley was the first area in the United States where land subsidence due to groundwater overdraft was recognized. As of 1971 subsidence has been stopped due to a reversal of the water-level decline.

The valley is a large structural trough filled with over 1,500 ft of alluvium, including, in the lower part, the semiconsolidated Santa Clara Formation of Pliocene and Pleistocene age (Figure 9.2). Sand and gravel aquifers predominate near the valley margins, but the major part of the alluvium is fine grained.

Below a depth of 200 ft the groundwater is confined by layers of clay, except near the margins. Initially wells as far south as Santa Clara flowed, but decline in head (water-pressure surface) occurred due to pumping for agricultural purposes (90 percent of use until the mid-1930s and 18 percent in 1967). By 1965 the artesian head had declined 150 to 200 ft within most of the confined area. The decline was not continuous; recharge of the aquifer occurred between 1938 and 1947, in part because of controlled infiltration from surface reservoirs.

Most of the wells tapping the artesian system range from 500 to 1,000 ft in depth, although a few reach 1,200 ft. Well yields in the valley range from 500 to 2,500 gal/min.

QUESTIONS (9, PART B)

Refer to Figures 9.2 and 9.3, showing land subsidence for two different periods, in answering the following questions.

1. According to Figure 9.2, what locations showed maximum subsidence in the Santa Clara Valley from 1934 to 1960?

2. a. What was the total subsidence from 1934 to 1960 in these areas?

b. What was the average annual rate in these areas for this period?

3. What geologic material is largely responsible for the subsidence?

4. Would subsidence be more noticeable at Alviso (on the Bay) or at Mountain View?

5. What types of damage and geomorphic changes would indicate subsidence?

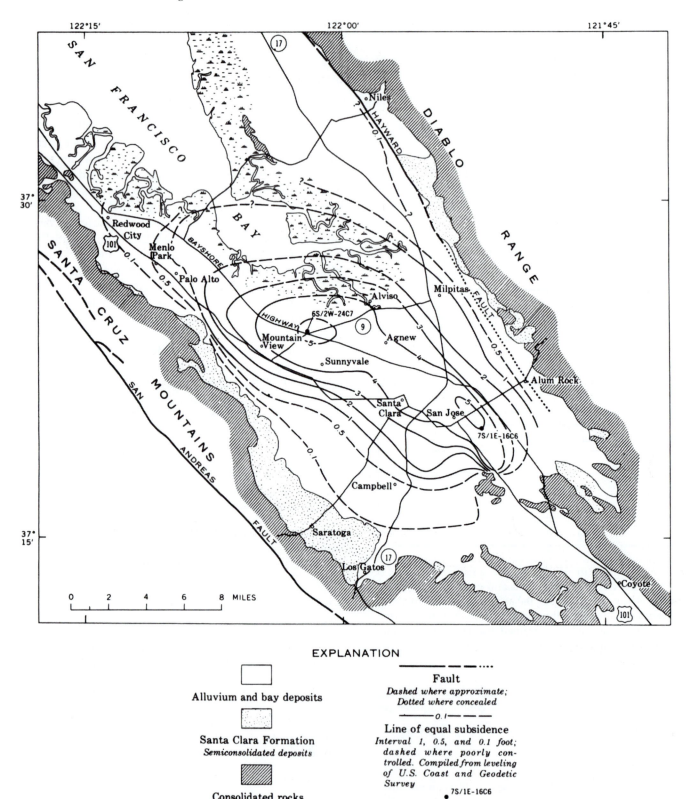

FIGURE 9.2 Land subsidence (in ft) in the Santa Clara Valley, 1934–60 (Johnson et al. 1968, p. A-7).

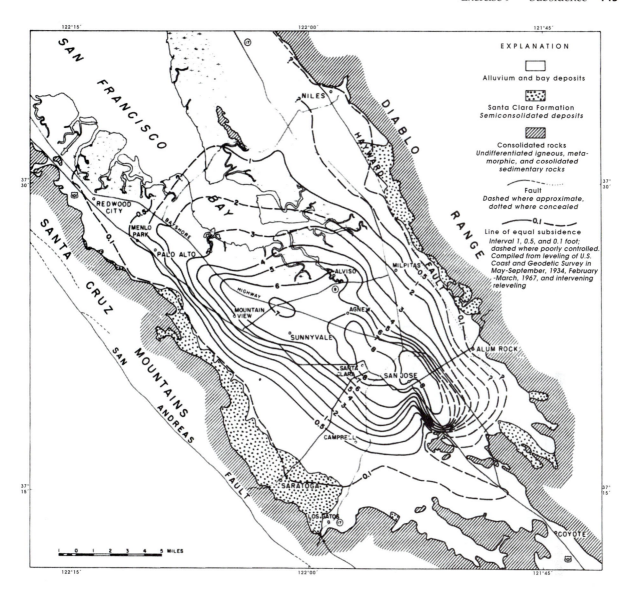

FIGURE 9.3   Land subsidence (in ft) in the Santa Clara Valley, 1934–67 (Poland, 1971).

6. Would you expect any subsidence in the areas marked with diagonal lines? _____ Explain.

7. Figure 9.3 shows the total subsidence for the central part of the valley from 1934 to 1967. What was the annual rate of subsidence for the area around the highway junction (Bayshore Highway and #9) north of Sunnyvale and east of Mountain View from 1960 to 1967? (First refer to Figure 9.2.)

8. Where had the maximum subsidence occurred by 1967?

9. What was the increase in subsidence between 1960 and 1967 on the Bayshore Highway south of Agnew?

Damage to well casings in the valley due to subsidence has amounted to over $4 million. Even greater damage occurred in the northern part of the valley because more than 17 square miles sank to below highest tide and protective works had to be installed. The fluctuations in water level in an 840-foot deep well in San Jose are shown in the hydrograph in Figure 9.4. This record shows the changes between 1915 and 1967.

10. What was the water level in the well in 1915? _____ In 1967? _____

11. During what period would this artesian well have been a flowing well?

12. a. How many highs and lows occur in the water level in any 5-year period? (Figure 9.4)

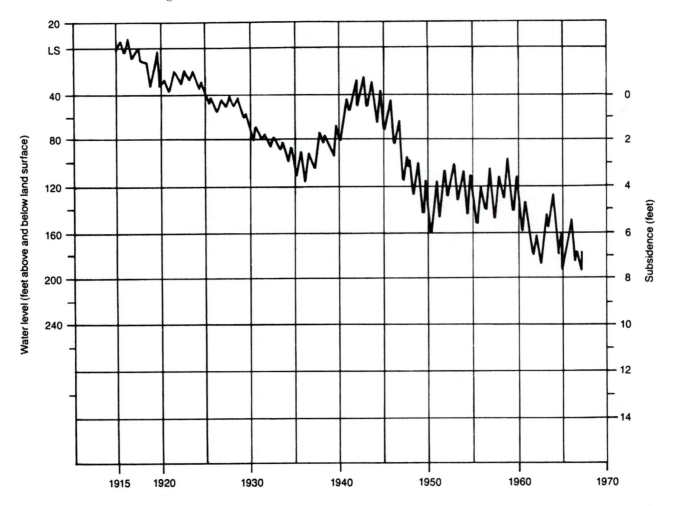

FIGURE 9.4 Change in water level in a well at San Jose, California (USGS by Poland, 1971, Open file Rept. is now listed as mF no. 336).

b. What is the cause of these minor fluctuations?

13. What may have caused the upward trend in the water-pressure surface between 1938 and 1944?

14. Using the data in Table 9.2, plot on Figure 9.4 (where possible) the decline of the land surface at a nearby bench mark (P7) in San Jose. Use the scale on the right side of Figure 9.4. Connect each point with a straight line.

15. In Figure 9.4 the slope of a straight line joining the subsidence in 1912 and that in 1967 would give the average rate of subsidence for this period. In general, how did the rate of subsidence occurring between 1935 and 1948 differ from earlier rates?

16. How is the trend of the water-pressure surface related to the change in subsidence rate?

17. When did subsidence cease according to data in Table 9.2?

18. In 1965 importation of water to the valley began. Some imported water was used to recharge the groundwater through stream channels. According to data in Figure 9.5:
a. What was the response in groundwater pumpage to the increased importation of water? When did this response occur? What was the pumpage rate in 1965 and in 1980?

b. What was the response of the water-pressure surface (artesian head) to the importation of water?

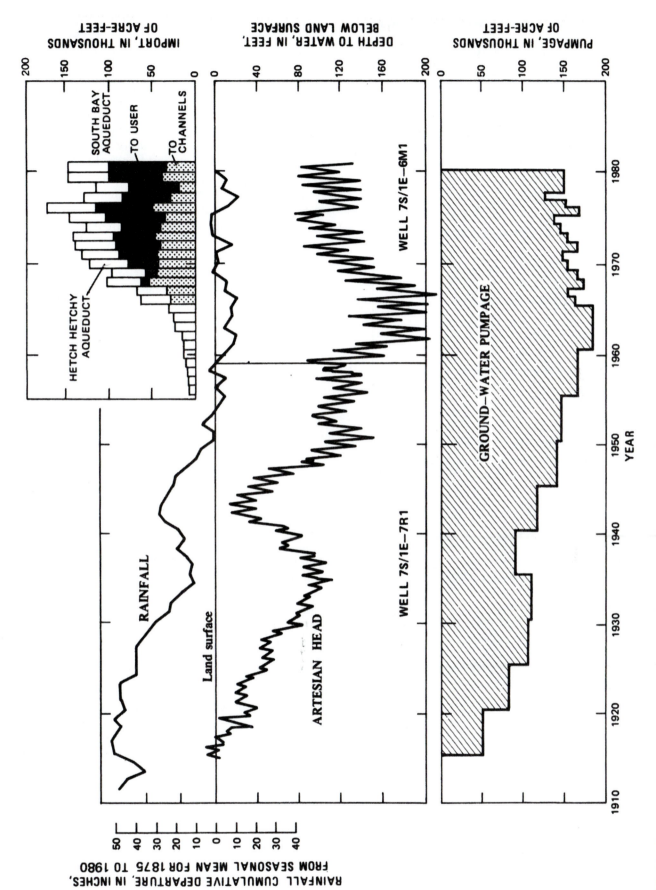

FIGURE 9.5 Artesian head change in response to rainfall, pumpage, and water imports. (From Poland and Ireland, 1988)

TABLE 9.2 Subsidence at Bench Mark P7 in San Jose

Year Leveled	Total Subsidence (ft)
1912	0.0
1920	0.3
1934	4.6
1935	5.0
1936	5.0
1937	5.2
1940	5.5
1948	5.8
1955	8.0
1960	9.0
1963	11.1
1967	12.7
1969	13.0
1988	13.0
1990	13.0

c. Did decreased pumpage, increased precipitation, or water imports have the greatest role in causing the water pressure surface (artesian head) to rise from the levels of the mid-1960s?

19. Between 1970 and 1980 the depth to water below land surface had risen to an average of 90 feet; between 1990 and 1995, the depth to water had risen to an average of 45 feet. What happened to the subsidence rate as a result of the change in the water-pressure surface?

20. Would subsidence resume if imports of water were reduced and pumping of groundwater was increased?

PART C. SULFUR EXTRACTION FROM ORCHARD SALT DOME, TEXAS

Natural and mining removal of salt from salt domes, sulfur from their cap rocks, and oil and gas from adjacent strata can result in subsidence features on the landscape. In this exercise we look at the collapse processes, landforms, and damages that occur with the extraction of sulfur from salt domes in the Houston dome area of Texas. Salt domes in this area are roughly cylindrical masses of rock salt (halite) formed by the upward migration of salt from Jurassic evaporites thousands of

feet below the surface. Flow of the salt (deformable and less dense than overlying sedimentary rock) occurs due to lithostatic pressure differences in the rock column. As a salt dome flows and punches upward through the overlying sediments over millions of years, it develops a cap rock of sulfate and carbonate minerals that forms from impurities in the salt. The cap rock is several hundred feet thick on top of the salt and drapes over the edge and the upper sides of the dome. Cap rocks may be covered by several hundred feet of sand and mud/shale layers in the Houston dome area.

Sulfur, formed from bacterial alteration of the sulfate minerals (anhydrite and gypsum), is a resource extracted from the cap rock by the Frasch process (sulfur is pumped out after being melted by superheated steam). In addition to the salt that is removed by solution mining from the dome, oil and natural gas are extracted from traps in steeply dipping strata that terminate at the edge of the salt dome. Salt domes are also used for storage of oil and gas in cavities that have been solution mined in the salt (e.g., U.S. Strategic Petroleum Reserve).

Extraction of sulfur from the cap rock has produced **trough subsidence** (gentle downwarping over the extraction zone) and **collapse sinkholes** (smaller, circular, steep-walled depressions). Natural subsidence over salt domes is indicated by depressions and saline lakes over dome crests. When a cavern is formed by extraction of salt or sulfur, the cavern roof may fail, with collapse of rock into the cavern. As more of the overlying rock and sediment fails, upward migration of the broken zone occurs.

Damage from subsidence following salt dome resource extraction includes loss of the flooded land, damage to roads and buildings, loss of well casings (greatest risk), and loss of drilling equipment. In one instance after sulfur extraction had been completed at a site, an attempt to recover some of the well casing caused the almost immediate collapse of the land surface to a depth of 70 feet and loss of the rig.

At the Orchard Salt Dome in Fort Bend County, Texas (Figure 9.6, topographic map), subsidence related to removal of sulfur has been documented (Mullican, 1988). The dome is about 7,000 feet in diameter, covering an area of 620 acres as shown by the occurrence of subsurface structural contours on the salt surface. Structural contours outline the shape of a rock unit in the subsurface (see Figure 9.6). Minimum depths to the cap rock and the salt are 285 feet and 375 feet, respectively. Production of sulfur was from the dome margin in the zone 1000–3000 feet below the surface, from 1938 to 1970. Maps and photo surveys document the change beginning with the first sinkhole (100 ft in diameter) that developed near the southwest (SW) margin in 1941. By 1952 subsidence troughs and circular sinkholes in the northeast part of the dome began, followed by other failures there (Figure 9.7). After sulfur extraction terminated in 1970, high-altitude U-2 photographs in 1979 showed additional subsidence in the northeast part of the dome.

FIGURE 9.6 Subsidence topography above Orchard Salt Dome, Texas. Structural contours are the heavy solid or dashed lines (in feet below sea level) on the surface of the salt dome, which is nearly vertical. "A" marks one of the open circles (a well). Spot elevations and land surface contours are in feet.

(Modified from Mullican, 1988; used with permission of the UT Bureau of Economic Geology.)

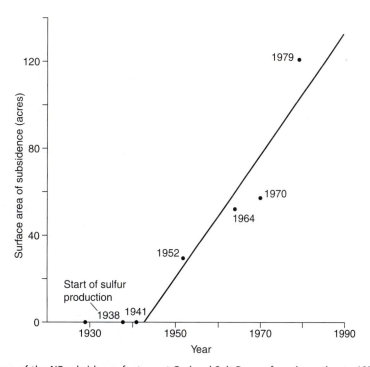

FIGURE 9.7 Change in surface area of the NE subsidence feature at Orchard Salt Dome, from inception to 1979. (Mullican, 1988, used with permission of the UT Bureau of Economic Geology.)

QUESTIONS (9, PART C)

1. What is the diameter and surface area of Orchard Dome?

2. a. At what depth is:

the top of the salt?

the top of the cap rock?

b. According to spot elevations and contours in Figure 9.6, what is the approximate elevation of the land surface above the dome (i.e., above sea level)?

3. a. In what unit or part of Orchard Dome is the sulfur that was extracted?

b. Describe how the sulfur was mined.

4. From the description of the Orchard Salt Dome and information in Figure 9.6, make a simple concept sketch of the salt dome in cross section showing the shape of the dome, the position of the cap rock above it, and the overlying rock and sediment. Label the rock units. Also mark and label the approximate elevation of the landscape, sea level, top of cap rock, and top of salt. Structural contours are shown and explained in Figure 9.6. (Your instructor may provide part of this concept sketch.)

5. How far below sea level is the lowest structural contour for the dome?

6. On Figure 9.6, identify and label the two types of subsidence features that are described in the introduction. The two types are:

7. Refer to Figure 9.7 to determine the surface area of the large northeastern subsidence feature in 1970, when sulfur extraction ended, and record answer here.

8. By 1979, the NE subsidence feature had extended to the SE beyond the area seen in Figure 9.6. It reached across a road almost to one of the small open circles (marked "A" in Figure 9.6) at the end of an unimproved road.

a. According to data in Figure 9.7, by how many acres had the flooded subsidence area increased between 1970 and 1979?

b. What do the small open circles on the map (Figure 9.6) represent?

c. Explain, with the aid of your salt dome cross section in Question 4, why these circles occur outside the salt dome. (Hint: Related to the mechanism of salt dome formation.)

9. What was the average rate of increase in subsidence area between 1941 and 1979 for this major subsidence trough?

10. A reconnaissance flight over the dome in 1985 documented an increase in the number of sinkholes from 16 to 22 in the central dome area since 1979. Investigators were not able to accurately determine if there were any increases in the NE subsidence trough in 1985.

a. What is the latitude and longitude marked near the NE edge of the dome in Figure 9.6?

b. Describe how you might go about determining the current number and distribution of sinkholes and troughs in Orchard Dome today? (Hint: The answer to Question 10.a should help.)

PART D. SUBSIDENCE ABOVE COAL MINES, ILLINOIS AND COLORADO

Coal-fired power plants provide the largest share of electric energy in the United States. Of major concern is their impact on air quality (sulfur and nitrogen oxides, mercury, and CO_2, a greenhouse gas). Carbon capture with sequestration is proposed to address the CO_2 problem and this technique will benefit from implementation of new, clean coal technologies for electric energy generation. Successful sequestration depends on our ability to find adequate subsurface space for injection of captured CO_2. Assuming we will be able to extract energy from coal in an environmentally responsible way and given that other fossil fuels have passed their peaks in the United States, coal mining will increase in importance.

The coal is mined in surface and underground operations and both processes have environmental impacts such as acid mine drainage, acid soils and waste piles, and disrupted groundwater systems. In the case of surface mining, the landscape and surface drainage is completely disrupted and in the past, erosion and stream sedimentation, landslides, and revegetation of the landscape presented problems. Fortunately, the Surface Mining Control and Reclamation Act of 1977 reduced the impact, but problems still exist, particularly in mountain top removal. In the case of underground mining of coal, surface subsidence is a process that continues to cause problems. Understanding the subsidence process and risk in coal mining is the objective of this part of the exercise.

The **room and pillar** method of underground coal mining results in voids (rooms and tunnels) separated by pillars that support overlying rock (sandstone, siltstone, and shale) and regolith. In this mining process ~50% or more of the coal is removed from the section of the coal seam being mined. Over a period of time (often years to many decades) after mining, when the overlying rock sags and breaks into the rooms, tunnels, and shafts, surface subsidence occurs and forms small **circular depressions or collapse sinkholes** (over shallow depth mines, usually within 100 feet or less of the surface) or larger **subsidence troughs.** Factors in the form, size, and amount of subsidence include void size, depth of mining, strength of rocks (including fractures and joints), age of timbers, soundness of the coal pillars, and weakening of the coal pillars due to fluctuations in groundwater (exposure to air promotes oxidation and a lowered water table reduces the buoyant force). The extent of missing or "pulled" pillars (at the very end of mining) also reduces roof support. In **longwall mining,** a more recent mining technique not considered here, almost 95% of the coal is removed in the mining process, which results in a more uniform lowering of the mined-out region as mining progresses. It can still have an impact on structures at the surface. **Surface mining,** in which the overlying regolith and bedrock are removed to expose coal, subsidence of structures is not a problem (but settlement and landslides may occur in the reclaimed landscape).

Part D of this subsidence exercise is based on reports from Illinois and Colorado, but the examples here would apply to most of the country where coal is mined by the room-and-pillar method. To find the general areas in the United States where subsidence from coal mining might be a problem, see the Coal Fields of the Conterminous United States map (USGS Open-File Report OF 96–92). To locate actual mined areas in these coal regions, one should check state

natural resources agencies or geological surveys for extent of mining maps and more detailed mine maps.

Extent-of-mining maps show the mined areas on topographic maps. More **detailed coal mine maps** show the actual rooms, pillars, tunnels, and adits (nearly horizontal passages into a mine); however, these maps might not be completely accurate or easily available. Any building that is close to a mine could be at risk in some areas, even if it is shown on a map to be somewhat beyond the mined area. For building-site selection purposes, drilling can reveal the existence and condition of rooms, pillars, previous failures, and any later filling of voids. It might not be possible to protect some existing buildings from collapse due to the difficulty of predicting where subsidence will occur and the cost of drilling and any void filling that is needed (Turney, 1985).

Roads, buildings, and utilities are subject to damages from subsidence. Concrete or asphalt roads may crack or sag, or completely collapse into circular pits. Building damage shows up as a sagging roofline, cracks in bricks, separation of steps, cracks in drywall, distorted windows, sticky doors, floor sags, and cracking noises in the home. Utilities also provide clues to subsidence, with cracked water and sewer pipes, water and natural gas leaks, and dirty tap water.

QUESTIONS (9, PART D)

1. a. Longwall coal mining is a more modern mining technique used in some areas, but the older type of underground coal mining used in many parts of the country is known as:

b. Make a labeled sketch below showing a map view of a subsurface coal mine in which about 50% of the coal has been removed by the room and pillar method (see the lowest part of the cross section in Figure 9.8 for a view of such a mine).

2. a. What are the two subsidence landforms common above these coal mines?

b. Which of these two subsidence landforms would likely develop over shallow-depth mines and might have a connection between the surface and the mined-out void?

3. List at least four factors that control subsidence over coal mines.

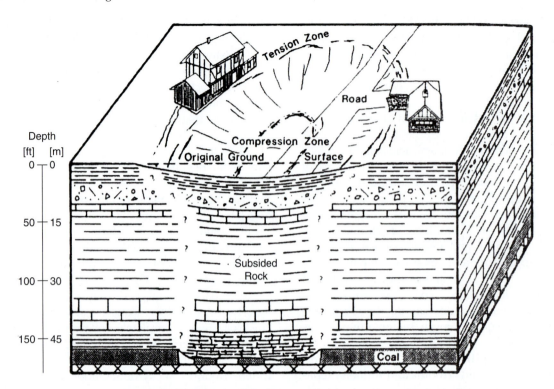

FIGURE 9.8   Trough or sag subsidence over a coal mine with collapsed pillars. Note the damage on the corner of each house over the fracture zone. The unit below the coal seam represents a clay paleosol (an old soil).

(Modified from Turney (1985) after DuMontelle and others, 1981, Illinois State Geological Survey; Environ. Geol. Notes, No. 99; used with permission.)

4. a. Circle the option below that would most likely be the setting for a pronounced subsidence feature at the surface, if the size of the void and the rock types were the same in both cases.

Void in a deep mine

Void in a shallow mine

b. Explain

5. Why might your house be at risk even if it were not directly above a void shown on a mine map? (Hint: Review Introduction.)

6. Over a period of several months, what damages, clues, or processes might you observe that help you to recognize that your house is subsiding?

7. In Figure 9.8, a block diagram of a subsidence event, what is the major type of subsidence feature shown?

8. In Figure 9.8, describe what is occurring:
   a. to the land and the buildings (visible damage) on the margins of the subsidence feature

   b. to the land in the central area of the subsidence feature.

9. a. In Figure 9.8, on the margins of the collapsed area above a depth of ~35 ft, draw (and identify) a sloping line to show the increased width of surface subsidence that begins at this depth.
   b. What does the line that you have drawn in this diagram suggest about the difficulty of determining if your house is "at risk"?

   c. The level above which the width of subsidence begins to increase also is where a major change in geologic materials occurs. Above is unconsolidated glacial till in the regolith; below is limestone bedrock. Why does the increase in width of the subsidence occur here?

10. Geologists use standard symbols for geologic materials in most cross section diagrams, as in Figure 9.8. Draw and identify below several of the geologic symbols for the rocks and/or unconsolidated sediments shown in the block

diagram (Figure 9.8). You may use those symbols identified in the diagram, its caption, the questions above, or other sources.

Bibliography

DuMontelle, P. B., Bradford, S. C., Bauer, R. A., and Killey, M. M., 1981, *Mine subsidence in Illinois: Facts for the homeowner considering insurance.* Illinois Geological Survey, Environmental Geology Notes 99, 24 p.

Hynes, J. L. ed., 1986, *Proceedings of the 1985 Conference on coal mine subsidence in the Rocky Mountain Region:* Colorado Geological Survey Special Publication 31, 316 p.

Ireland, R. L., Poland, J. F., and Riley, F. S., 1984, *Land subsidence in the San Joaquin Valley, California, as of 1980:* U.S. Geological Survey Professional Paper 437-1, 93 p.

Johnson, A. I., Moston, R. P., and Morris, D. A., 1968, *Physical and hydrologic properties of water-bearing deposits in subsiding areas in central California in mechanics of aquifer systems:* U.S. Geological Survey Professional Paper 497-A, 71 p.

Mullican, W. F. III, 1988, *Subsidence and collapse at Texas salt domes:* The University of Texas at Austin Bureau of Economic Geology, Geological Circular 88-2, 35 p.

Poland, J. F., 1970, Status of present knowledge and needs for additional research on compaction of aquifer systems. In *Land Subsidence.* Proceedings of Tokyo Symposium, Sept. 1969, LASH/AIHS UNESCO, p. 11–21.

Poland, J. F., 1971, *Land subsidence in the Santa Clara Valley, Alameda, San Mateo, and Santa Clara counties, California:* U.S. Geological Survey Open File Report (mF no. 336).

Poland, J. F., and Green, J. H., 1962, *Subsidence in the Santa Clara Valley, California: A progress report:* U.S. Geological Survey Water-Supply Paper 1619-C, 16 p.

Poland, J. F., and Ireland, R. L., 1988, *Land subsidence in the Santa Clara Valley, California, as of 1982:* U. S. Geological Survey Professional Paper 497-F, 61 p.

Turney, J. E., 1985, *Subsidence above inactive coal mines: Information for the homeowner.* Colorado Geological Survey Special Publication 26, 32 p.

USGS, 1996, *Coal fields of the conterminous United States:* USGS Open-File Report OF 96-92. http://pubs.usgs.gov/of/1996/of96-092/doc.htm

Wilson, A. M. and Gorelick, S., 1996, The effects of pulsed pumping on land subsidence in the Santa Clara Valley, California: *Journal of Hydrology,* v. 174, p. 375–396.

River Floods

"We are presently in a period of marked global increases in damage to life and property from floods and from other naturally catastrophic phenomena. The problem may be partly exacerbated by anthropogenic climatic change ("global warming"), but it is certainly tied to the human propensity to build in hazard-prone areas."

—BAKER, 1994

"The reason they call them floodplains is that it is plain that they flood."

—GORE, 1995

INTRODUCTION

Tens of billions of dollars have been spent by the federal government, mainly since the 1930s, on flood control projects such as dams, levees, and channel improvements. Nevertheless, damages from flooding have continued to rise in the United States.

A flood occurs when bodies of water flow over land that is not usually submerged. Although we commonly think of floods as being caused by a river that overflows its banks during periods of excess precipitation or snowmelt, floods can also be caused by dam failures, high tides along seashores, high water levels in lakes, or high groundwater.

River floods occur when the water height (known as *stage,* and recorded as number of feet above a local base height) passes an arbitrary level. This level usually is the bank-full stage, or when a river is completely filling its channel. When a stream channel can no longer hold the increased water (its *discharge,* or total amount of water flowing past a site) and overflows its banks, then the river is said to be in flood. Water then flows on valley floors or floodplain adjacent to the normal river channel. Of course, higher discharges mean higher stages or levels of water within and outside a river channel.

Floods are a natural characteristic of rivers. They occur in response to the interaction of hydrologic, meteorologic, and topographic factors. Human actions also impact floods, for example building dams or levees, channelizing rivers, developing urban, suburban, or agricultural areas, and deforestation.

River floods are not the only types of floods. The term flood also is associated with the inundation of lake and marine coasts and alluvial fans. Even relatively flat and poorly drained upland regions are subject to flooding. In the latter case, water that would normally infiltrate to become groundwater or be carried away in poorly defined channels, sewers, or ditches remains on the land during excessive rainfall. When this water enters basements or garages, or otherwise makes life uncomfortable for the inhabitants of the area, the term flood is applied even though the area is not a floodplain, a coastal region, or an alluvial fan.

This exercise is designed to acquaint you with the nature of river floods, the problems that are created by flooding, and potential solutions. Part A analyzes flood losses in the United States, introduces the concepts of flood frequencies or recurrence intervals, magnitudes, and the variations of discharge with drainage basin area, which we consider here as intensity.

Parts B and C provide specific cases of floods. You will apply concepts and, using information from other sources, explore several solutions to flooding including those in the broad categories of (1) land use regulation and (2) engineering projects on rivers and watersheds.

Part D explores the identification of flood plains on topographic maps and the nature and use of Flood Insurance Rate Maps (FIRMs).

PART A. FLOOD FREQUENCY AND MAJOR FLOODS

Flood Discharge

The *discharge* of a stream is the volume of flow that passes a specific location in a given period of time. Discharge rates are usually expressed in cubic feet/second (cfs) or cubic meters/second (m^3/s). If the area of the wetted channel cross section (the measured width of the channel at a site multiplied by the depth of the water) and the velocity of the stream are known, then the discharge can be determined by the following formula:

$$Q = A \times V$$

where Q is the discharge, in cfs or m^3/s, A is the cross-sectional area of the stream in square feet or square meters, and V is the velocity, in ft/s or m/s.

Flood Frequency

Where flood records are available, computations of flood frequency are based on peak annual floods (the maximum discharge for the year at a specific station). Flood frequency is expressed as a recurrence interval (or return period), which is the average time interval (in years) between the occurrence of two similar floods, with the same water levels. The recurrence interval (T, in years) for a flood of a given discharge is determined by this formula:

$$T = (n + 1)/m$$

where n equals the number of years of record, and m is the rank or order of the annual flood discharges from the greatest (1), to the smallest for the number of years of record.

To understand better the impact and frequency of floods and to assist in flood prediction, one method is to plot flood data on a flood frequency graph with the discharge plotted on the vertical axis and recurrence interval on the horizontal axis. A straight "best-fit" line is drawn to join the points for each year. The average number of years that will elapse until a given magnitude flood occurs again can be estimated from this line. In other words, based on the line there is a certain probability that a flood of a given magnitude can be expected to occur x times within a fixed time interval. For example, if the recurrence interval for an annual maximum discharge of 500 cfs is found to be 4 years, then we would expect to see a discharge of 500 cfs or greater five times during a 20-year period (20 years divided by the 4-year recurrence interval).

The longer the number of years of flood records, the higher the probability is that very large floods with very large discharges have been recorded. Estimates of 100-year floods are better if they are based on 100 or more years of record. Where there are few years of flood data, such as in Table 10.1, reliable estimates of 100-year floods may be difficult.

The annual probability of exceedence, P, is the reciprocal of T. Written as a formula

$$P = 1/T$$

This is the probability or chance that in a single year the annual maximum flood will equal or exceed a given discharge. A flood having a recurrence interval of 10 years is one that has a 10 percent chance of recurring in any year; a 100-year flood has a 1 percent chance of recurring in any year.

The 100-year flood, as defined statistically, is a legal definition of areas that are likely to be flooded. If, in the United States, someone chooses to purchase a home in the 100-year floodplain, they must obtain flood insurance.

SOMEWHERE IN THE UNITED STATES, YEAR 2000 PLUS OR MINUS. Nature takes its inexorable toll. Thousand-year flood causes untold damage and staggering loss of life. Engineers and meteorologists believe that present storm and flood resulted from ··· conditions [that] ··· occur once in a millennium. Reservoirs, levees, and other control works which have proved effective for a century, and are still effective up to their design capacity, are unable to cope with enormous volumes of water. This catastrophe brings home the lesson that protection from floods is only a relative matter, and that eventually nature demands its toll from those who occupy flood plains. (Hoyt and Langbein, 1955)

QUESTIONS (10, PART A)

Flood Frequency in the Seattle/Tacoma area

Questions 1-11 investigate recurrence intervals, 100-year floods, and changing flood frequencies for two watersheds in the state of Washington. These data are slightly modified from U.S. Geological Survey information. Population in the Puget Sound area is growing rapidly, and humans have made many changes to rivers and drainages.

Pick **one** of the four data sets (Mercer Creek 1, Mercer Creek 2, Green River 1, or Green River 2) in Table 10.1. Be sure, if you are working in a laboratory or class, that one or more students select each of the four data sets. Each data set spans 11 years of record along Mercer Creek or the Green River. Use your chosen data set to estimate the likely discharge for a 100-year flood. Follow the steps below.

1. Rank the peak flood discharges for the data set you have chosen in order of magnitude, starting with 1 for the largest and ending with 11 for the smallest. Write these results in the "Rank" column.

TABLE 10.1 Flood Data for Mercer Creek and the Green River. Discharge Data in cfs.

	Mercer Creek-Data Set 1				Mercer Creek-Data Set 2		
Year	Peak Flood Discharge	Rank (1 is greatest)	Recurrence interval	Year	Peak Flood Discharge	Rank (1 is greatest)	Recurrence interval
1957	180			1979	518		
1958	238			1980	414		
1959	220			1981	670		
1960	210			1982	612		
1961	192			1983	404		
1962	168			1984	353		
1963	150			1985	832		
1964	224			1986	504		
1965	193			1987	331		
1966	187			1988	228		
1967	254			1989	664		
	Green River-Data Set 1				Green River-Data Set 2		
Year	Peak Flood Discharge	Rank (1 is greatest)	Recurrence interval	Year	Peak Flood Discharge	Rank (1 is greatest)	Recurrence interval
1941	9310			1976	4490		
1942	10900			1977	9920		
1943	12900			1978	6450		
1944	13600			1979	8730		
1945	12800			1980	5200		
1946	22000			1981	9300		
1947	9990			1982	10800		
1948	6420			1983	9140		
1949	9810			1984	10900		
1950	11800			1985	7030		
1951	18400			1986	11800		

2. Use the formula $T = (n+1)/m$ and determine the recurrence interval of each of the 11 floods. Write the results for each year in the "recurrence interval" column.

3. Determine an appropriate vertical scale for your discharge data. The vertical scale should be chosen such that the numbers you plot from the data above fill about one-half or slightly more of the length of the scale. Write the appropriate numbers for your discharge along the left edge of the scale.

4. Plot the discharge and recurrence interval for each of your 11 floods.

5. Draw a best-fit straight line, not a dot-to-dot curve, through the data points. Extend your line to the right side of the graph.

6. Based on your data, what is the predicted discharge for a 100-year flood?

7. Either find someone who has plotted the second set of data for your stream, or repeat steps 1 through 7 to determine the predicted discharge for a 100-year flood, using the second set of data for your stream. (You may plot the second set of data on Figure 10.1; note that depending on your choice of numbers for the first plot, you may need to have a second set of values on the vertical axis.) How does your prediction made in Question 6 compare with the answer from the other set of data for the river you plotted?

8. Suggest possible human activities in the watershed that could have caused the differences in predicted floods that result from the two sets of data.

9. When you have completed interpretation of the stream you selected, find students who have done the other stream. How do their data compare with yours? What human activities did they suggest for the changes in flood predictions they discovered?

10. Based on the flood predictions for all four data sets, what does the contrast in predicted flood discharges imply about the usefulness of the 100-year flood as a legal designation for these two streams?

11. What information do you need to know if you are about to buy a house that is located adjacent to, but just outside of, the 100-year floodplain?

Large Floods in the United States

Data in Table 10.2, partly from the U.S. Army Corps of Engineers, show the damages suffered and deaths due to flooding. The U.S. Army Corps of Engineers (2000) provides data on *damages avoided* by flood projects and by emergency activities of the Corps. For example, in 1999 flood projects reduced potential damages by $2.8 billion (75% by reservoirs and 25% by levees), while emergency activities of sandbagging and technical assistance saved $48 million in losses, a much smaller amount.
Do Questions 12–16, which refer to Table 10.2.

12. On Figure 10.2, place a point for each decade to show monetary flood loss (in billions of dollars) for each decade. What is the general trend in flood loss in the United States between 1900 and 2000 as determined from Table 10.2 and your graph?

13. Now place an open circle for each decade to show the U.S. population in millions of people at the end of each decade.

14. a. For flood damage losses in the 20th century in Figure 10.2, describe and explain the trend.

b. What is the role, if any, of growth in population and rising flood losses in the 20th century?

c. What other factors contribute to increased losses?

15. Discuss the effectiveness of flood mitigation in the 20th century with your lab group. Are flood control systems effective?

16. According to the Corps of Engineers (2000), for the decade of the 1990s, the average damage loss per year was about $5 billion. For the same period the average value of flood damage reduction by projects per year is estimated at $20 billion. Using this information and that in the graphs you plotted discuss the statement "Flood-control dams and levees have been effective in reducing flood damage" and suggest additional measures for reducing flood damage.

The Discharge/Area Ratio

Questions 17 and 18 investigate ranges in flood discharge (Q) with drainage basin size (A) for several different sizes of rivers. This **flood intensity** comparison is based on data in Table 10.3, which will be plotted in Figure 10.3.

17. Use the flood records in Table 10.3 to calculate the discharge/area ratio for each river. Record your calculation in the column on the table. Plot the calculated ratio against drainage-basin area in Figure 10.3. Identify the six rivers that you are able to plot.

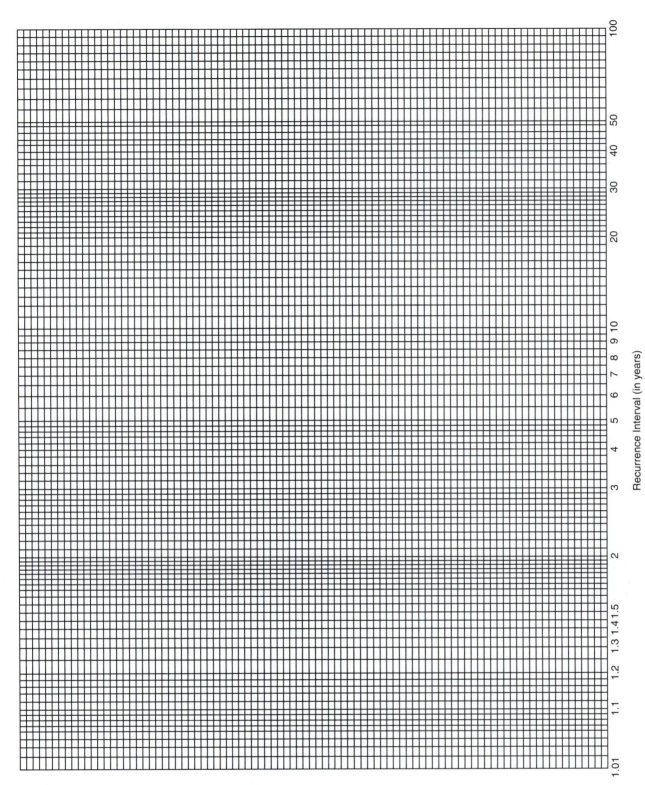

Recurrence Interval (in years)

FIGURE 10.1 Flood frequency curve based on data from Table 10.1.

TABLE 10.2 Flood Damages in the United States. Monetary losses for years prior to 1993 have been adjusted to constant 1993 dollars.

Decade Ending	Flood Damages, in Billion $	Deaths	U.S. Population, in Millions
1909	7.75	204	90
1919	18.28	958	105
1929	16.31	1048	122
1939	24.07	929	131
1949	19.45	619	149
1959	28.04	791	177
1969	17.19	754	201
1979	39.88	1806	227
1989	27.49	1119	246
1999	49.56	956	273

(Flood data from Richards, 1997 and U.S. Army Corps of Engineers, 2000) population data from U.S. Census Bureau (2000).

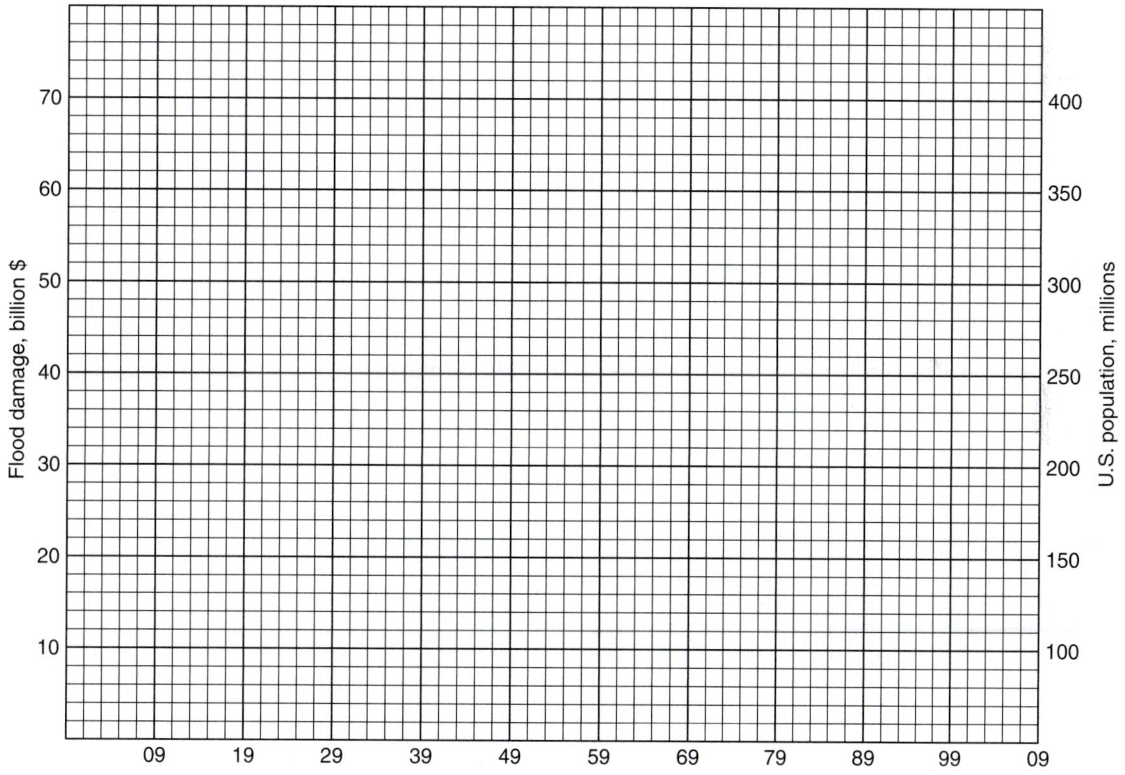

FIGURE 10.2 U.S. population and flood impacts in billions of dollars, by decade.

18. From your calculations and plots determine the general relationship between Q/A ratio and drainage basin area. With increasing area of drainage basin, is there an increase or decrease in the Q/A ratio? Briefly explain the reason for this.

PART B. ZANESVILLE, OHIO FLOOD

Flood of 1959 in Zanesville, Ohio

The Zanesville flood of January 21–24, 1959, was one of the most devastating floods since 1913 in a widespread area of Ohio. Sixteen lives were lost, damage exceeded $100 million, and 17,000 buildings were flooded. A storm on January 14–17 saturated the ground, which then partly froze and later was covered

TABLE 10.3 Discharge/Area Ratios for 9 Rivers

River		Drainage Basin Area (km²)	Maximum Discharge, Q (m³/s)	Q/A Ratio (m³/s)/km²
Name	Location			
Woallva	Hawaii	58	2,470	
Yaté	New Caledonia	435	5,700	
Pioneer	Australia	1,490	9,840	
Tam Shul	Taiwan	2,110	16,700	
Eel Scotia	California	8,060	21,300	
Han Koan	South Korea	23,880	37,000	
Cheng Jiang	China	1,010,000	110,000	
Lena	USSR	2,430,000	190,000	
Amazon	Brazil	4,640,000	350,000	

with snow. In a northeastward band extending from Cincinnati through Columbus, 6 inches of rain fell on January 20–21; more than half of the state received at least 3 inches of rain (Edelen et al., 1964).

Flood-control reservoirs in some basins reduced peak flows of some streams and prevented flood damage. The Licking River, which discharges into the Muskingum River at Zanesville (see Figure 10.4), exhibited higher flood levels in 1959 than during the 1913 flood, although 5 miles upstream from their confluence the stage of the 1913 flood was 4.5 ft higher than that of the 1959 flood.

A stream-gaging station lies on the bank of the Licking River 3.65 miles upstream from its mouth and at an elevation of 683.7 ft (Figure 10.5 map in the back of the book). The maximum stage of the 1959 flood at this gage was 32.46 ft, which means that the elevation of the flood at this point was 716.2 ft. Any structure in the immediate area of the gaging station and lower than this elevation would have suffered flood damage.

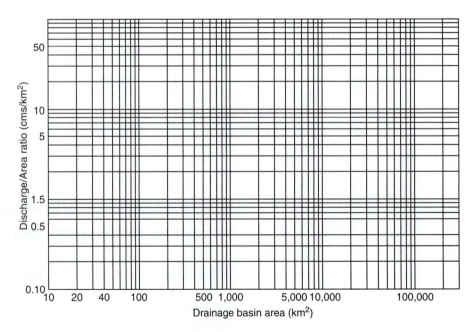

FIGURE 10.3 Variation in discharge/area ratio or flood intensity with drainage-basin area.

FIGURE 10.4 Confluence of Licking (left) and Muskingum (right) rivers at Zanesville during 1959 flood. (USGS, from the *Zanesville Signal*)

In this exercise we examine the extent, frequency, and causes of some river floods, and ways to reduce loss of life and property.

QUESTIONS 10, PART B

1. Using the data in Table 10.4 and the graph paper (Figure 10.6) plot the data and draw a straight line for frequency of floods at Dillon Falls, Ohio. Use a dashed line beyond the 40-year recurrence interval.

2. What are the expected elevations of the following floods at the Dillon gaging station as determined from Figure 10.6?

  a. 30 years:

  b. 50 years:

  c. 100 years:

TABLE 10.4 Recurrence Interval, Stage, and Discharge at Dillon Falls, Ohio

Recurrence Interval at the Gaging Station (Yr)	Elevation Above Mean Sea Level, Stage (ft)	Discharge (cfs)
40	710.2	27,400
20	708.7	24,300
10	707.0	21,100
5	705.1	18,000
3	703.5	15,600

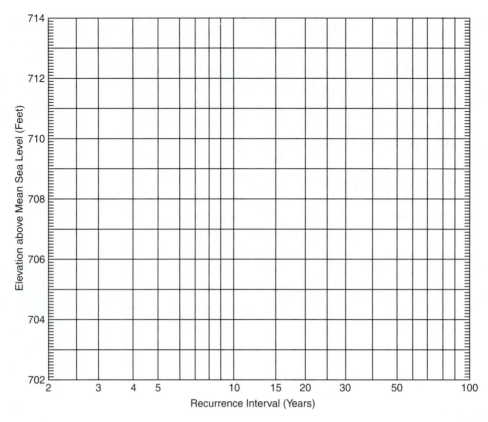

FIGURE 10.6   Frequency of floods above the 703-feet elevation on Licking River at Dillon Falls, Ohio.

3. Using the data in Table 10.4 construct a graph of elevation versus discharge (rating curve) in Figure 10.7.

4. Discharge can be obtained for a flood of a given frequency by using information on the flood frequency–elevation curve (Figure 10.6) and the rating curve (Figure 10.7). What is the expected discharge at Dillon, Ohio
   a. for a 30-year flood?

   b. for a 50-year flood?

5. If you visit a flood area after the water recedes, what indications might there be on buildings, trees, and land surfaces of the level of the flood maximum?

6. Figure 10.8 is a flood profile of the Licking River for the 1959 flood at Zanesville. Elevations are plotted against the distance upstream from the mouth (in miles along the centerline of the river). The maximum elevation of the 1959 flood at the gage given in the introduction. Profiles of higher or lower floods can be plotted on this diagram, although

backwater effects from the Muskingum River and channel constrictions may affect the profile.

   Using the 1959 flood profile (Figure 10.8) as a guide and the 40-year flood elevation at the gaging station as given in Table 10.4, determine the expected maximum stage of a 40-year flood at the floodmark elevation 2.9 miles upstream from the river mouth (Figure 10.8). Assume a similar drop in flood level (from the 1959 flood level) at the gaging station <u>and</u> at this floodmark elevation.

7. List several approaches that have been applied to reduce losses from river floods. (Consult your notes, textbooks, or references, if necessary.) Group the approaches as either control of the river and/or control of land use or human activity.

PART C. CENTENNIAL FLOOD ON BIG THOMPSON RIVER, COLORADO

Flash floods and associated erosion, deposition, and mass movement devastated parts of north-central Colorado in the Front Range on the night of July 31–August 1, 1976. This event caused 139 deaths and $35 million damage, mainly on the Big Thompson

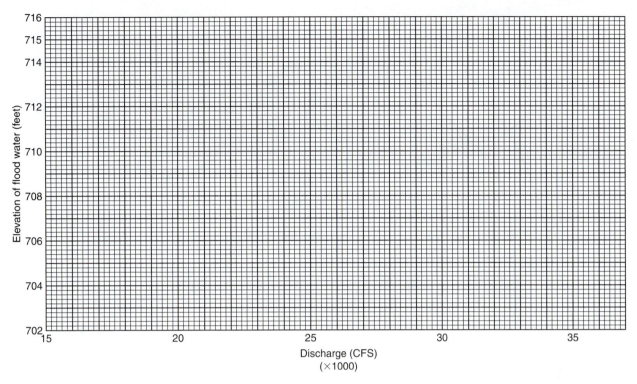

FIGURE 10.7   Rating curve for Licking River at Dillon Falls, Ohio.

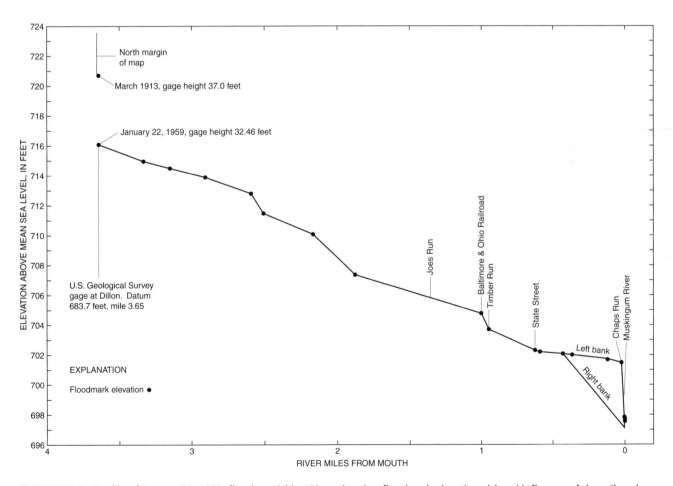

FIGURE 10.8   Profile of January 22, 1959, flood on Licking River, showing floodmark elevations (•) and influence of the railroad.

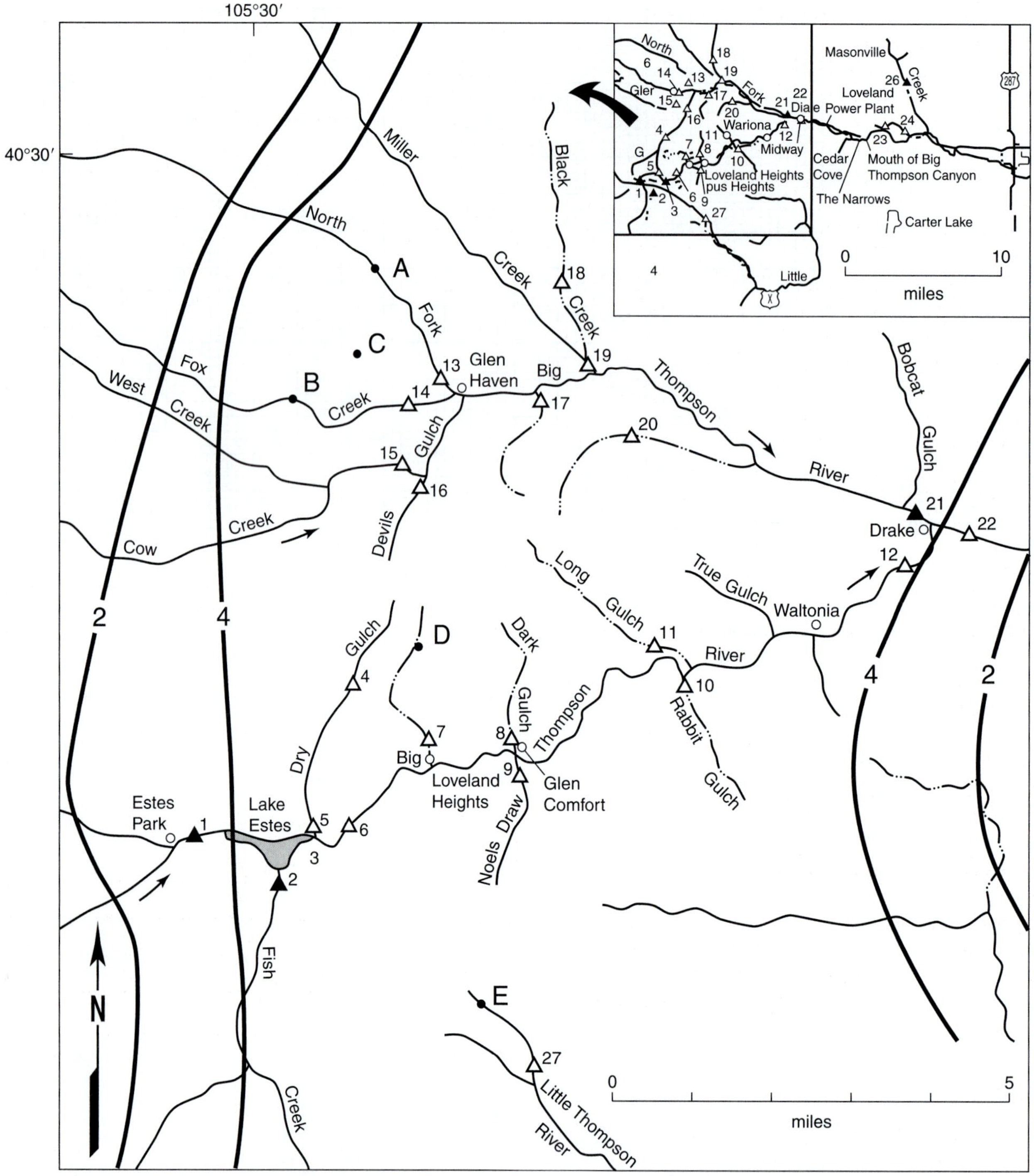

FIGURE 10.9 Location map and distribution of precipitation, in the mountains of the Big Thompson River Basin, July 31–August 2, 1976. Open and shaded triangles are discharge and precipitation sites; other precipitation sites represented by shaded circles. Isohyets are in inches. Arrows by rivers indicate flow direction.

(Base map from McCain et al., 1979; isohyets modified from Shroba et al., 1979)

River between Estes Park and the mouth of Big Thompson Canyon, 9 miles west of Loveland, Colorado (Figure 10.9).

The storms developed when moist, conditionally unstable air was pushed steadily westward up the slopes of the Front Range. By 1830 MDT (6:30 PM Mountain Daylight Time; times in this section are given in a 24-hour system, as this is how they were originally

reported) a N–S line of strong thunderstorms had developed on the foothills. These nearly stationary storms intensified as additional low-level moist air flowed westward from the plains. This system locally released more than 7 in. of precipitation between 1930 and 2040 MDT, July 31. The high precipitation rate and the steep terrain caused rapid surface runoff and the Big Thompson River quickly reached flood stage.

Peak discharges exceeded the 100-year flood discharge at several sites, although the rainfall, flood discharges, unit discharges, and dollar losses were not unprecedented for areas along the eastern foothills and plains of Colorado (McCain et al., 1979). The major component of the dollar losses was damage to U.S. Highway 34; losses not counted included tax receipts, tourist business, water wells, septic tanks, and property devaluation because of location in a designated flood plain.

Significant geomorphic changes occurred in the steep terrain with relief of more than 3,000 ft and canyon walls as steep as 80 percent. These included channel scouring for most of the river, where the stream gradient was greater than 2 percent, on the outside of meander bends and in narrow reaches. In reaches where the gradient flattened to less than 2 percent and where the stream channel widened, boulders (some with an intermediate diameter of 7 ft) and finer sediment were deposited on point and channel bars and on flood-plains. Overbank sedimentation was the primary impact downstream of Big Thompson Canyon.

On the tributaries, sheetfloods transported boulders onto lesser slopes of alluvial and debris fans and eroded unprotected slopes. Debris avalanches, slides, and flows were generated by the saturated soils and provided debris that destroyed some buildings.

Both historic and geologic evidence suggests that other basins along the Front Range are vulnerable to flood events of magnitude similar to that of the one that struck the Big Thompson Basin on the eve of Centennial Sunday (Shroba et al., 1979).

The primary objective of this exercise is to understand the nature and causes of flash flooding in mountain terrains. The secondary objective is to investigate the impacts of the flooding and associated water erosion and sediment deposition. For more information on the Centennial Flood, see www.coloradoan.com/news/thompson/.

TABLE 10.5 Cumulative Precipitation, July 31–August 2, 1976, for Selected Sites in the Big Thompson River Basin. Mean value of precipitation for July here is 1.5–2.1 inches. The 2-and 4-inch rainfall contours (isohyets) are shown in Figure 10.9.

Site	Cumulative Precipitation (in.)
2	5.2
3	6.0
4	8.1
6	9.9
7	10.5
8	10.4
9	10.6
10	8.0
11	10.0
13	10.0
14	9.9
15	6.3
16	8.0
17	8.8
18	8.1
19	8.2
20	8.0
27	7.9
A	6.3
B	5.8
C	10.0
D	10.1
E	9.8

QUESTIONS (10, PART C)

1. Examine Table 10.5. Plot the cumulative rainfall for the sites in this table on Figure 10.9 and draw in the lines of equal precipitation (isohyets) for 6, 8, and 10 inches.

2. Where were the centers of precipitation?

3. What is the explanation for the intense, localized nature of this precipitation event?

4. From this map, where would you expect most flood and mass-movement damage from the storm to occur?

5. Refer to discharge hydrographs (Figure 10.10) for selected sites on the Big Thompson River and the North Fork Big Thompson River. Compare sites 1, 21, and 23. Do you think the flooding at Estes Park (site 1) was serious? Explain.

6. Do the hydrographs support the precipitation data in indicating the area(s) of greatest runoff and potential damage? Explain.

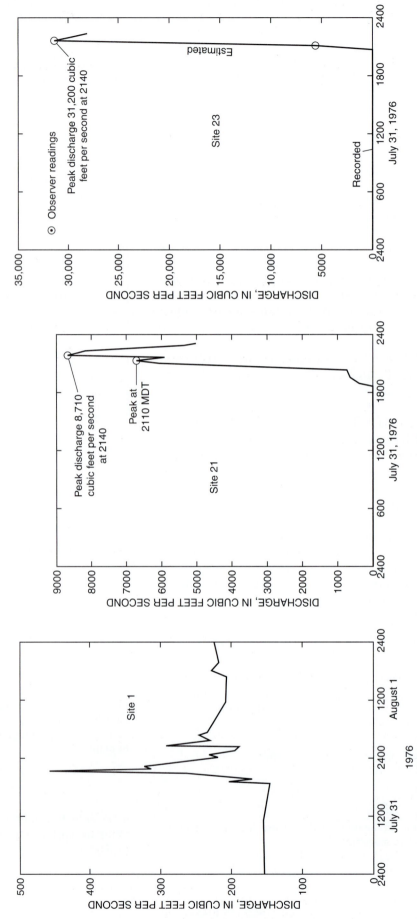

FIGURE 10.10 Hydrographs for Big Thompson River (Sites 1 and 23) and North Fork Big Thompson River (Site 21). Bottom scale is in hours, using a 24-hour clock. See Figure 10.9 for site locations (McCain et al., 1979).

7. There are two major rivers in the Big Thompson River system: The Big Thompson (BT) in the south and the North Fork Big Thompson (NFBT) in the north. The flood peak at site 22, east of Drake, occurred at the same time as the minor peak at site 21 (2110 MDT) on the NFBT (see Figure 10.9).

    a. How did the flood crest moving down the BT impact the water level on the NFBT at Site 21?

    b. Why was there a second (main) peak at Site 21 on the NFBT?

8. If the main peak at site 21 had occurred 30 to 35 minutes earlier than shown in Figure 10.10, what impact would this have had on the size of the flood at site 23 at the mouth of Big Thompson Canyon (Figure 10.9, Inset map)?

9. The flood crest reached Site 23 at the mouth of Big Thompson Canyon, 7.7 miles east of Site 22, 30 minutes after the crest at Site 22. What was the average speed of the flood crest in miles per hour and feet per second? Could you outrun this flood crest?

10. Refer to Table 10.6

    a. At what three sites were the unit discharges or discharge/area ratios of drainage basins (cfs/mi^2) the greatest?

TABLE 10.6 Hydrologic Data for Selected Flood Sites in Figure 10.9

Site No.	Stream and Location	Drainage Area (mi^2)	Peak Discharge (cfs)	Unit Discharge (cfs/mi^2)	Average Velocity (ft/s)	Average Depth (ft)
4	Dry Gulch	2.00	3,210	1,600	12	3.3
6	Big Thompson R., Lake Estes	9.00[a]	4,330	481	8	4.6
7	tributary near Loveland Heights	1.37	8,700	6,350	26	5.5
8	Dark Gulch	1.00	7,210	7,210	28	5.1
9	Noels Draw	3.37	6,910	2,050	21	5.7
10	Rabbit Gulch	3.41	3,540	1,040	13	4.7
11	Long Gulch	1.99	5,500	2,760	19	5.8
12	Big Thompson R. above Drake	34.0[a]	28,200	829	22	8.3
14	Fox Creek	7.18[b]	1,300		9	2.8
16	Devils Gulch	0.91	2,810	3,090	12	2.1
17	Tributary near Glen Haven	1.38	9,670	7,010	29	5.6
18	Black Creek	3.17	1,990	628	11	4.5
20	Tributary west of Drake	1.26	3,240	2,570	18	3.0
21	North Fork Big Thompson R. at Drake	85.1[b]	8,710		12	5.2
22	Big Thompson R. below Drake	276.[b]	30,100		16	10.3
23	Big Thompson R. at mouth of canyon	305.[b]	31,200		26	10.6

a. Approximate contributing drainage area during flood of July 31–August 1, 1976.

b. Contributing drainage area for flood of July 31–August 1, 1976, unknown.

(from McCain et al., 1979)

b. How do the locations of these sites compare with the distribution of maximum precipitation for the storm (see Figure 10.9)?

11. Plot the peak discharge and drainage area for each of the stations with the three highest unit discharges for the 1976 storm on the peak discharge/drainage area diagram for eastern Colorado (Figure 10.11). Unit discharges provide a measure of the relative size of a flood event in cubic feet per second per square mile (cfs/mi^2). Considering the three data points plotted for the 1976 flood, do you think that even more outstanding floods will occur on Big Thompson River, or is this the largest that can be expected? Explain.

12. Explain why unit discharge for Site 23 (see inset map, Figure 10.9) could not be plotted on this graph (Figure 10.11; see Table 10.6.)

13. From Table 10.6, which site had the highest average velocity of stream flow?

14. Given that the previous peak discharge at Site 21 was 1290 cfs in the 1965 flood, how many times larger was the 1976 flood than the earlier flood at this site?

15. The ratio of the peak discharge of the 1976 storm to the discharge for the 100-year flood was determined for selected stations as follows:

Site	Peak Discharge, 1976 (cfs)	Ratio
12 (BT)	28,200	3.8:1
21 (NFBT)	8,710	1.4:1
22 (BT)	30,100	2.9:1
23 (BT)	31,200	1.8:1

For the drainage basins upstream of Drake, was this flood greater on the Big Thompson River (BT, Site 12) or on the North Fork Big Thompson River (NFBT, Site 21)?

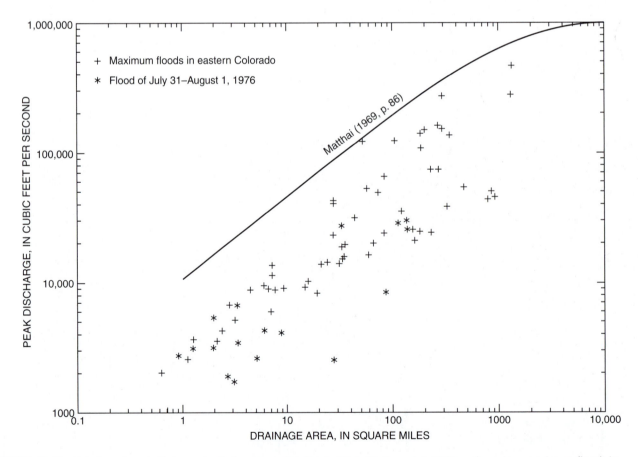

FIGURE 10.11   Relation of peak discharge to drainage area for flood of July 31–August 1, 1976, and previous maximum floods in eastern Colorado. The curve was drawn through the average of the eleven largest floods in the United States between 1890 and 1969. The empirical equation for the curve below the 200 mi² area is Q = 4000 AO.61, where Q = discharge and A = drainage area.

(Modified from McCain et al., 1979; after Matthai, 1969)

16. Using the information in Question 15, what is the expected discharge for the 100-year flood? (Hint: The ratio shown is: discharge for 1976 storm/discharge for 100-year flood.)

17. Study Figure 10.12 showing the Waltonia area before and after the flood. Describe the changes in buildings, roads, vegetation, and stream bed.

18. At other sites along the river, significant changes (erosion and deposition) also occurred on the floodplain and on alluvial and debris fans. Large accumulations of boulders and debris occurred on some of the fans; other fans underwent erosion. Study Figure 10.13. Would it have been safer to reside during the 1976 storm on the inside or outside of a meander? Explain.

FIGURE 10.12   Aerial photos of Waltonia, 1.5 miles southwest of Site 12 before (A) and after (B) the flood. Approximate scale: 1 in. = 200 ft. Most of the community is on a debris fan. Highway 34 and two large motels are shown in the floodplain in A. Big Thompson River is immediately behind (north of) the motels (Shroba et al., 1979).

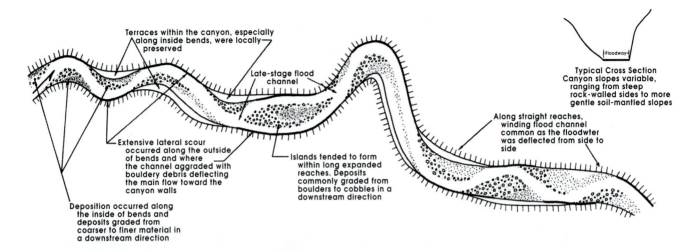

FIGURE 10.13 Pattern of scour and deposition within canyons of Big Thompson and North Fork Big Thompson rivers. Floodway is about 200 ft wide with gradients of 2–4 percent (Shroba et al., 1979).

19. The heaviest precipitation from this flood occurred between 1930 and 2040 hours MDT at Glen Comfort and 1930 and 2200 hours at Glen Haven. There was little opportunity to warn anyone. Use this information with the other information in this exercise to prepare two paragraphs, one arguing for and the other against purchase of a waterfront lot for a house in one of the many river valleys or canyons similar to Big Thompson along the Front Range of the Rockies. Use a separate sheet of paper for your answer.

PART D. FLOOD PLAINS AND FLOOD INSURANCE RATE MAPS

Meandering Rivers and Identification of Flood Plains on Topographic Maps

The first step in avoiding flood hazards is to avoid building in rivers. Some geoscientists suggest that the area known as the flood plain should instead just be called part of the river. The implication of this is important: Flood plains are natural parts of river systems that do not happen to be covered by water all of the time. When flood plains are covered by water, however, the damage to unprepared (and in some cases even prepared) people and communities can be disastrous.

In this part of the exercise we are going to use several different topographic maps in the colored plates section of the book, and identify different flood plains on the maps. First, refer to Figure 10.14, which is a sketch of typical topographic features seen along a flood plain.

Note on this figure that natural rivers meander (bend) a lot. The floodplain is the topographically low area adjacent to the river. Although when we look at a river it may appear to have a permanent channel, meandering rivers will change their channels and erode and redeposit over their entire floodplain, given

enough time. The river determines how much time is "enough"; it may be a few years, a few decades, or a few centuries. But over time, rivers will occupy and modify their floodplains.

On a shorter time period, floods will cover parts of the river valley. How much gets covered is determined by how big the floods are.

QUESTIONS (10, PART D)

Refer to Figure 2.10 in Chapter 2, Bloomington, Indiana.
1. Sketch a topographic profile of the valley of Griffy Creek, from the 750' contour in contact with the "L" in Bloomington in the west to the zero in the 750' contour above the "16" in Township 16 in the East. Mark the floodplain on the profile with a different color.

2. Is the Drive-in Theater (Southwest of the dam) in a floodplain? Explain.

3. Is Payne Cemetery (east-central part of the map) in a floodplain? Explain.

Refer to Figure 6.10 in Chapter 6, Draper, Utah.
4. Shade on this map the floodplain of Little Cottonwood Creek in the area of Glacio Park, near the eastern edge of the map.

5. Do you think the floodplain west of Beaver Ponds Springs is broad or narrow? What evidence do you use to support your answer?

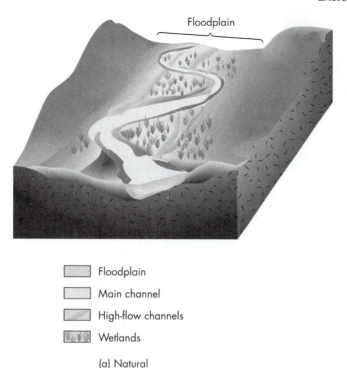

Floodplain

☐ Floodplain

☐ Main channel

☐ High-flow channels

☐ Wetlands

(a) Natural

FIGURE 10.14 Components of a floodplain. (Modified from Keller, 2002).

Refer to Figure 8.23 in Chapter 8, Athens, Ohio.
6. Shade on this map (1975), the area of the Hocking River floodplain from the Fairgrounds to the Radio Tower.

7. Are any buildings located on the floodplain? Identify three of the major ones.

8. What has happened to the course of the river through time in the area south of the sewage disposal facility?

Flood Insurance Rate Maps

In 1968, Congress created the National Flood Insurance Program. Flood Insurance Rate Maps (FIRMs) are part of this program. According to the FEMA website (http://www.fema.gov/pdf/fhm/ot_frmsb.pdf), flood information on FIRMs is based on historic, meteorological, hydrologic, and hydraulic data, as well as the amount and types of open space, the existence or absence of flood control works, and the status of development. FIRM are used to determine if homeowners need to obtain flood insurance. It is less expensive to own a home if you do not have to pay for flood insurance.

It is important to note that FIRMs represent a "snapshot" in time, as they reflect the conditions in existence at the time of the creation of the map. Should weather patterns change, perhaps as local or regional manifestations of global climate change, should the amount of open space change, should flood control works be installed or removed, or should the status of development change with population growth, FIRMs may quickly become outdated.

For example, if an area upstream from a site along a river is changed from forest to urban, the amount of water that runs off (goes rapidly to streams rather than soaking into the ground or being used by plants) increases dramatically. In wooded or rural areas, typically 5–30 percent of the rainfall runs off, while in business and industrial areas, 50–90 percent of the water runs off and in residential areas 25–70 percent of the water runs off. The amount of runoff will also depend on other factors such as the density of development in an area, the types of soils, the angle of hillslopes, and the intensity and duration of individual storms. In general, however, development leads to a 2–3 times increase in the amount of runoff (Iowa Department of Transportation, 2004).

Washington, DC

Figure 10.15 is a FIRM map of part of Washington, DC. The left side of the map is the Potomac River. Rock Creek flows into the Potomac from the east. The circular "Zone C" on the east side of the river is centered on the Lincoln Memorial.

9. Let's first look at the background (nonflood) information on the map. What types of background data are shown on the map?

10. What types of background data are missing from the map that would be helpful in identifying flood hazards?

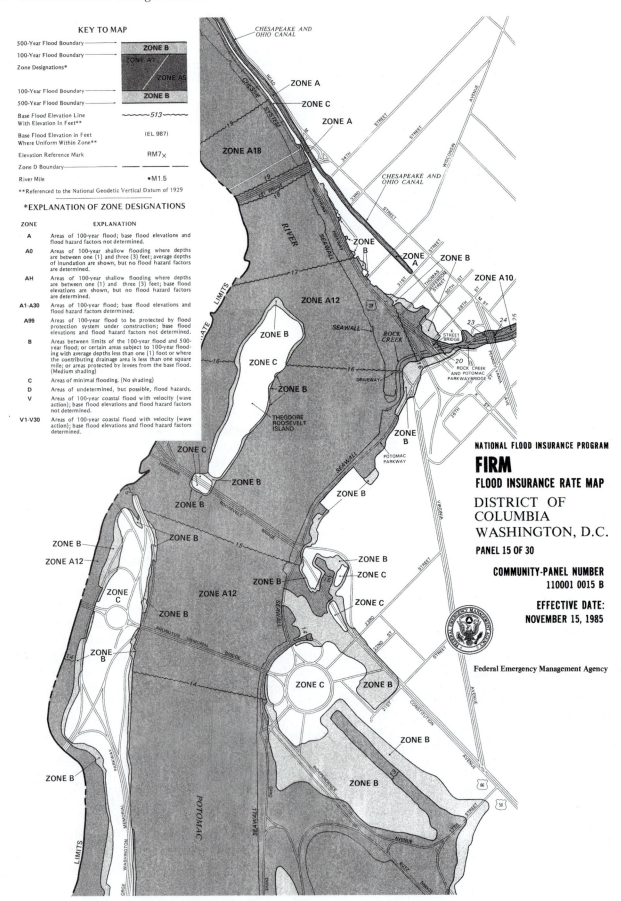

KEY TO MAP

500-Year Flood Boundary	
100-Year Flood Boundary	
Zone Designations*	
100-Year Flood Boundary	
500-Year Flood Boundary	
Base Flood Elevation Line With Elevation In Feet**	~513~
Base Flood Elevation in Feet Where Uniform Within Zone**	(EL 987)
Elevation Reference Mark	RM7×
Zone D Boundary	
River Mile	●M1.5

**Referenced to the National Geodetic Vertical Datum of 1929

***EXPLANATION OF ZONE DESIGNATIONS**

ZONE	EXPLANATION
A	Areas of 100-year flood; base flood elevations and flood hazard factors not determined.
AO	Areas of 100-year shallow flooding where depths are between one (1) and three (3) feet; average depths of inundation are shown, but no flood hazard factors are determined.
AH	Areas of 100-year shallow flooding where depths are between one (1) and three (3) feet; base flood elevations are shown, but no flood hazard factors are determined.
A1-A30	Areas of 100-year flood; base flood elevations and flood hazard factors determined.
A99	Areas of 100-year flood to be protected by flood protection system under construction; base flood elevations and flood hazard factors not determined.
B	Areas between limits of the 100-year flood and 500-year flood; or certain areas subject to 100-year flooding with average depths less than one (1) foot or where the contributing drainage area is less than one square mile; or areas protected by levees from the base flood. (Medium shading)
C	Areas of minimal flooding. (No shading)
D	Areas of undetermined, but possible, flood hazards.
V	Areas of 100-year coastal flood with velocity (wave action); base flood elevations and flood hazard factors not determined.
V1-V30	Areas of 100-year coastal flood with velocity (wave action); base flood elevations and flood hazard factors determined.

NATIONAL FLOOD INSURANCE PROGRAM

FIRM

FLOOD INSURANCE RATE MAP

DISTRICT OF COLUMBIA WASHINGTON, D.C.

PANEL 15 OF 30

COMMUNITY-PANEL NUMBER
110001 0015 B

EFFECTIVE DATE:
NOVEMBER 15, 1985

Federal Emergency Management Agency

FIGURE 10.15 Flood Insurance rate map of part of Washington, DC.

11. What hazard zones are identified on the map? How often are floods expected in zone A? In zone B?

12. Are areas in zone C are completely free from flood hazards?

13. Are all the hazard zones on the map the result of natural processes? Look carefully and explain.

14. Although the data that comprise flood hazard depictions on this map were collected in 1974 and revised in 1975, what year was this map published? What changes might have taken place in this area between the time the data were collected and the map was published?

15. What year is it now? What additional changes do you think might have taken place in the drainage basin of the Potomac River? What will be the likely impact of these changes on flood hazards that have been mapped?

Bibliography

Baker, V. R., Kochel, R. C., and Patton, P. C., 1988, *Flood geomorphology:* John Wiley, New York, 503 p.

Baker, V. R., 1994, Geomorphological understanding of floods: *Geomorphology*, v. 10, p. 139–156.

Cross, W. P., and Brooks, H. P., 1959, *Floods of January–February 1959 in Ohio:* U.S. Geological Survey Circular 418, 54 p.

Dunne, T., and Leopold, L. B., 1978, *Water in environmental planning:* San Francisco, W. H. Freeman, 818 p.

Edelen, G. W., Jr., Ruggles, F. H., Jr., and Cross, W. P., 1964, *Floods at Zanesville, Ohio:* U.S. Geological Survey Hydrologic Investigations Atlas HA-46.

Gore, A., 1995, speech, cited in Hulsey, Brett, 1997, *$ubsidizing disaster:* Madison, WI, Sierra Club Midwest Office, p. 7.

Hansen, W. R., 1973, *Effects of the May 5–6, 1973 storm in the Greater Denver Area, Colorado:* U.S. Geological Survey Circular 689, 20 p.

Hays, W. W., ed., 1981, Facing *geologic and hydrologic hazards—earth science considerations:* U.S. Geological Survey Professional Paper 1240 B, 109 p.

Hoyt, W. G., and Langbein, W. B., 1955, *Floods:* Princeton, NJ, Princeton University Press.

Hulsey, Brett, 1997, *$ubsidizing disaster:* Madison, WI, Sierra Club Midwest Office, 18 p.

Iowa Department of Transportation, 2004, Design Manual. Retrieved March 27, 2005, from ftp://165.206.203.34/design/dmanual/04a-04.pdf

Keller, E. A., 2002, Introduction to Environmental Geology: Second edition Upper Saddle River, N.J., Prenticehall, 563 p.

Larimer, O. J., 1973, *Flood of June 9–10, 1972, at Rapid City, South Dakota:* U.S. Geological Survey Hydrologic Investigations Atlas HA-511.

Leopold, L. B., 1974, *Water: a primer:* San Francisco, W. H. Freeman, 172 p.

Leopold, L. B., 1994, *A view of the river:* Cambridge, MA, Harvard University Press, 298 p.

Malamud, B. D., Turcotte, D. L., and Barton, C. C., 1996, The 1993 Mississippi River flood: A one-hundred or a one thousand year event?: *Environmental and Engineering Geoscience*, v. II, no. 4, p. 479–486.

Matthai, H. F., 1969, *Floods of June 1965, in the South Platte River basin, Colorado:* U.S. Geological Survey Water-Supply Paper 1850-B, 64 p.

McCain, J. F., Hoxit, L. R., Maddox, R. A., Chappell, C. F., and Caracena, F., 1979, *Storm and flood of July 31–August 1, 1976, in Big Thompson River and Cache la Poudre River basins, Larimer and Weld counties, Colorado: Part A. Meteorology and Hydrology in Big Thompson River and Cache la Poudre River basins:* U. S. Geological Survey Professional Paper 1115, p. 1–85.

Morgan, A. E., 1971, *Dams and other disasters:* Boston, Porter Sargent, 422 p.

Mount, J. F., 1995, *California rivers and streams, the conflict between fluvial process and land use:* Berkeley, CA, University of California Press, 359 pp.

Newson, M., 1994, Hydrology and the river environment: New York, Oxford University Press, 221 p.

Rahn, P. H., 1986, *Engineering geology: An environmental approach:* New York, Elsevier, 589 p.

Rahn, P. H., 1996, *Engineering geology: An environmental approach:* Upper Saddle River, NJ, Prentice-Hall, 657 p.

Rantz, S. E., 1970, *Urban sprawl and flooding in southern California:* U.S. Geological Survey Circular 601-B, 11 p.

Ritter, D. F., 1978, *Process geomorphology:* Dubuque, IA, W. C. Brown, 603 p.

Shroba, R. R., Schmidt, P. W., Crosby, E. J., and Hansen, W. R., 1979, *Storm and flood of July 31–August 1, 1976, in Big Thompson River and Cache la Poudre River basins, Larimer and Weld counties, Colorado: Part B. Geologic and Geomorphic effects in Big Thompson Canyon area, Larimer County:* U. S. Geological Survey Professional Paper 1115, p. 87–152.

Soule, J. M., Rogers, W. P., and Shelton, D. C., 1976, Geologic hazards, geomorphic features, and land-use implications in the area of the Big Thompson flood, Larimer County, Colorado: 1976 *Colorado Geological Survey Environmental Geology no. 10.*

U.S. Army Corps of Engineers, 1975, Guidelines for reducing flood damage. In *Man and his physical environment*, 2nd edition, eds. G. D. McKenzie and R. O. Utgard, Minneapolis, Burgess, p. 53–57.

U.S. Army Corps of Engineers, 2000, Annual flood damage reduction report to Congress for Fiscal Year 2000 (in cooperation with the National Weather Service Office of Hydrology and the Climate Prediction Center). Retrieved February 8, 2004, from http://www.usace.army.mil/inet/functions/cw/cecwe/flood00/index.htm

U.S. Census Bureau, 2000, The population profile of the United States 1999. Retrieved February 8, 2004, from http://www.census.gov/population/www/pop-profile/profile1999.html

U.S. Water Resources Council, 1976, A uniform technique for determining flood flow frequencies. *WRC Bulletin no. 17*, 26 p.

Coastal Hazards

INTRODUCTION

Coastal zones are among the most highly populated regions on Earth. In the United States, it has been estimated that about 80 percent of the population lives within 50 miles of a coast. Worldwide, the number is probably 50–60 percent. Coasts, therefore, have great potential for either good or bad interactions between natural processes and human activities. The number of people living near coasts will rise in the future, both as world population rises and as more people move to cities, many of which are near coasts.

Processes in coastal zones are distinct from other geologic processes such as earthquakes, volcanoes, floods, or landslides, in that coastal processes are always active. Between earthquakes there is no shaking. Between coastal storms, however, both water, in the form of waves and currents, and sediments, in the form of sand (if we are looking at a typical beach), are continuously in motion. Indeed, coastal zones, like rivers, can be thought of as integrated systems of both water movement and sediment movement. Coasts, of course, can also be impacted by other geologic processes such as landslides, river and coastal floods, earthquakes, and volcanic eruptions.

Thirty of the 50 states in the United States are considered coastal states; that is, they have either marine or Great Lakes shorelines. Nearly all of these states have some coastal areas that are experiencing moderate to severe erosion problems (Williams et al., 1990), as shown in Figure 11.1.

Coasts are highly energetic zones, where dynamic processes related to water and wind interact with land-related processes. Natural and human activities impact both water and land processes. Table 11.1 summarizes the causes of coastal land loss. Note that sediment transport, and, in some cases, volcanic eruptions, can also lead to coastal land gain.

Long-term observation of the interactions of people and coasts has led to the development of several principles, which are listed in Table 11.2.

The nature of coastal issues also depends on whether a coast is predominantly broad sand beaches or narrow beaches below steep cliffs. A broad sand beach will be more vulnerable to high waves, storm surges (water pushed toward shore by the force of winds), and tsunamis, as it will let the waves travel further inland. A steep cliff will break up waves before they can travel inland. In some areas, one type of coast will dominate for many miles; in other areas coasts are diverse assemblages of rocky cliffs, regolith bluffs, and small pocket beaches of sand or gravel.

Worldwide sea-level rise, attributed partly to global warming, has magnified the problems of the ocean coasts in recent years. Natural long-term fluctuations in lake levels have caused problems for coastal installations and transportation systems on large lakes.

In this exercise, we look at shoreline erosion (Part A) and lake level trends (Part B) in the Great Lakes, tsunami hazards along the west coast (Part C), and storm and/or hurricane-related geologic hazards along the east and south coasts (Part D). There is a wide range of vulnerability of oceanic coasts to sea-level rise. Throughout these exercises, look for causes listed in Table 11.1 and applications of the principles in Table 11.2. Note that the cases in this exercise are general examples. Should you ever decide to live along a coast, site-specific investigation of water, beach, and land processes will help you assess the specific risks you will face.

PART A. COASTAL EROSION, LAKE ERIE

In many areas along the shores of the Great Lakes, shore and bluff erosion present serious hazards to structures. In this part of the exercise we examine recession rates on the south shore of Lake Erie. (Refer to Figure 11.1 for the location of Lake Erie.)

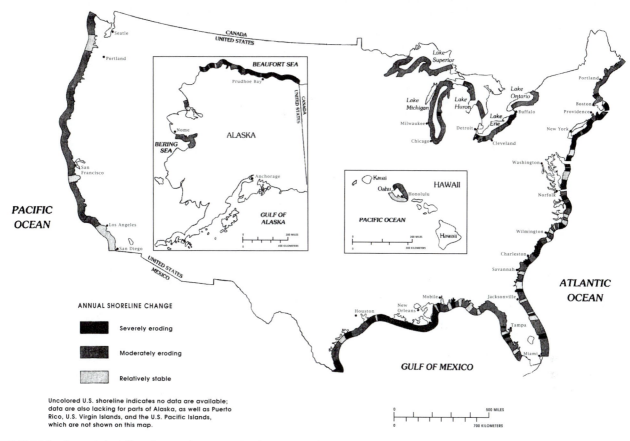

FIGURE 11.1 Annual shoreline change due to erosion in 30 states (Williams et al., 1990, and others).

TABLE 11.1 Primary Causes of Coastal Land Loss	
Natural Processes	
Agent	**Examples**
Erosion	waves and currents storms landslides
Sediment Reduction	climate change stream course changes source depletion
Submergence	land subsidence sea-level rise
Wetland Deterioration	herbivory freezes fires saltwater intrusion
Human Activities	
Agent	**Examples**
Transportation Coastal Construction	boat wakes, altered water circulation sediment deprivation (bluff retention) coastal structures (jetties, groins, seawalls)
River Modification	control and diversion (dams, levees)
Fluid Extraction	water, oil, gas, sulfur
Climate Alteration	global warming and ocean expansion increased frequency and intensity of storms

TABLE 11.1 Primary Causes of Coastal Land Loss (continued)

Excavation	dredging (canal, pipelines, drainage)
	mineral extraction (sand, shell, minerals)
Wetland Destruction	pollutant discharge
	traffic
	pumping-induced saltwater intrusion
	failed reclamation
	burning

Modified from http://pubs.usgs.gov/of/2003/of03-337/intro.html (downloaded September 17, 2007)

TABLE 11.2 Principles of Interaction of People with Coasts (Keller, 2000)

- Coastal erosion is a natural process that becomes a hazard when people build along the coast.
- Any shoreline construction causes change.
- Stabilization of the coastal zone by engineering protects the property of a few people at the cost to many (through state and federal taxes and emergency assistance).
- Engineering structures that are designed to protect a beach can instead, given enough time, contribute to the destruction of the beach instead.
- Once engineering structures are built, it can be very difficult to reverse their impacts.

Over 3 million people live in the Ohio counties bordering Lake Erie. Land loss due to shoreline erosion is a critical problem for those living immediately adjacent to the lake, with millions of dollars of property damage occurring annually.

The 220 miles of Ohio shore is made up largely of regolith comprised of glacial till, clay, and sand, plus resistant limestone and dolostone bedrock exposures in the Catawba and Marblehead peninsula area at the western end of the lake and shale exposures in the Lorain and Cleveland area in the west central part of the lake. Wetlands, low bluffs, and gently sloping shores characterize the western one-third of the coast. Sandy beaches generally overlie clay deposits along the shore in this area.

The low relief and location of the western shoreline make it susceptible to flooding because of wind setup during storms out of the northeast. This wind setup, usually associated with low atmospheric pressure events (in which the water level rises), leads to a phenomenon known as a *seiche*, where water that has piled up at one end of the lake then sloshes back and forth in the elongated basin of Lake Erie. The out-of-phase water levels between Toledo and Buffalo (Figure 11.2) are an example of this phenomenon.

Most of the eastern third of Ohio's Lake Erie shoreline consists of nonresistant lake clay, sand, and glacial till, which form bluffs 10–60 feet high. In this section shoreline erosion has historically been a problem, just as it is today. Whittlesey (1838) reported that most of the Ohio shoreline from the Pennsylvania line to Marblehead had lost an average of 130 feet of land to the lake between 1796 and 1838. Although natural processes such as wave erosion and mass wasting

were almost solely responsible for the erosion recorded in the earlier years, structures such as groins, breakwaters, and dikes have had an influence in more recent times (Carter, 1973).

The amount of land lost and rate of shoreline recession over nearly a century (1876–1973) in an area in Lake County, Ohio, east of Cleveland, is shown in a diagram of shoreline locations taken from topographic maps of the area (Figure 11.3). Aerial photographs

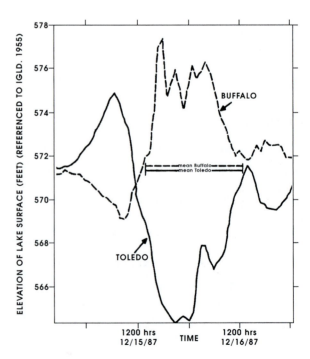

FIGURE 11.2 Water-level curves for Buffalo and Toledo during the storm of December 15–16, 1987 (Fuller, 1988).

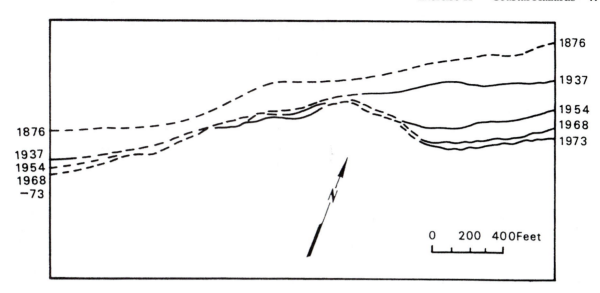

FIGURE 11.3 Shoreline retreat, 1876–1973, Lake County, Ohio (Carter, 1976).

taken at two different times, in 1954 and in 1973 after 19 years of erosion (Figure 11.4), show the changes for the same area of the coast.

Between 1954 and 1973 a portion of the shore eroded away. Shore erosion occurs through the combined effects of waves and currents. Waves, especially during storms, attack the bluffs along the shore, causing them to fail. (See Exercise 8.) Waves and wave-formed currents transport the sand along the shore. These longshore currents flow in a predominant direction. Any obstruction traps the sediment carried by the currents on its up-current side (the side from which the current is coming). On the down-current side of the obstruction the currents pick up a new load of sand. Because sand, in the form of a wide beach, is an effective buffer against storm waves, groins cause deposition of sand that slows bluff erosion on the up-current side. On the down-current side additional erosion occurs.

QUESTIONS 11, PART A

Refer to Figure 11.1 to answer Questions 1 and 2. For Questions 7 and 8 you will need a sheet of tracing paper the size of the photos (about 4″ by 8″). (Questions 7–18 are modified from Kennedy and Mayer, 1979.)

1. Which coast of the United States (north, south, east, or west) has the least problem with severe erosion?

2. According to Figure 11.1, which Great Lake has the highest percentage of U.S. shore in the severe erosion category?

3. Many factors affect rates of coastal erosion (see Part B). List three.

4. Refer to Figure 11.2. What was the maximum change in feet from high to low water level at Toledo during the December 1987 storm?

5. If the water-level differences between Buffalo and Toledo were due mainly to atmospheric pressure differences, over which end of the lake would the low-pressure area have been at midnight on December 15, 1987?

6. Refer to Figure 11.3 and determine average annual rates of coastal retreat between 1876 and 1937. Measure distances perpendicular to the coast and use the scale given. Record measured distances, too.
 a. Half-inch from the east margin of the diagram? Rate = _____ ft/yr
 b. Half-inch from west margin of the diagram? Rate = _____ ft/yr
 c. Coast with the least change? Rate = _____ ft/yr

7. Compare the 1954 and 1973 photos (Figure 11.4). Describe the physical and cultural changes that have occurred along the coast. (The photos are almost the same scale; Figure 11.8 shows lake level change.)

8. The straight objects jutting into the water are *groins*, structures designed to protect the bluffs by trapping sand and gravel. What changes have occurred from 1954 to 1973 in the number of groins?

9. List the physical processes involved in the changes that have occurred on this coast. Our next task is to compare the 1954 and 1973 shorelines and bluff bases by sketching on tracing paper these and other coastal features. (Another approach is to make a transparency copy of the 1954 photo, overlay it on the 1973 photo, and mark in pen the changes as described for the tracing paper technique.)

10. Cover the 1954 photo with a half sheet of tracing paper and secure it with paper clips or removable masking or clear tape.

    a. Outline major streets within an inch of the east–west road along the top of the bluff, to provide reference marks.

b. With a sharp pencil on the tracing paper, trace the base of the bluff with a solid line and the shoreline with short dashes. Label the figure "1954."

c. Draw and label the groins.

d. Now construct six parallel reference lines approximately perpendicular to the coast on the tracing paper so that we can measure the actual change in the position of the bluff

FIGURE 11.4 Aerial photographs for 1954 (top) and 1973 (bottom) of Painesville Township Park, Lake County, Ohio (Ohio Department of Transportation).

base. Draw the first line parallel to and along the largest groin. Mark this line C. Draw two more parallel lines (A and B) one-inch apart to the west. Draw and label three lines (D, E, F), again parallel to line C and one-inch apart, to the east.

e. Remove the 1954 tracing paper and place it over the 1973 photo. Align the streets where you can, so the tracing and the photo can be accurately compared.

f. Using a colored pencil, trace the base of the bluff with a solid line and the shoreline with short dashes. Label the colored-pencil lines "1973."

11. What is the maximum width of beach lost between 1954 and 1973?

12. Shade the area between the 1954 and 1973 solid lines representing the base of the bluff. What does the shaded area represent?

13. Compare and explain the changes in the coast west and east of the groins.

14. From your observations of the beach in the top photo and the net change in the coast, what direction does the dominant longshore current flow?

15. What really protects the bluffs along the coast from the energy of the waves when groins are installed?

In this part of the exercise we will use the changes in the position of the base of the bluff, as measured along the six reference lines, to determine the rate of recession of the coast and to predict the future position of part of the coast.

16. At each of the reference lines (A–F) on the tracing paper or overlay, measure the distance between the solid lines (base of the bluff) you drew for 1954 and the lines you drew for 1973. Use the scale given in the photos to measure the distances. Record the distances in the appropriate boxes below.

West lines			East lines		
Profile line	Distance between 1954 and 1973 bluffs, in ft	Rate of bluff change, in ft/yr	Profile line	Distance between 1954 and 1973 bluffs, in ft	Rate of bluff change, in ft/yr
A			D		
B			E		
C			F		
Average			Average		

17. Divide each measured distance between 1954 and 1973 bluffs by 19 years (the time between 1954 and 1973) to obtain the average annual rate of change in the position of the bluff for each profile line. Record these data in the indicated spots in the table in Question 16.

18. Determine the average distance and rate for the western and eastern part of the coast. What geologic processes can explain the differences in the average rates?

19. Assuming the annual rates determined at the two reference lines with the most rapid recession, calculate how much recession would be expected at these lines by 1993.
 a. At line _____, the recession in feet expected between 1973 and 1993 is _____

b. At line _____, the recession in feet expected between 1973 and 1993 is _____

20. Plot on your tracing paper map, using long dashes, the expected 1993 coastline between the two lines with the most rapid recession. You have now used your understanding of past geologic processes to predict future changes. Since 1993 is now past and a photograph is available (Figure 11.5), you can determine how close your prediction is to the actual change that occurred.

21. Place your tracing paper map over the 1993 photograph (which is approximately the same scale), and use another color to draw a solid line to mark the location of the base of the bluff in 1993.

FIGURE 11.5 Aerial photograph taken in 1993 at Painesville Township Park, Lake County, Ohio (Ohio Department of Natural Resources).

22. Describe the differences between the actual recession and the recession that you predicted. What are possible reasons for the differences, if any?

23. What were the average annual recession rates at the two lines that had the most rapid retreat between 1973 and 1993?
   a. At line _____, the recession rate in feet/year between 1973 and 1993 was _____

   b. At line _____, the recession rate in feet/year between 1973 and 1993 was _____

24. Compare the position and number of houses in the 1954, 1973, and 1993 photographs. What changes occurred?

25. How do the average annual rates for recession of the bluffs for the periods 1954–1973 and 1973–1993 compare with the rate of recession from 1876–1937? What factors might explain the differences?

26. If installation of the groins was a factor in the different rates, when do you think they were installed? _____ Explain. Refer to Figure 11.3.

27. Many coastal experts have suggested that structures that interfere with longshore drift should not be build along shorelines because they produce net erosion. Do you agree or disagree with the experts? Explain.

28. As a consulting coastal geoscientist, you are asked by the residents of the area for help in solving their problem. What advice do you give them?

PART B. ANNUAL AND LONG-TERM WATER-LEVEL FLUCTUATIONS ON THE GREAT LAKES

Many factors control erosion rates on coasts. In addition to engineered structures, earth material, and wave energy, among others, water level is also important. High-water and low-water levels on the Great Lakes cause problems for coastal communities and transportation systems. The problems associated with high water generally cause the greatest losses for individual homeowners.

   Flooding of docks, farmland, roads, and residential areas is the result of high water due to long-term (climatic), annual (seasonal), or short-term (meteorological) conditions. Increased erosion or the potential for it leads to lost property or expensive engineered shore-protection systems. Even recreation is impacted

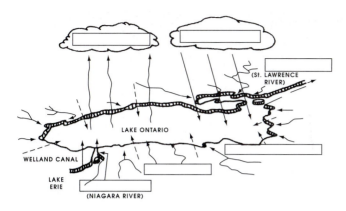

FIGURE 11.6 Factors affecting lake levels. Boxes represent processes described in the text. Dashed lines represent ground-water discharge (Freeman and Haras, n.d.).

by fluctuations in water level. High water reduces clearances under bridges, covers beaches, and reduces harbor protection; low water results in damage to boats, makes harbors inaccessible, reduces shipping capacity, and impacts some domestic water systems. Both high- and low-water conditions can impact wetlands and fishing.

Long-term fluctuations in lake levels are controlled by the climate over the Great Lakes basin. In the period of continuous records for Lake Erie, the variation in lake levels has been slightly more than 5 feet (Fuller, 1987). Annual fluctuations are due to yearly variations in precipitation, evaporation (and temperature), surface runoff, groundwater discharge, inflow, and outflow (Figure 11.6). If losses and gains to the lake do not balance, then water-volume and lake-level changes occur. In the case of Lake Erie, the much higher than normal levels of the 1970s and 1980s (Figure 11.7) are primarily due to above-normal precipitation in the Great Lakes basin.

QUESTIONS 11, PART B

1. Read the introduction to Part B and enter the processes controlling lake levels in the appropriate boxes in Figure 11.6.

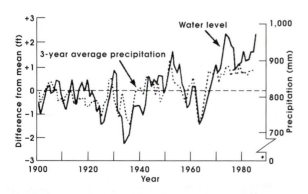

FIGURE 11.7 Annual Lake Erie water levels, 1900–1987, and average precipitation for the Great Lakes Basin (Fuller, 1987, after NOAA).

2. (Web research question) Lakes Huron and Michigan are at low levels. One factor appears to be the enhanced scour at the head of the St. Clair River at Port Huron and Sarnia. Search online for the latest developments and summarize them.

3. In Figure 11.7, the lowest level shown occurred at the time of the "dust bowl" on the western plains. When was this low level of Lake Erie? From the figure, is there any evidence that the evaporation rate from Lake Erie must have been high?

4. On an annual basis, in what months is Lake Erie (and the other Great Lakes) likely to be highest? When is it likely to be lowest?

5. Study Figure 11.8. In what periods (by beginning and ending years) since 1860 has Lake Erie been unusually high? Assume a mean elevation of about 570.4 feet.

6. In what years has Lake Erie been unusually low?

7. What are the highest and lowest annual mean elevations for Lake Erie?

8. What are the hazards associated with high lake levels?

9. Are there any hazards or impacts on society of low lake levels? Explain.

10. From the information gathered in Parts A and B of this exercise, we have seen that the water levels in the Great Lakes are likely to continue fluctuating. Although global warming could change averages and ranges of water levels, in any case erosion is expected to be a problem in some areas. To reduce losses to individual coastal residents and to taxpayers (who pay for roads, water supplies, insurance subsidies, tax deductions for lost buildings, and rescue costs in the coastal zone), expanded land-use planning and zoning of coastal areas by state, provincial, or municipal governments may be warranted. On a separate sheet of paper:

   a. Discuss the advantages and disadvantages of a 50-year building-setback zone. With such a setback zone, no permanent residential structures may be installed on the coast within the expected 50-year erosion zone.

   b. Also discuss any alternative that would minimize interference with the physical and biological processes along the coast, resource loss, and costs to individuals and to society.

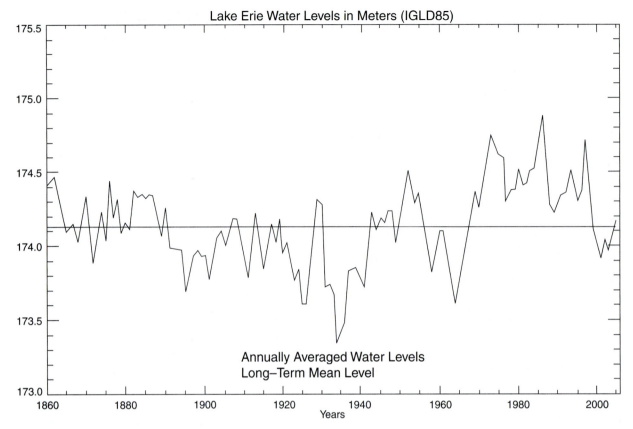

FIGURE 11.8 Lake Erie annual mean water levels, in meters, 1860–2005.

(Modified from NOAA)

PART C. TSUNAMI HAZARDS ON THE WEST COAST

Tsunamis

The tsunami of December 2004, in the Indian Ocean, reminded the world about the destructive impacts these waves can have. This tsunami, triggered by a magnitude 9.1 earthquake, killed an estimated 200,000 people. The waves reached heights of 30 meters (approximately 100 feet), and destroyed many buildings and communities.

The size of the rupture zone that was associated with the Indian Ocean earthquake has been approximated by study of the distribution of aftershocks. This rupture zone, when overlaid on a map of the Pacific Northwest (Figure 11.9), shows that a similar earthquake could be anticipated adjacent to the United States. This local earthquake could also trigger a local tsunami. In January 1700, the last time a major earthquake shook the northwest, a tsunami traveled across the Pacific Ocean and was recorded in Japan.

The Pacific coast of the United States is also subject to tsunamis that have been generated by earthquakes elsewhere in the Pacific Ocean. A tsunami that was triggered by the 1964 Alaskan earthquake caused damage along the west coast of the United States. In addition to the amount of energy released by the earthquake and the proximity to the coast, the height of a tsunami at a specific place is influenced by the tidal stage when the waves hit, tidal currents, topography of the land (how much low-lying land is along the shore), changes in topography due to earthquake-related uplift or subsidence, and the bathymetry (depth and shape) of the seafloor immediately offshore.

Coastal Processes

When tsunamis impact barrier islands, spits, and broad, low-sloping coasts, or where they travel far in kind along rivers the destruction of engineered structures and modifications of landforms can be rapid and very significant. Tsunamis can be survived, if individuals are educated about warning signs and actions to take (Atwata and others, 1999). Communities must also be prepared (Samant and others, 2008), for example by having clearly marked evacuation routes.

The plate tectonic setting of the coast of Washington, Oregon, and northern California is generally similar to the plate tectonic setting of the Sumatra earthquake of 2004. In the area of Sumatra, the India Plate is subducting beneath the Burma microplate. In the Pacific Northwest, the Juan de Fuca plate is subducting beneath the North America plate. The Sumatra earthquake had a magnitude of 9.1, and it is anticipated that the Pacific Northwest could have a similar size event, which could also unleash a major tsunami.

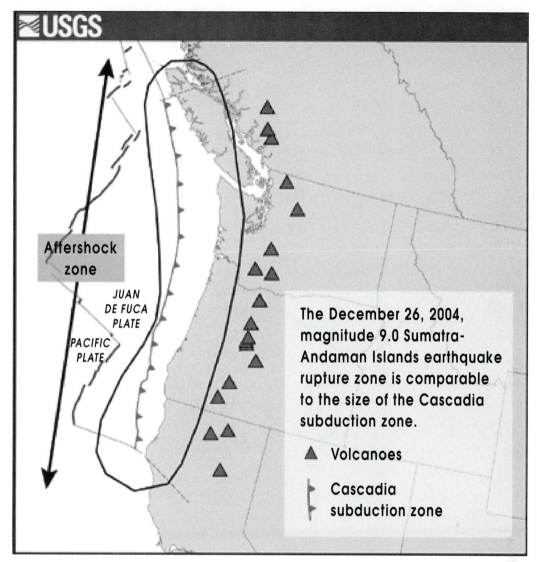

FIGURE 11.9 Sausage-shaped region depicts the rupture zone size of the Sumatra earthquake of December 2004, overlaid on a map of the Pacific northwest. Aftershocks occurred throughout the entire rupture zone. (http://soundwaves.usgs.gov/2005/03/cascadiaLG.gif)

Ocean Shores, Washington, is a popular vacation destination for people from the Tacoma–Seattle metropolitan area. Its population of slightly more than 4,000 residents can increase ten-fold on busy summer weekends. Ocean Shores is proud of its designation as "tsunami aware." It has signs that remind visitors of tsunami hazards, and pamphlets and signs that illustrate tsunami escape routes. The tidal range in this area can be more than 3 m (12 feet).

QUESTIONS 11, PART C

Refer to Figures 11.10 and 11.11, which are two maps of the Washington coast. Figure 11.10 is a 1915 map that shows the Ocean Shores area, and Figure 11.11 shows the same area in 1994. The original maps were different scales, and they have been reduced for use in this exercise.

1. What changes have taken place between 1915 and 1994 in the shape of the coast near Ocean Shores? It is especially important to look in the area of the two jetties at the entrance to Gray's Harbor.

2. Based on these maps, has deposition or erosion been dominant between 1915 and 1994 near Ocean Shores? Near Westport? Explain your evidence.

3. What changes in land use have taken place between 1915 and 1994 in the area shown on these maps?

4. What is the highest elevation of land in Ocean Shores? Westport?

5. What do the changes in land use imply about the risks from storm waves or tsunamis in this area?

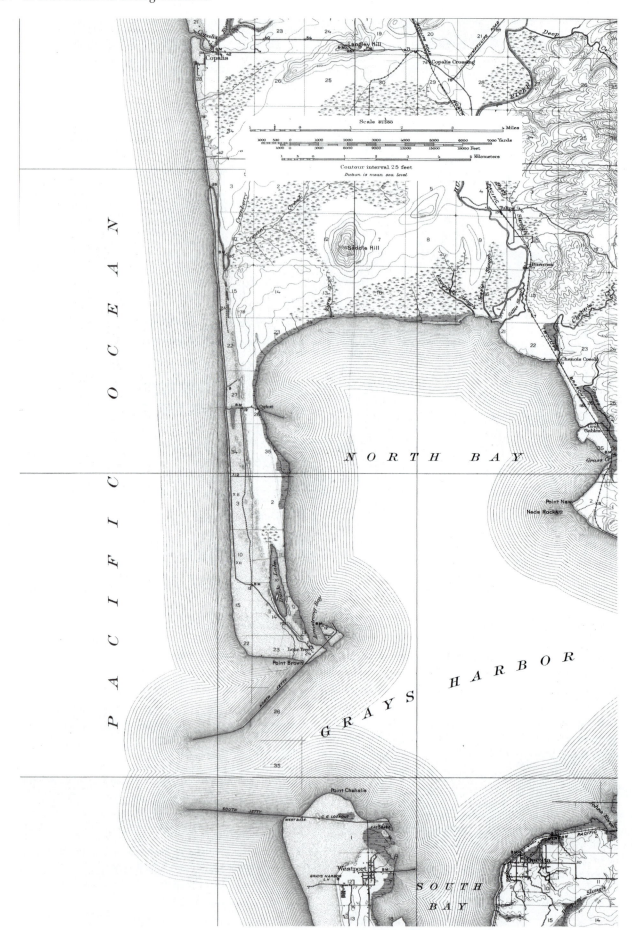

FIGURE 11.10 Ocosta, Washington (now better known as the Ocean Shores area), topographic map from 1915.

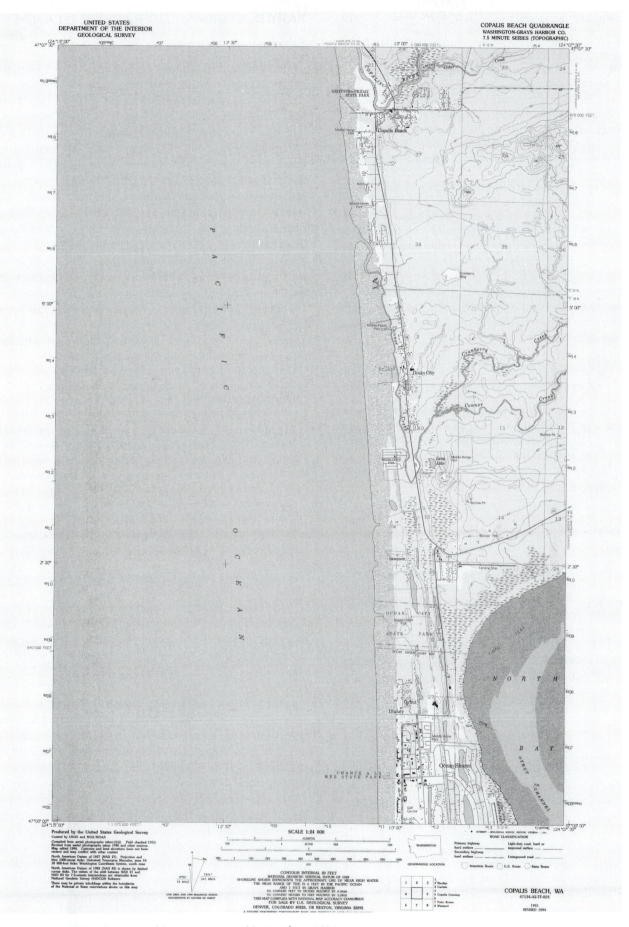

FIGURE 11.11 Ocean Shores, Washington, topographic map from 1994.

Tsunami Wave Heights

There are two different common origins for tsunamis that can strike the coast of the Pacific Northwest. The first are tsunamis that are generated by distant geologic events, such as the Alaska earthquake of 1964. The second are tsunamis that are generated by local subduction zone earthquakes.

After the 1964 Alaskan earthquake, tsunami waves in this area reached 9.7 feet above the water level of the tides at Ocean Shores, and 14.7 feet above tide at Wreck Creek, which is about 20 miles north of Ocean Shores.

6. Study the 1994 topographic map (Figure 11.11) to determine areas that would be impacted by a 10-feet rise in water. Use a colored pencil and mark these areas on the map. Assume that the 10-foot contour is approximately halfway between the edge of the water and the 20-foot contour.

7. How much more land would be impacted by a 20-foot rise in water? Use a different color, and mark these areas on the map as well.

8. After the Sumatra earthquake, tsunami wave heights reached 30 m (100 feet). Recent studies suggest that a local magnitude 9+ earthquake could generate a 20 m (65 foot) wave height at the Ocean Shores, Washington, area of the coast. Use a third colored pencil, and indicate on the map areas that would *not* be inundated by a 20 m wave.

Escape?

9. Use the 1994 topographic map (Figure 11.11) and suggest a route for escape from the peninsula that includes Ocean Shores. Sketch your route below or clearly indicate it on the map.

10. If a distant earthquake generates the tsunami, there may be several hours before the tsunami hits. Will your suggested escape route likely work with several hours of advance warning?

11. If a local major earthquake occurs, it may be less than 30 minutes before a tsunami hits. Will your suggested escape route likely work with less than an hour of advance warning?

12. Given your analyses in Questions 10 and 11 above, are alternative escape routes needed and if so, what alternatives do you suggest for people living, visiting, or working in the Ocean Shores area?

PART D. STORMS AND BARRIER ISLANDS

Storms

The most dramatic changes along coastlines take place during storms. Although hurricanes are dramatic in their power, other storms can also cause much damage. These include storms such as the "pineapple express" of coastal Oregon and Washington and "northeasters" along the East Coast. In both hurricanes and these other storms, strong winds can do much damage to structures. The combination of waves and a rise in water level known as storm surge also has major impacts along the coast. Storm surges during Hurricane Katrina reached nearly 30 feet; when combined with high tides, the runup of water can be even higher. The amount of storm surge is dependent upon air pressure, wind speed, and the near-shore shape of the ocean bottom. Low air pressure and high winds let the water build up higher. Where the sea bottom has a gentle slope, storm surges can build up quite high. Where the sea bottom has an abrupt drop-off, the surges are not likely to be as great.

Think About It

The force of moving water is tremendous. One gallon of water weighs about 8.35 pounds. There are about 7.48 gallons in a cubic foot. This means that each cubic foot of water weighs about 62.5 pounds. A cubic yard of water weighs just slightly less than 1,700 pounds. This means that a wave that is 3 feet high by 3 feet wide, and 100 yards long, will weigh 170,000 pounds. Add the force of motion, and it is easy to see how even smaller waves can pack the punch to knock down houses and easily erode beaches.

Figure 11.12 shows a typical low-lying coast with barrier islands that lie slightly offshore. Although this example is based on Texas, the same processes can occur along other low-lying coasts, such as much of the Gulf of Mexico and Atlantic coasts of the United States.

It is important to note that storms such as hurricanes have winds that rotate in a counterclockwise direction. This means that as a storm approaches a coast, winds in the 12 o'clock to 3 o'clock quadrant of the storm will blow toward the shore, while winds in the 6 o'clock to 9 o'clock quadrant will be blowing offshore. The onshore winds can pile up water into a storm surge, while the offshore winds will relatively lower the water level. So the same storm, where sites are even only a few miles or tens of miles apart, can have dramatically different impacts from surge, waves, and coastal flooding. Study this figure carefully, as it provides the context for the questions that follow.

Hurricanes are rated on the Saffir–Simpson scale. Table 11.3 shows the wind velocities and typical storm surges associated with each magnitude of storm.

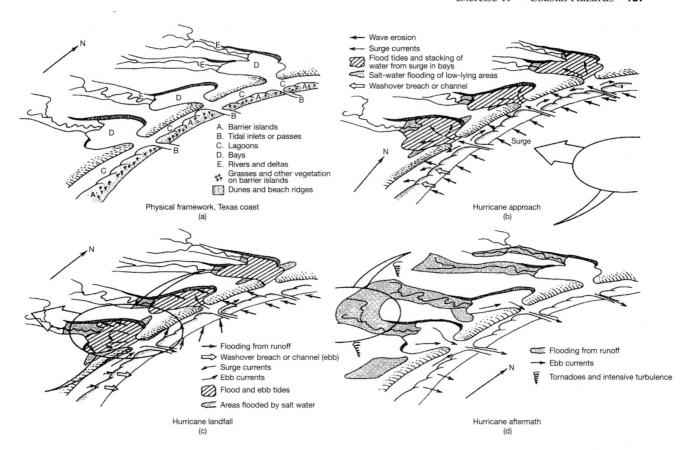

FIGURE 11.12 Impacts of a hurricane as it makes landfall. Figure (a) shows the morphology of the coast. Note that there are no towns shown in this drawing. Figure (b) shows the approach of the storm. The greatest storm surge will be on the north side of the storm in this drawing, as this is where the counterclockwise winds will pile up water the most. Figure (c) depicts landfall conditions. Tremendous amounts of wave erosion continue along the coast, while south of the storm path (in this drawing) ebb currents push water back out to sea. Any houses built along the barrier island, therefore, must be protected not only from waves that come from the sea, but also from currents that flow to the sea from land. Figure (d) shows residual flooding and ebb currents.

(From McGowen et al., 1970)

TABLE 11.3 Saffir–Simpson Hurricane Scale (from http://www.nhc.noaa.gov/aboutsshs.shtml).

Storm Category	Wind Speed Range (in miles per hour)	Typical Storm Surge (in ft)
1	74–95	4–5
2	96–110	6–8
3	111–130	9–12
4	131–155	13–18
5	>155	>18

Barrier Islands

Barrier islands are elongate ridges consisting mainly of sand that extend a few hundred meters to many kilometers along a coast. They have a beach on the ocean side and commonly are capped by sand dunes. Backed by a lagoon, marsh, or tidal flat, these islands are separated by inlets that probably formed during a storm in a wash-over erosion event. Barrier islands are part of a dynamic longshore transport system, migrating by erosion and deposition in response to natural and human factors. In response to changing sediment supply and sealevel rise, barrier islands retreat or decrease in size as their sand is lost to the lagoon or into deeper water offshore. Major changes occur during storm surges that wash over much of an island, causing large structural losses on heavily developed islands. Such losses are not new as revealed by two cases: Hog Island, Virginia, and Isles Dernieres, Louisiana.

In the late 1800s 10-km long Hog Island had a population of about 300, mainly in the town of Broadwater, where there was a "lavish hunting and fishing club … and 50 houses, a lighthouse, a school, and a church and cemetery." Much of the town was destroyed by a hurricane in 1933, which flooded the island and killed "the protective pine forest." All inhabitants had left by the early 1940s and the town site is now about 400 meters offshore under several meters of water (Williams et al., 1990).

About 120 km southwest of New Orleans, Isles Dernieres is a 32-km long barrier island chain that is "one of the most rapidly eroding shorelines in the world" (Williams et al., 1990). About 500 years ago this area was part of the now abandoned Lafourche delta complex of the Mississippi River. By the mid-1800s the Isles Dernieres were a single wide barrier island, also with a resort community and mature forest. A hurricane destroyed the resort and killed hundreds of people in 1856. Since then the single barrier island was cut by tidal inlets into five islands, which have continued to erode and retreat because of storm erosion and a rise of >1 m in relative sea level (some subsidence occurs). With low relief, some islands are now overwashed by storms and high tides six to eight times per year. Thus they now offer less protection to the wetlands and estuaries behind them than a healthy barrier island would. Between 1887 and 1988, the average erosion rate was 11.1 m/year. Today, the islands continue to decrease in area and migrate landward. These changes help to support the suggestion that Louisiana is losing an acre of coastal land every 24 minutes (O'Malley, 1999). Research continues on this gulfside coastal retreat, which in 1992, the year of Hurricane Andrew, reached 59 m/year.

In the questions for Part D we explore the nature and impacts of hurricanes in the Gulf of Mexico, with examples from the coasts of Texas and New Orleans (Questions 1 and 2) and offshore Louisiana (Questions 3 to 11). At Isles Dernieres, a chain of barrier islands, the focus is on changes to Raccoon Island located at 29° 03' N, 90° 56' W, between 1935 and 1994.

QUESTIONS 11, PART D

1. Use Figure 11.13 below, and the data shown in Figure 11.12 and Table 11.3, to analyze the likely impacts of a hurricane on this area.

  a. while a category 5 hurricane is still offshore

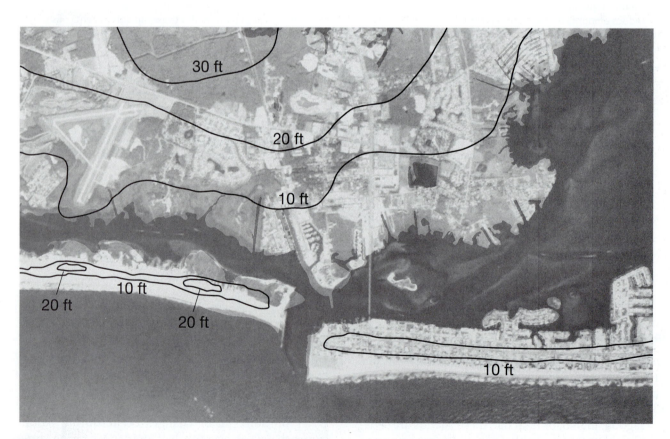

FIGURE 11.13 Hypothetical map of a community along the coast.

b. if a category 2 hurricane hits land east of the town

c. if a category 4 hurricane hits land west of the town

d. after the storm center has passed

2. By 2100, it has been estimated that sea level may rise between 1 and 3 feet.

a. Refer to Figure 11.14 and identify on it additional areas of New Orleans that are likely to be below sea level if a 3-foot rise takes place.

b. What likely impacts from hurricanes could be more severe if a 3-foot rise in sea level takes place?

Examine Figure 11.15a, a map showing changes in Isles Dernieres between 1887 and 1996 (and read the introduction to the barrier islands part of this exercise) to help answer the following. (Note: The Gulf of Mexico is on the south side of the island.)

3. a. How many named islands are there shown at Isles Derniers in 1996? _____ What is the most westerly island?

b. How many major or large islands are shown on Figure 11.15a in 1887?

c. Was Isles Dernieres at one time a single long island?

4. Which of the named islands appear to have lost the most area since 1887?

5. a. In Figure 11.15b (aerial photo mosaic, 1996), identify the following by marking them on the figure:
>   a sand beach
>   wetland
>   spit
>   shoal

b. What is a pass or coupe?

Figure 11.16 is a compilation of four different editions of the West Derniere, Louisiana, quadrangle, beginning with one

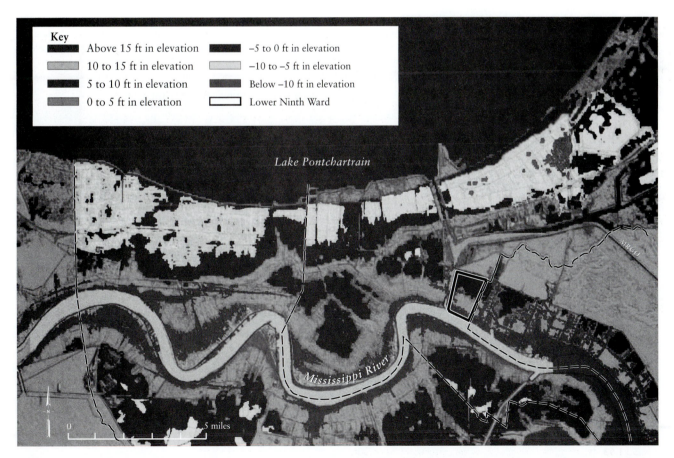

Key

▮ Above 15 ft in elevation	▮ −5 to 0 ft in elevation
▮ 10 to 15 ft in elevation	▮ −10 to −5 ft in elevation
▮ 5 to 10 ft in elevation	▮ Below −10 ft in elevation
▮ 0 to 5 ft in elevation	▭ Lower Ninth Ward

Lake Pontchartrain

MRGO

Mississippi River

N

0   5 miles

FIGURE 11.14 Topographic map of the New Orleans area. Note that much of the town lies below sea level, and that many of the higher elevations are along levees next to the Mississippi River (From McCulloh, Heinrich, and Good, 2006).

a.

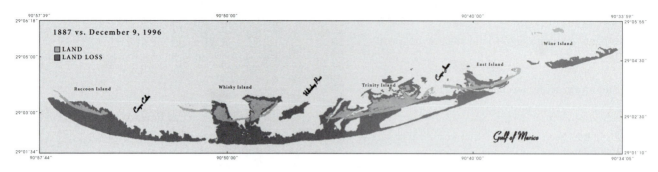

b.

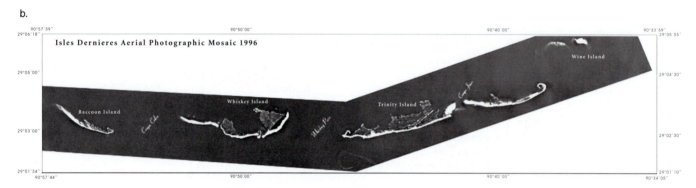

FIGURE 11.15   (a) Coastal change of Isles Dernieres, Lousiana, between 1887 and December 9, 1996, showing land loss and breakup of the island; (b) Aerial photographic mosaic, 1996.

(Adapted from Penland et al., 2003, http://pubs.usgs.gov/of/2003/of03-398/posters/pdf/cont_pdf/id_atlas.pdf)

published in 1935. Note the dashed lines in the 1935 quadrangle that could form a set of equal squares if extended. Each square on the map is numbered and is known as a section. With this set of numbered sections (squares) we have a very good reference system that we can use to trace the changes in this barrier island. Also on this map are named bench marks.

6. a. In the 1935 map (Figure 11.16), what is the section number that contains the word ISLES?

b. What sections contain the word DERNIERES?

c. What is the name of the triangulation point on the west end of the island?

d. What is a section?

e. What is the area of a section? (See the map exercise of this manual.)

f. Write out the abbreviations for T 23 S and R 15 E. (Hint: See the map exercise of this manual.)

7. a. In 1935, what was the maximum length of this island <u>in feet</u> (not including small eastern islands)? (Use a paper edge to measure the map distance and compare with the map scale.)

b. In 1935, what was the maximum width?

c. In 1994, what was the maximum length of this island <u>in feet</u> (not including small eastern islands)?

d. In 1994, what was the maximum width?

8. a. Measured along the boundary between Sections 33 and 32, how many feet has the Gulf side of the island retreated between 1935 and 1994?

b. What is the average rate of retreat? (Show your work.)

9. a. What part of the bayside (N side) of the island showed the most change between 1935 and 1994? (Indicate by Section number and mark on Figure 11.16.)

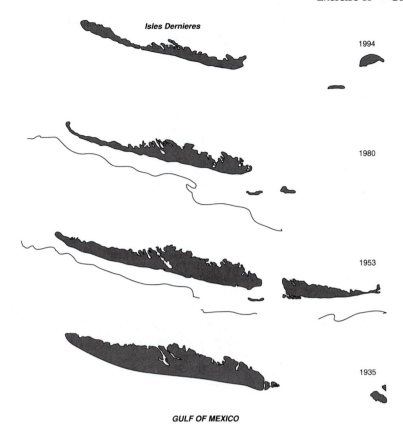

FIGURE 11.16 Compilation from the West Derniere, Louisiana, quadrangles: 1935, 1953, 1980, and 1994 (adjusted to the same scale).

b. What processes contributed to the change?

10. In the 1994 map, what are the materials shown on the island?

11. a. What do you think will eventually happen to the island?

b. Will this have any impact on other areas? Explain your answer.

Bibliography

Atwater, B. F., Cisternas V., M., Bourgeois, J., Dudley, W. C., Hendley 11, J. W., and Staofter, P. H., 1999 *Surviving a Tsunami—Lessons from Chile, Hawaii, and Japan: U.S. Geological Survey Circular* 1187, 18 p.

Carter, C. H., 1973, *Natural and manmade features affecting the Ohio shore of Lake Erie*: Ohio Department of Natural Resources, Division of Geological Survey, Guidebook no. 1, 34 p.

Carter, C. H., 1976, *Lake Erie shore erosion, Lake County, Ohio: Setting, processes, and recession rates from 1876 to 1973*: Ohio Department of Natural Resources, Division of Geological Survey, Report of Investigations, no. 99, 105 p.

Carter, C. H., Neal, W. J., Haras, W. S., and Pilkey, O. H., 1987, *Living with the Lake Erie shore*: Durham, NC, Duke University Press, 263 p.

Freeman, N. G., and Haras, W. S., n.d., *What you always wanted to know about Great Lakes levels and didn't know whom to ask*: Ottawa. Environment Canada, 28 p.

Fuller, J. A., 1987, High water creates problems on Lake Erie shore: *Ohio Geology Newsletter*, Ohio Department of Natural Resources, Division of Geological Survey, Spring, 1987, 5–6. p.

Fuller, J. A., 1988, Storm-induced water-level changes in Lake Erie: *Ohio Geology Newsletter*, Ohio Department of Natural Resources, Division of Geological Survey, Spring, 1988, 6 p.

Keller, E. D., 2000, Environmental geology: 8th edition, Upper Saddle River, NJ, Prentice Hall, 562 p.

Kennedy, B. A., and Mayer, V. J., 1979, *Erosion along Lake Erie*: The Ohio State University Research Foundation, Great Lakes Schools Investigation 6, 8 p.

McCulloh, R. P., Heinrich, P. V., and Good, B., 2006, *Geology and hurricane protection strategies in the greater New Orleans area*: Louisiana Geological Survey, Public Information Series 11, 31 p.

McGowen, J. H., Groat, C. G., Brown, L. F., Jr., Fisher, W. L., and Scott, A. J., 1970, *Effects of hurricane Celia—A focus on environmental geologic problems of the Texas coastal zone*: Bureau of Economic Geology, University of Texas, Austin, Circular 70–3, 35 p.

Platt, R. H., Beatley, T., and Miller, C., 1991, The folly at Folly Beach and other failings of U.S. coastal erosion policy: *Environment*, v. 33, no. 9, 6–9, 25–32 p.

Penland, S., et al., 2003, Shoreline changes in the Isles Dernieres Barrier Island Arc: 1887–1996 Terrebonne Parish, Louisiana. http://pubs.usgs.gov/of/2003/of03-398/posters /pdf/cont_pdf/id_atlas.pdf

O'Malley, P., 1999, Emergency erosion control: *Erosion Control*, v. 6, no. 9, 61–66 p.

Rosenbaum, J. G., 1983, Shoreline structures as a cause of shoreline erosion: A review. In *Environmental Geology*, ed. R. W. Tank. New York, Oxford University Press, 198–210 p.

Samant, L. D., Tobin, L. T., and Tirker, B., 2008, Preparing Your Community for Tsunamis: A Guidebook for Local Advocates: Geohazards International, 58 p., downloaded from http://www.geshaz.org/contents/publications/ PreparingYourCommunityforTsunamisVersion2-1.pdf, May 14, 2008.

Thieler, E. R., and Bush, D. M., 1991, Hurricanes Gilbert and Hugo send powerful messages for coastal development: *Journal of Geological Education*, v. 39, no. 4, 291–298 p.

Whittlesey, C., 1838, Geological report. In *Second Annual Report of the Ohio Geological Survey*, 41–71 p.

Williams, S. J., Dodd, K., and Gohn, K. K., 1991, *Coasts in crisis*: USGS Circular 1075, 3

Williams, S. J., Dodd, K., and Gohn, K. K., 1990, *Coasts in crisis*: U.S. Geological Survey Circular 1075, 32 p.

III. Introduction to Water Resources and Contamination

INTRODUCTION

"Water is, in a sense, both artery and vein to urban life."

—SCHNEIDER, RICKERT, AND SPIEKER, 1973

"It seems apparent that water will become much more expensive in the future . . ."

—KELLER, 2000

Lack of concern for some geological resources is frequently based on faith in economics and an option of substitution. Water is the one geological resource for which there is no substitute. Shortages of high-quality water degrade human health, reduce biological productivity of a region, lower the quality of life, and ultimately may produce conflict between individuals and countries.

SOURCES OF WATER

Fortunately the Earth's water cycles (it is a renewable resource) through evaporation, precipitation, surface runoff, and groundwater flow. We depend on our disruption of the hydrologic cycle (Figure III.1) to support our society. With increasing population in the United States and globally, we need increased efficiency of use, improved adaptation to seasonality of precipitation and runoff, options for responding to local climate changes

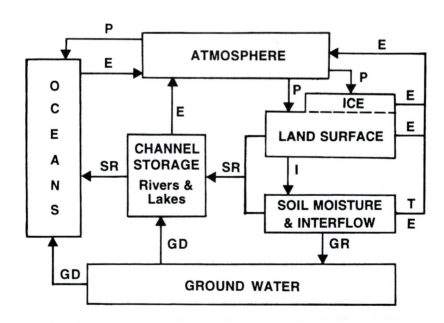

FIGURE III.1 The Earth's hydrologic cycle. (See Question 1 for explanation of abbreviations.)

due to global environmental change, and increased consideration of the environmental benefits and needs of other components of the hydrologic cycle such as wetlands. We also must remember that there is a connection between surface water and groundwater in this cycle; most rivers flow because of groundwater discharging into them.

Society's sources of water include rivers, lakes, groundwater (fresh and saline), and the oceans. Some cultures collect precipitation directly; precipitation also is used directly by agriculture. Most of the water that is withdrawn for use is freshwater (<1000 mg/L of dissolved solids) obtained from surface water (75%) and groundwater (25%). Some saline water (>1000 mg/L dissolved solids) is also withdrawn. Treated water released from wastewater treatment facilities is a source for irrigation of parks and golf courses and groundwater replenishment in a growing number of aquifer storage and recovery systems. In regions where water supply is a problem because of climate and/or increasing population, pollution, environmental concern, and capital shortages, the traditional "supply management" approach to managing water resources is changing to "water-demand management," which includes conservation programs.

AVAILABILITY OF WATER

On a global scale water abundance is not a problem. It is the availability of water in the right form in the right place at the right time that is a problem. Thus, the water available for human use is the key. Although more than 97 percent of the total water on Earth is in the oceans, this water is too salty for most uses. Less than 1 percent of the Earth's water is available for human use. The other 99 percent is unavailable or unsuitable because of its salinity (oceans and deeper groundwater) or its location and form (glaciers).

The annual runoff (surface and groundwater) per person has been used as a measure of the availability of water resources for a region or country. Although there may be local shortages and excesses of freshwater, the annual *water yield or runoff* from rivers and groundwater on a per capita basis is the critical figure in determining the water status in a region. According to Falkenmark and Widstrand (1992), when annual runoff is less than $1600 \, \text{m}^3$ per person water stress begins, and when it reaches $1000 \, \text{m}^3$ per person, chronic shortages occur. At less than $500 \, \text{m}^3/\text{person}$, there is extreme scarcity. As population increases or runoff decreases in countries and regions, more areas with some level of water stress are to be expected.

In 2000, the global total offstream use of water was about $6000 \, \text{km}^3/\text{yr}$. This figure is significant in comparison to the average annual global runoff of $47,000 \, \text{km}^3$, which includes water that cannot be diverted for human use because of location or environmental needs or lack of storage. When we realize the importance of total freshwater runoff and how much there is, we begin to focus on our options for water management. Do we increase conservation and wastewater reuse, increase desalination (with associated increases in costs), move people to environments with surplus water, increase interbasin water transfers, increase domestic precipitation collection and storage systems, or all of these? Maybe we will also begin to reflect on a possible planetary capacity for humans. On a global basis currently the per capita use of water is about $700 \, \text{m}^3$ per year; for the United States it is about $1800 \, \text{m}^3$ per year.

HUMAN USE

Humans only require a gallon of water a day for drinking, but domestic use in most developed areas of the world includes drinking, cooking, bathing (30% of household use), washing dishes, laundry (10%), flushing toilets (40%), watering lawns and gardens, and sometimes washing cars and filling swimming pools. In the United States, this household total averages about 80 gallons per person per day. Water delivered by public water supply systems, which includes household (about 60%), commercial, and industrial deliveries, and supplier uses and losses in delivery is 180 gallons per person per day (Solley et al., 1993). However, per capita public-supply use varies with region. For example, when measured in gallons per day (gpd), Nevada uses 344 gpd/person and Utah 308 gpd/person, while Massachusetts uses 130 gpd/person and Vermont 116 gpd/person. Groundwater accounts for about 40 percent and surface water 60 percent of water in public supply systems.

Private water supply systems also withdraw water from surface and groundwater sources for commercial, industrial, irrigation, livestock, mining, and thermoelectric plant cooling uses. Most is freshwater, but for some industrial, mining, and thermoelectric needs it is saline. Irrigation is the largest category of freshwater use.

If we apportion all the freshwater withdrawals (not just for public water supply) to individuals in the nation, then the United States withdraws about 1,340 gpd/person for all these uses. (If we include both fresh and saline water it is 1,620 gpd/person.) The range of use by state or district is as low as 256 gpd/person in Washington, DC, where use is primarily domestic and commercial, to 19,600 gpd/person in Idaho, where large quantities are used for irrigation and there is a small population.

All of the above uses are known as *offstream* use because the water was withdrawn or diverted from the river (or lake, reservoir, or groundwater reservoir!) to a place of use. *Instream* use of water refers to uses within the channel. This includes hydroelectric generation and such difficult to quantify uses as navigation, pollution abatement, recreation, and fish and wildlife habitat. All of these uses provide benefits to humans in the Earth system.

Water use is considered to be *consumptive* when it is withdrawn and does not go directly back to the immediate water supply system. It includes water that is evaporated, transpired, incorporated in products and crops, and consumed by humans and livestock.

Although we don't use all of our share of the nation's water supply directly in our homes, we consume goods that require water to produce. When we buy a car, each pound of steel in it required 60,000 gallons of water; production of a pound of beef may have required more than 2,000 gallons of water. Shortages of water for irrigation of crops can translate to a shortage of food, forcing up food prices globally (Brown, 2006). In this way a water shortage on the other side of the world may impact you here in North America; a local drought can become a "global drought."

WATER QUALITY

Natural water is not necessarily "pure" or distilled; even water from glacier ice has ions precipitated from the atmosphere. Natural water contains minerals and elements from the atmosphere and the enclosing river channel or groundwater aquifer (calcite, silica, calcium, sodium, nitrates, iron, etc.), dissolved gases (oxygen, CO_2, etc.), organic wastes from plants and animals, and microbes (viruses, bacteria, and parasites such as *Giardia* and *Cryptosporidium*). A quick measure of the inorganic content of water is the total dissolved solids (TDS) which can be obtained using a conductivity meter. Evaporation of seawater provides a good demonstration of the mineral matter than can be dissolved in clear water.

The content of dissolved organic and inorganic material in water is usually given as ppm or even ppb (parts per billion). One ppm is 1 drop in 13 gallons of water. A ppb is a very small quantity. It is equal to: 1 inch in 16,000 miles; 1 second in 32 years; and 1 square foot in 36 square miles. Yet these ppb and ppm quantities are important in water quality as contaminants in water at these levels can cause health problems.

In discussions of water quality criteria, one must specify the use of the water. Water that is good for irrigation, may not be good for other uses such as drinking, boiler feed, bathing, or food processing. For swimming the important factor is not dissolved solids but fecal coliform bacteria, which indicate sewage pollution—a health hazard—and closure of beaches. For fish habitat, the dissolved oxygen level is important. For public water supplies in the United States, the USEPA has Primary Regulations for Human Health and Secondary Regulations based on aesthetics. For example, the Secondary Maximum Contaminant Level (SMCL) for Total Dissolved Solids (TDS) is 500 mg/L.

Waterborne diseases in the developed world are much reduced because of water treatment systems and public understanding of contamination. Still 20% of U.S. community drinking water systems violate some public safety requirement and in 1995 fish advisories for consumption covered all of the Great Lakes, (Great Lakes Information Network, 2005) and many rivers and smaller lakes. About half the population in 45 least developed countries lack access to safe drinking water. Understanding water contamination and remediation is essential in improving the quality of life for hundreds of millions of humans.

CONTAMINATION OF GROUND AND SURFACE WATERS

Contamination of water occurs from domestic, industrial, and agriculture wastes produced by society. Some contamination is from systems designed to place waste on the land or in the water: septic tanks with drain fields (nitrates might be the world's number one contaminant), land disposal of sludge, and disposal wells. Some contamination occurs because the wastes are accidentally discharged: animal feedlots, acid mine drainage, and landfills. In addition, sometimes nonwastes contaminate water: accidental spills from truck and train accidents, highway salt, fertilizers and pesticides, and leaky underground storage tanks (USTs). The types of pollutants include inorganic solids, minerals, metal ions, salts, bacteria and viruses, undecomposed organic matter such as grease, volatile organic compounds (VOC), radioactive waste, and silt. Of the VOC, the chlorinated solvents (TCE, PCE, and TCA)[1] are widespread as they were used as cleaning agents at many sites and have produced plumes of contaminated water far from the source.

Contamination of slow-moving groundwater may last for centuries and be difficult to clean up. Pulses of recontamination of a groundwater aquifer occur from contaminated soils above the water table during recharge events. We need proper waste handling systems to prevent groundwater and surface water contamination, and we need to protect the zones of groundwater recharge from contamination.

Addressing past contamination is a more difficult job. How do we clean up sites? Options include removal and reburial or decontamination of the fluid or the soils; on-site destruction of the hazardous material by biological, mechanical, and chemical techniques; or sealing off the site from further use and letting natural long-term remediation occur. These and other options are topics of research on clean-up of contaminated groundwater aquifers.

Finally, in discussions of potential pollution by human activity, we must always consider *background or baseline levels* in a watershed. Simply finding high levels of dissolved solids, for example, in a basin does not mean that humans are responsible. Some bodies of

---

[1] TCE, Trichloroethelyne; PCE, Tetrachloroethylene; and TCA, Trichloroethane.

water are naturally unfit for human consumption. While distilled water has 0 ppm of dissolved solids, the value for rainwater is 10 ppm, Lake Michigan is 170 ppm, the Missouri River is 360 ppm, the Pecos River is 2600 ppm, the ocean is 35,000 ppm (35 kg, or 77 pounds, of salt per cubic meter), and the Dead Sea is 250,000 ppm. Nonetheless, these environments provide useful habitats for certain organisms and in some cases mineral resources for humans.

In this section on Water Resources and Contamination, we focus on groundwater, how it moves, its quantity, and its quality. We also look at contamination of surface water and groundwater, groundwater overdraft, and saltwater intrusion of aquifers.

QUESTIONS III

1. The Earth's hydrologic or water cycle consists of several reservoirs for H_2O connected by flows of water or water vapor. The typical textbook representation of the hydrologic cycle usually is a cross-section diagram of a small part of the Earth's surface with an ocean, landmass, and clouds. A box model (Figure III.1) is another representation of the cycle that emphasizes the connections (arrows) and the importance of both groundwater and the cryosphere (ice). List and identify by name the different symbols in Figure III.1. One, the letter P, is provided as an example below.

P—Precipitation

2. Determine how many gallons of water you use in a day. Consider a rainy and cold Saturday, when you spend most of the day in your house or dormitory. To determine this amount, go to the USGS website that has a calculator for determining water usage on a per-capita basis: http://ga.water.usgs.gov/edu/sq3.html. Record your answers in Table III.1 below and be ready to explain in class why your "Saturday per-capita value" is higher or lower than the average of the class.

3. List any other water uses or factors that might increase or decrease the amount of water used in a single-family home.

4. According to SIWI–IWMI (2004), between 300 and 3,000 liters of water are needed to produce a kilogram of grain. Each one of us, depending on our diet, is responsible for the 2,000–5,000 liters of water used to produce the food we consume each day. This water is "hidden"; it does not show up in our household water calculations. If we are seeking to understand water use in a region, this agricultural use must be taken into consideration.

a. Given that 1 gallon (U.S.) of water equals 3.78 liters and assuming the food in your daily diet required 4,000 liters of water to produce, how many "food gallons" of water per day are you using?

b. How many times greater is the quantity of "food gallons" than the value of your standard household gallons calculated in Question 2 (above)?

5. Table III.2 provides average water requirements for production of food items. Use it in answering this question.

In Table 2, the water requirement given for cereals is up to 3 m^3 per kilogram. Below, in a *conversion example*, this quantity (3 m^3/kg) is converted to gallons (U.S.) per pound.

water used in growing a pound of cereal:

$$\frac{3 \text{ m}^3 \times 264.2 \text{ gallons}/1 \text{ m}^3}{1 \text{ kg} \times 2.2 \text{ pounds}/1 \text{ kg}} = 792.6/2.2 = 360 \text{ gallons/pound}$$

a. From data in Table III.2 and the example above, how many gallons of water are used to produce a pound of grain-fed beef? (Hint: Simply consider how many times greater is the use of water for beef than the upper limit of water for cereal.)

TABLE III.1 Determining Your Daily Water Use from USGS (2004) Website[*]

Number of	Number of	Number of
Baths:	Showers: (Length =)	Clothes wash loads:
Brushings (teeth):	Hand/face wash:	Shaves (face/legs):
Dishwasher:	Dishes (hand):	
Flushes:	Drinks (8oz):	
Your water use =	**gallons**	

[*] http://ga.water.usgs.gov/edu/sq3.html

TABLE III.2 Water Required to Produce Selected Foods

Selected Food Items	Water Required		My estimated weekly intake of selected foods in kg *or* lbs	Estimated weekly water use in food production
	m³/kg	gal/lb		
Citrus fruits	1	120		
Cereals	0.4–3	48–360		
Poultry	6			
Roots and tubers	11			
Grain-fed beef	15			
My total water use for food production				

(Modified from Table 2 of 2004 Stockholm International Water Institute report, Water—More Nutrition per Drop[*])

[*] http://www.siwi.org/downloads/More_Nutrition_Per_Drop.pdf

b. Complete column 2 in Table III.2 with the equivalent number of gallons per pound.

c. Estimate your weekly intake of selected foods in pounds (or kilograms) and record in Table III.2.

d. Determine your weekly "food gallons" *or* "food liters" of water per week and record in column 4 ($1\,m^3 = 1000$ liters).

6. As human population increases and changes in diet (more protein) continue, more food will be needed. At the same time, groundwater depletion due to irrigation threatens the needed increase in food production. What options would you suggest for addressing this problem?

Bibliography

Brown, L. R., 2006, *Plan B 2.0: Rescuing a planet under stress and a civilization in trouble:* New York, W. W. Norton.

Falkenmark, M., and Widstrand, C., 1992, Population pressure on water resources: *Population Bulletin,* v. 47 (3), p. 19–27.

Gleick, P. H., ed., 1993, *Water in crisis: A guide to the world's fresh water resources:* New York, Oxford University Press, 497 p.

Great Lakes Information Network, 2005, Fish Consumption in the Great Lakes. Retrieved May 27, 2008, from http://www.glin.net/humanhealth/fish/index.html

Keller, E. A., 2000, Environmental geology, 8th edition: Upper Saddle River, NJ, Prentice-Hall, 562 p.

Schneider, R. A., Rickert, D. A., and Spieker, A. M., 1973, *Role of water in urban planning and management:* U.S. Geological Survey Circular 601H, 10 p.

SIWI–IWMI, 2004, *Water—More nutrition per drop:* Stockholm International Water Institute, Stockholm, 36 p. Retrieved August 17, 2007, from http://www.siwi.org/downloads/More_Nutrition_Per_Drop.pdf

Solley, W.B., Pierce, R. R., and Perlman, H.A., 1993, *Estimated use of water in the United States in 1990:* U. S. Geological Survey Circular 1081, 76 p.

USGS, 2004, *Water Science for Schools Questionnaire #3.* Retrieved August 17, 2007, from http://ga.water.usgs.gov/edu/sq3.html

Groundwater Hydrology

INTRODUCTION

Groundwater is an important component of the hydrologic cycle. It feeds lakes, rivers, wetlands, and reservoirs; it supplies water for domestic, municipal, agricultural, and heating and cooling systems. Groundwater resources at a site vary with natural and artificial recharge and discharge conditions. Because we dispose of wastes improperly or mishandle materials on the land surface, we pollute some groundwater reservoirs. For resource planning and waste management, it is essential that we understand the quantity, quality, and movement of water in bedrock and regolith or surficial aquifers. This exercise is an introduction to the basics of groundwater hydrology (hydrogeology) and to interpretation of the subsurface with geologic cross sections.

PART A. GROUNDWATER

Part of the water that reaches the land in the form of precipitation infiltrates to become groundwater. Groundwater occurs in openings in rocks and unconsolidated materials (Figure 12.1) and moves under the influence of gravity or pressure. An *aquifer* or groundwater reservoir is a water-saturated geologic unit composed of rock or unconsolidated materials that yields water to wells or springs. Generally, unconsolidated materials such as sand and gravel have more spaces than solid rock; the openings are due to incomplete cementation of the grains or to fracturing or partial solution of the rock. Openings in igneous and metamorphic rocks are generally due to fractures and joints. The ratio of the open spaces relative to the rock or regolith volume is called *porosity* (*n*), which is expressed as a percentage (Table 12.1). Porosity is a storage factor.

Not all the pores of a rock or regolith are available for flow of water. Some water does not flow because of molecular forces, surface tension, and dead-end pores. *Effective porosity* (n_e) is the amount of pore space that is available for transmitting water. (In some coarse-grained materials effective porosity approximates specific yield and gravity drainage.) Effective porosity is difficult to measure and is often approximated from total porosity and lab test data. Effective porosity is a factor in velocity of groundwater flow and is expressed as a percentage (Table 12.1).

The movement of water through a rock is also controlled by its permeability or hydraulic conductivity. The term *hydraulic conductivity* (*K*) is used in hydrogeology to describe the ease with which water can move through a formation and is often measured in units of length/time. Both fine grained and poorly sorted materials have low *K* values. Values for *K* are obtained in the lab and in the field. Representative values for different rock and unconsolidated materials are given in Table 12.1 in ft/day or ft d^{-1}. Sometimes values for *K* are given in m/d, m/s, or gal per day/ft^2. Some materials, such as clay and silt, may have a high porosity and hold much water; however, they have low effective porosity and low hydraulic conductivity because the openings are very small or not connected. Such units are aquitards because they retard the flow of water. Aquifers that have high hydraulic conductivity provide large quantities of water to wells.

In some aquifers groundwater occurs under *water table* or unconfined conditions (Figure 12.2). In this case the water table is the boundary between the *zones of aeration and saturation*. Where the water table intersects the land surface, *springs*, seeps, streams, and lakes are formed. The position of the water table can be determined by measuring the depth to water in a well tapping an *unconfined aquifer*.

Layers of low permeability confine many aquifers, and water in them is stored under pressure (Figure 12.2). When a well is drilled into such a *confined or artesian aquifer*, water rises in the well to some level above the base of the confining bed. In some cases the well may even flow at land surface. The water level (also known as the *potentiometric, piezometric, or water-pressure surface*) represents the artesian pressure in the confined aquifer.

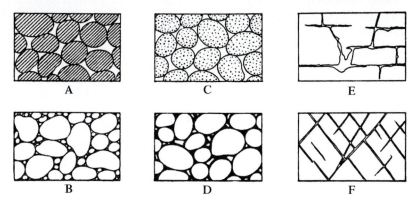

FIGURE 12.1 Types of primary (A–D) and secondary (E, F) porosity. A, well-sorted sedimentary deposit having high porosity; B, poorly sorted sedimentary deposit having low porosity; C, well-sorted sedimentary deposit consisting of pebbles that are themselves porous, so that the deposit as a whole has a very high porosity; D, well-sorted sedimentary deposit whose porosity has been diminished by the deposition of mineral matter in the interstices; E, rock rendered porous by solution; F, rock rendered porous by fracturing (Meinzer, 1923, p. 3).

TABLE 12.1 Range in Hydrologic Properties of Selected Geologic Materials

Material (rock or regolith)	Porosity (n) %	Effective Porosity (n_e) (%)	Hydraulic Conductivity (K) (ft/day)
Gravel	25–40	15–30	100–10000
Sand	30–40	10–30	0.1–1500
Clay, silt	45–60	1–10	10^{-7}–10
Till	20–40	6–16	10^{-7}–0.1
Sandstone	10–30	5–15	10^{-4}–1
Shale	1–10	0.5–5	10^{-8}–10^{-3}
Limestone	1–20	0.5–5	10^{-3}–10^4
Igneous rocks	0–40	0–30	10^{-8}–10
Metamorphic rocks	0–40	0–30	10^{-8}–10

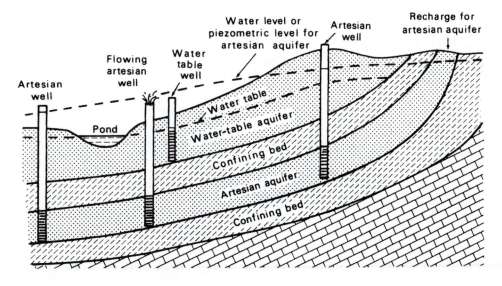

FIGURE 12.2 Schematic diagram of artesian and water table aquifers. Horizontal lines show well screen.

Most commonly the water table forms a gently sloping surface that follows the land surface (i.e., higher under hills than adjacent valleys). The water-pressure surface in artesian systems also generally follows topographic contours but in a more subdued manner. When the water table is at or near the land surface, groundwater may evaporate or be transpired by plants in large quantities and thus returned to the atmosphere.

The water table or water-pressure surface controls the direction of groundwater flow and can be mapped in a manner similar to contouring surface topography (Figure 12.3). In this case, however, control points are water elevations in wells, springs, lakes, or streams. Groundwater flows in the direction of decreasing head, which means that it flows from high to low pressure in a groundwater system. The high-pressure areas are where the water table is high or the water-pressure surface has a high value. On the contour map of the water table the flow lines cross the contour lines at right or 90° angles; the flow of groundwater effectively moves down slope or down gradient. Note how the flow lines curve to maintain the 90° crossing of each contour line in Figure 12.3.

The *hydraulic gradient* (*I*) is the difference in water level per unit of distance in a given direction. It can be measured directly from water-level maps in feet per foot or feet per mile. It is the slope of the water table surface or the water-pressure surface. (See "Slope or Gradient" in Part B of Exercise 3).

By using water-level maps in conjunction with topographic maps, the depth to the water table or water-pressure surface can be determined. This depth will vary with time depending on the season and the amount of recharge supplied by precipitation infiltrating the aquifer and the amount of discharge by pumping and by natural outflow to springs and streams. If discharge exceeds the rate of recharge to the aquifer, the water level in the aquifer will decline, and some wells could become dry.

The rate of groundwater flow generally ranges from 5 ft/day to 5 ft/year. It is usually less than 1 ft/day, but velocities greater than 400 ft/day have been measured. Groundwater *velocity* (*v*) depends on *hydraulic conductivity* (*K*), the hydraulic gradient (*I*), and the *effective porosity* (n_e). Sometimes permeability (*P*) and specific yield (*a*) are used in place of hydraulic conductivity and effective porosity, respectively. The

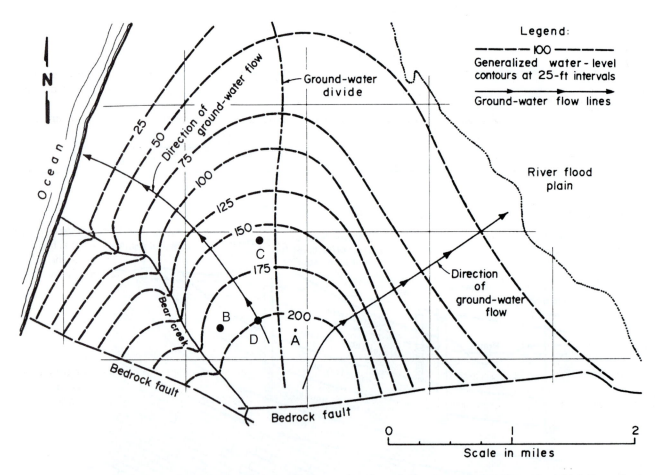

FIGURE 12.3 Water-level contour map showing elevation of the upper surface of the saturated zone (the water table). Groundwater flows down-gradient at right angles to the contours, as shown by the two flow lines that have been added to the map. One mile = 5,280 feet.

(Modified from Johnson, 1966, p. 40)

following formula is used to determine groundwater velocity in ft/day, where n_e is unitless and given as a decimal (i.e., 10 percent = 0.10), and I is in ft/ft.

$$v \text{ (ft/day)} = [K \text{ (ft/day)} \, I \text{ (ft/ft)}]/n_e$$

Knowing the amount of groundwater moving in an aquifer under a property may be of interest for resource development. The quantity of groundwater (Q), in cubic feet per day (ft^3/day) or cubic feet per second (cfs, if multiplied by 0.0000115), that passes through a cross-sectional area of an aquifer can be determined by means of *Darcy's Law*:

$$Q \text{ (ft}^3/\text{day)} = K \text{ (ft/day)} \, A \text{ (ft}^2) \, I \text{ (ft/ft)}$$

where A, the cross-sectional area through which flow occurs in ft^2, is equal to the width of the aquifer times its saturated thickness. Darcy's Law shows that the quantity of flow increases with an increase in K, A, or I.

These two equations, for the velocity and quantity of groundwater flow, are useful for estimating the movement and potential availability of water in an aquifer. More sophisticated computer models, which account for geologic variations in the subsurface, are used by professionals to predict groundwater flow. The velocity equation, based on Darcy's Law, may not apply where: (1) groundwater flows through low hydraulic conductivity materials at low gradient, and (2) turbulent flow occurs through large fractures or openings. The following questions are based on the conditions shown in Figure 12.3.

QUESTIONS 12, PART A

1. What is the average water-level gradient or slope along the eastern flow line of Figure 12.3 between the 200-ft and 50-ft contour? Give your answer in ft/ft and ft/mi and show your work.

2. If an aquifer near the floodplain in the eastern part of Figure 12.3 is 30 ft thick and has a porosity of 10 percent, how much water is *stored* in a 0.1 mile by 0.1 mile area of the aquifer? Give your answer in cubic feet and gallons; show your work (see Appendix A for Conversions).

3. If the hydraulic conductivity (K) of the aquifer is 150 ft/day, and the effective porosity (n_e) is 15%, what is the

average groundwater velocity in the vicinity of the eastern flow line! Show your work.

4. On Figure 12.3, construct a groundwater flow line downslope from each of sites A, B, and C. (See the instructor for other possible sites).

5. a. If gasoline were spilled at A, would it discharge with groundwater directly into the ocean? Explain.

b. If gasoline were spilled at B, would it discharge with groundwater directly into the ocean? Explain.

c. If gasoline were spilled at C, would it discharge with groundwater directly into the ocean? Explain.

6. In the western part of the aquifer, at and down gradient from D, the hydraulic conductivity is 100 feet per day and the effective porosity is 30 percent.
a. What is the average velocity along the flow line from D? Show your work.

b. If the velocity of the groundwater is assumed to also represent the movement of the contaminant, what is the time required for gasoline spilled at site D to travel to the end of the flow path? Show your work (time-distance/velocity D).

7. Using Darcy's Law, what is the quantity (Q) of groundwater flowing horizontally through a 2ft × 2ft square of an aquifer with $K = 180$ ft/day and an hydraulic gradient of 1 ft/1000 ft?

PART B. SUBSURFACE GEOLOGY

In addition to looking at the flow of groundwater, it is important to have a mental understanding of what the distribution of subsurface geologic units may be like. Different geological units will allow different amounts of water to flow through them depending on their hydraulic conductivity (Table 12.1). Understanding the relationships among subsurface geologic units is very important in developing an understanding of how groundwater moves in complex subsurface environments. Although there is much difference in the geology of subsurface units in different geological settings, the example below illustrates how sedimentary units might be related in many parts of the northern United States where glaciation has modified or formed the regolith.

When geologists begin to investigate the subsurface geology of an area, they often are limited to data gathered from existing wells. These data are in the form of well logs, which are a driller's or a geologist's written records of the different types of rock materials encountered as a well was drilled. By using logs from one well, a graph can be drawn that shows units encountered, as is illustrated in Figure 12.4. The depths at which different geologic units are encountered are shown. Depths below ground surface often are converted to elevations, so multiple wells can easily be plotted and compared. The transition from one geologic unit to another is called a contact.

Data from multiple wells can be connected (correlated) to form a geologic cross section. A cross section illustrates the subsurface relationships among different geologic units.

Where data are available from multiple wells, lines linking the geology in one well with the geology in another well can be sketched between the wells. Although there are no data between wells to suggest at what depth a particular geological unit may be encountered, it is reasonable to sketch a straight line where units can reasonably be connected (wells are vertical lines in Figure 12.5). If a unit is encountered in one well, but not in the next, the section must be drawn to indicate that it has not been encountered in the latter well. This is illustrated in Figure 12.5. Note that the sand unit at 760 ft was only found in one well on the right. Either it was not deposited elsewhere, or it has been eroded.

QUESTIONS 12, PART B

Your assignment is to interpret the well logs in Table 12.2, draw a topographic map, draw a geologic map, draw cross sections that illustrate the subsurface geology, and answer related questions about how water may move in the subsurface. The distribution of wells is shown on the map in Figure 12.6.

1. List the five different types of sediment or rock found in the wells (see Table 12.2).

2. Which sediment has the highest hydraulic conductivity? Which two have the lowest? (See Table 12.1 for hints.)

3. In Table 12.2, complete the data for *Depth* (to the bottom of each unit) and *Elevation* (height above sea level for the bottom of each unit) for wells 4–8. Wells 1–3 are completed in the table as guides.

4. On Figure 12.6, put the land elevation beside each well. Draw topographic contours showing the configuration of the land surface. Use a contour interval of 20 feet for your *topographic map*. The 260-foot contour is given on the map.

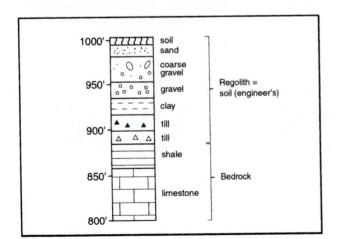

FIGURE 12.4 Typical log from a well, showing geologic materials and contacts.

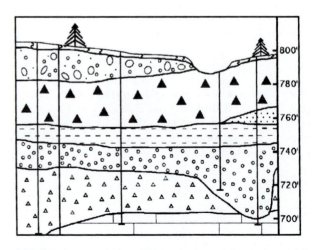

FIGURE 12.5 Geologic cross section, showing the correlation (connection) of geologic units in the subsurface between wells.

TABLE 12.2 Darcyville Well Logs. All measurements are in feet. Material is the type of geologic deposits (sediments). Thickness records how many feet of a particular unit were encountered when the well was drilled. Depth is how far down it is from the surface to the bottom contact of the specific type of geologic deposit. *Elevation is the height above sea level for the bottom of the geologic deposit.* Data for wells 1, 2 and 3 are filled in. Till is a poorly sorted silt and clay-rich sediment deposited by a glacier.

Well 1; Darcyville (land elev. 265)

Material	Thickness	Depth	Elevation
coarse gravel	40	40	225
till	45	85	180
clay	10	95	170

Well 2; Darcyville (land elev. 275)

Material	Thickness	Depth	Elevation
till	15	15	260
coarse gravel	25	40	235
till	15	55	220
fine sand	10	65	210
till	30	95	180
clay	25	120	155

Well 3; Darcyville (land elev. 290)

Material	Thickness	Depth	Elevation
till	60	60	230
fine sand	25	85	205
till	25	110	180
clay	25	135	155

Well 4; Darcyville (land elev. 305)

Material	Thickness	Depth	Elevation
coarse gravel	45		
till	20		
fine sand	45		
till	15		
clay	10		

Well 5; Darcyville (land elev. 310)

Material	Thickness	Depth	Elevation
coarse gravel	15		
till	40		
medium sand	40		
till	35		
clay	10		

Well 6; Darcyville (land elev. 290)

Material	Thickness	Depth	Elevation
till	70		
fine sand	15		
till	25		
clay	4		

Well 7; Darcyville (land elev. 265)

Material	Thickness	Depth	Elevation
coarse gravel	15		
till	35		
fine sand	10		
till	25		
clay	15		

Well 8; Darcyville (land elev. 255)

Material	Thickness	Depth	Elevation
coarse gravel	40		
till	35		
clay	10		

5. On Figure 12.6, beside each well, place the name of the material that is found at the land surface. These materials are those found in the top unit for each well.

6. Make a *geologic map* of the area in Figure 12.6 by interpreting the distribution of sediment (material) types and grouping any areas with similar materials. Do this by drawing a line to show the approximate contact between any two different materials at the surface. Without knowing the exact location of contacts, there will be more than one way to show the extent of materials. Your geologic map provides the distribution of sediments at the surface in this map area.

7. Draw two *geologic cross sections* (x to x', and y to y', for the map on Figure 12.6). The first cross section is through wells 1, 2, 3, and 4 (x–x'). The second cross section is through wells 5, 6, 7, and 8 (y–y'). Construct the cross sections on Figure 12.7 following the instructions below.

a. Draw the *profile* of the land surface; the profile for y–y' is given as an example. On the upper diagram in Figure 12.7 at the locations of the wells on the lower axis, draw a light line vertically above each well to the top of

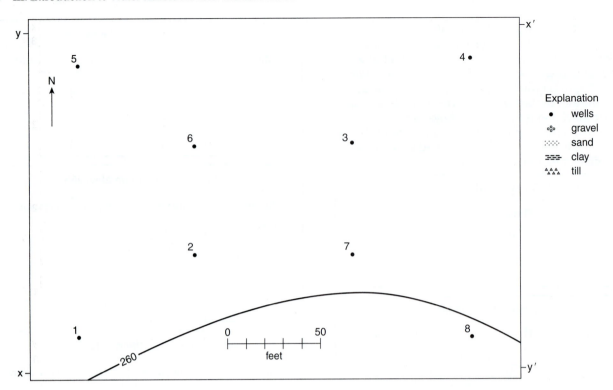

FIGURE 12.6 Map showing well locations near Darcyville, for completion of topographic and geologic maps. The 260-foot contour line is given.

the diagram (the line for well 1 is given). Mark the land elevation at the top of each well on this line with a short bar and label the elevation. Obtain the land elevation value for each well from Table 12.2.

Note: Optional: You could check agreement between your profile and a profile made from the topographic map constructed in Question 4, but this is optional. The points on profile y–y′ between wells were obtained from the contour map.

b. Add the stratigraphy for each well (wells 2–4 and 5–8). Starting from the topographic profile, place a tick mark on the vertical line for each well corresponding to the elevations where the geologic units change (i.e., the contacts of the units). The contacts for well 1 are entered on Figure 12.7.

c. Label each layer of each well as done for well 1.

d. Now complete the *geologic cross sections*. Look at the labeled geologic units in each well. Draw lines between wells to connect the same contacts separating geologic units. Not all units can be connected because some units may not occur in adjacent wells. Such units must "pinch out" before reaching an adjacent well when you construct your cross section.

8. Where is the best location to drill a water well for a house (domestic use well)? Briefly explain your choice based on the cross sections and geologic map that you constructed.

9. Assuming that all wells obtain water from every sand or gravel unit they intersect, which wells have the greatest potential for pollution from a nearby spill of toxic liquids? Explain your choices on the basis of the geology in your cross sections (Figure 12.7).

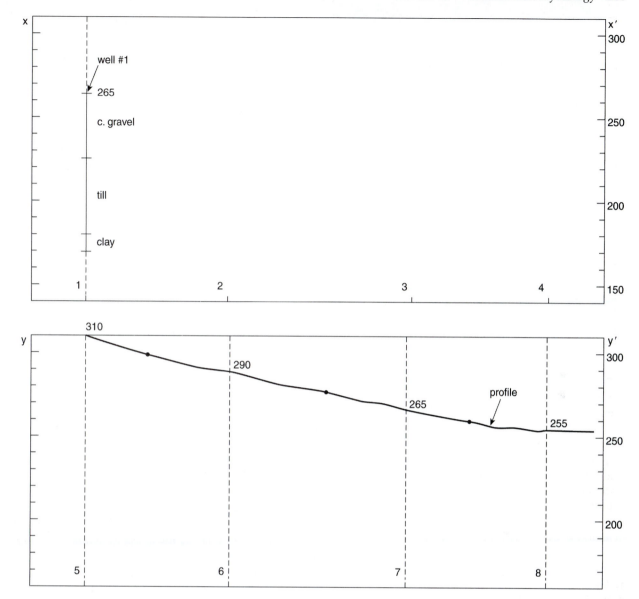

FIGURE 12.7   Geologic profiles and cross sections of the subsurface near Darcyville (x–x′ and y–y′). See Figure 12.6 for location. Interpretation based on wells 1–8; elevation of land surface given in feet.

Bibliography

Clark, L., 1988, *Field guide to water wells and boreholes*: New York, John Wiley and Sons, 155 p.

Driscoll, F. G., 1986, *Groundwater and wells*: St. Paul, MN, Johnson Division, 1089 p.

Fetter, C. W., 1994, *Hydrogeology*: New York, Macmillan, 691 p.

Freeze, R. A., and Cherry, J. A., 1979, *Groundwater*: Englewood Cliffs, NJ, Prentice Hall, 604 p.

Heath, R. C., 1983, *Basic ground-water hydrology*: U.S. Geological Survey Water-Supply Paper 2220, 84 p.

Johnson, E. E., 1966, *Groundwater and wells*: St. Paul, MN, Edward E. Johnson, Inc.

Manning, J. C., 1987, *Applied principles of hydrology*: Columbus, OH, Merrill, 278 p.

McGuinness, C. L., 1963, *The role of groundwater in the national water situation*: U.S. Geological Survey Water-Supply Paper 1800, 1121 p.

Meinzer, O. E., 1923, *The occurrence of groundwater in the United States*: U.S. Geological Survey Water Supply Paper 489, 321 p.

Ohio EPA, 1995, *Technical guidance manual for hydrogeologic investigations and ground water monitoring*: Columbus, OH Environmental Protection Agency.

Price, M., 1985, *Introducing groundwater*: Boston, Allen and Unwin, 195 p.

U.S.EPA, 1990, *Handbook, ground water, volume 1: Groundwater and contamination*, EPA/625/6–90/016a.

Water Quality Data and Pollution Sources

INTRODUCTION

Water quality in a stream, lake, well, or spring is a product of the quality of the precipitation and any changes that have occurred to the water at or below the surface of the earth. As water moves through the hydrologic cycle, it changes chemically, physically, and biologically (Hem, 1985). Some changes cause the quality to deteriorate and if the inputs are significant from human sources, the water is considered to be contaminated.

Major sources of contamination include industrial, municipal, agricultural, and mining activities. The contaminants may be physical, chemical, or biological. Some constituents affect taste, others may be corrosive, and a few are health hazards. All natural water, however, contains some impurities. These naturally occurring background concentrations reflect the soluble products in the soil and rocks through which the water moved.

Under the 1986 Amendments to the Safe Drinking Water Act, the U.S. Environmental Protection Agency (USEPA) established *National Primary Drinking Water Regulations* for specific contaminants that may have any adverse effect on human health and could be expected in public water systems (Table 13.1). These regulations indicate MCLGs and MCLs. MCLGs (Maximum Contaminant Level Goals) are recommended maximum limits of contaminants in drinking water. They are nonenforceable health goals of contaminant levels at which no known or anticipated adverse health effects occur and that allow an adequate margin of safety. MCLs (Maximum Contaminant Levels) are the maximum permissible levels. They are set as close to the MCLGs as is feasible using the best technology and techniques available. Additional organic chemicals and their potential health effects from exposure above the MCLs are available online from the USEPA.

In addition to health concerns addressed under the primary regulations, the USEPA has developed *National Secondary Drinking Water Regulations* that cover aesthetic qualities, such as taste and odor, of drinking water (Table 13.2). Under this section, Secondary Maximum Contaminant Levels (SMCLs) have been set for contaminants. The SMCLs are federally nonenforceable and establish limits for contaminants in drinking water that may affect aesthetic qualities and public acceptance of drinking water. Standards for drinking water in the United States continue to evolve.

The concentrations of substances dissolved in water are normally reported as milligrams per liter (mg/L), parts per million (ppm), parts per billion (ppb), or grains per gallon (gpg). Milligrams per liter and parts per million are practically the same for waters with less than 7,000 ppm total dissolved solids (TDS). The two terms are often used interchangeably. One mg/L means that a 1-liter sample of water contains 1 milligram of substance dissolved in it, or that in a million total units 1 unit consists of a dissolved substance. To correctly interpret the significance of water-quality data, it is necessary to understand something about the origin and reactions of certain compounds.

In this exercise we investigate the importance of selected inorganic chemicals in water and their potential sources. An estimate of the quantity of inorganic chemicals in water is provided by specific conductance. The capacity for water to conduct an electric current is determined by the temperature and the degree of ionization of the elements and compounds in the water. This capacity is known as *specific conductance,* and by measuring it we can determine the quantity of total dissolved solids (TDS) in the water. The concentration of dissolved solids (in ppm) is usually estimated by multiplying the specific conductance, which is given in μmhos/cm, or microsiemens, by 0.67. A mho, which is a unit for conductivity, is the reverse of an ohm, the unit for resistivity.

TABLE 13.1 National Primary Drinking Water Standards for Selected Contaminants (USEPA, 2007 http://www.epa.gov/safewater/contaminants/index.html#listmcl). See text for MCLG and MCL.

Contaminant	MCLG[1]	MCL[1]	Potential Health Effects	Contaminant Sources
Microbiological				
Coliforms (total)	0	5.0%[2,3]	Gastrointestinal Illness (GII)	human/animal fecal waste
Cryptosporidium	0	TT[3]	GII (diarrhea, vomiting, cramps)	human/animal fecal waste
Giardia lamblia	0	TT[3]	GII (diarrhea, vomiting, cramps)	human/animal fecal waste
Legionella	0	TT[3]	Legionnaire's Disease	multiplies in heating sys.
Turbidity	—	TT[3]	assoc. with microorganisms	soil runoff
Viruses (enteric)	0	TT[3]	GII (diarrhea, vomiting, cramps)	human/animal fecal waste
Inorganics				
Antimony	0.006	0.006	incr. blood cholesterol, decr. sugar	oil ref.; fire retard.; ceramics; solder
Arsenic	—[4]	0.010	skin, circulatory, cancer	nde,[5] orchards, electronics
Asbestos	7 MFL	7MFL[6]	benign intestinal polyps	asbestos cement in water mains
Barium	2	2	incr. blood pressure	drilling wastes; metal refin.; nde
Beryllium	0.004	0.004	Intestinal lesions	metal refin., defense industries.
Cadmium	0.005	0.005	kidney	galvanize pipes, battery waste
Chromium	0.1	0.1	dermatitis	steel and pulp mills, nde
Copper	1.3	(1.3)TT[7]	GII, liver, kidney	plumbing, nde
Cyanide (free)	0.2	0.2	nerve, thyroid problems	steel/metal/plastic/fertilizer fact.
Fluoride	4.0	4.0	bone disease, mottled teeth	additive, nde, aluminum waste
Lead	0	(0.015) TT[7]	mental dev, kidney, blood pres.	plumbing, nde
Mercury (inorganic)	0.002	0.002	kidney	factories, landfills, farms, nde
Nitrate (as N)	10	10	<6 mos, illness/death, "blue-baby"	fertilizer, septic tanks, nde
Nitrite	1	1	<6 mos, illness/death, "blue-baby"	fertilizer, septic tanks, nde
Selenium	0.05	0.05	hair, fingernail loss, numbness, circ.	oil ref., mines, nde
Thallium	0.0005	0.002	hair loss; blood, kidney, liver	ore-proc., electron, glass and fact.
Radionuclides				
Alpha particles	—[4]	15 pCi/L	cancer risk	nde of radioactive mines
Beta particles	—[4]	4 mrem/yr	cancer risk	decay radioactive mins.
Radium 226 + 228	—[4]	5 pCi/L	cancer risk	nde
Uranium	0	30 ug/L	cancer risk; kidney toxicity	nde
Organic Chemicals				
Atrizine	0.003	0.003	cardiovasc., repro. problems	herbicide runoff
Carbofuran	0.04	0.04	blood, nervous syst., repro.	soil fumigant (rice, alfalfa)
Chlordane	0	0.002	liver, nervous syst., cancer	residue of banned termiticide
2,4-D	0.07	0.07	liver, kidney, adrenal glands	herbicide runoff
Diquat	0.02	0.02	cataracts	herbicide runoff

TABLE 13.1 National Primary Drinking Water Standards for Selected Contaminants (USEPA, 2007 http://www.epa.gov/safewater/contaminants/index.html#listmcl). See text for MCLG and MCL. (continued)

Contaminant	MCLG[1]	MCL[1]	Potential Health Effects	Contaminant Sources
Lindane	0.0002	0.0002	liver, kidney	insecticide, cattle, lumber, gardens
PCBs	0	0.0005	anemia, blood, cancer risk	factories, dry cleaners
Benezene	0	0.005	anemia, blood, cancer risk	factories, gas storage, landfills
Carbon tetrachloride	0	0.005	liver, cancer risk	chemical plants
1,2 Dichloroethane	0	0.005	cancer risk	industrial chemical factories
Ethlbenzene	0.7	0.7	liver, kidney	petroleum refineries
Styrene	0.1	0.1	liver, kidney, circ. syst.	rubber/plastic factories, landfills
Tetrachloroethylene	0	0.005	liver, cancer risk	factories, dry cleaners
Vinyl chloride	0	0.002	cancer risk	PVC pipes, plastic factories
Dioxin (2,3,7,8-TCDD)	0	0.00000003	reproduction, cancer risk	waste incin. chemical factories
Disinfectants and By-products				
Chlorine (Cl_2)	MRDLG[8]=4	MRDL[8]=4.0	eye/nose irritation; stomach	add to disinfect drinking water
Total trihalomethanes (TTHMs)	—[4]	0.080	liver, kidney, nervous, cancer	byproduct of water disinfection
Bromate	0	0.010	cancer risk	byproduct of water disinfection.

[1] In milligrams per liter (mg/L) unless otherwise noted

[2] More than 5.0% samples total coliform-positive in month, then tests for fecal coliforms or E. coli. Fecal coliform and E. coli are bacteria that indicate water may be contaminated with human/animal wastes. Disease-causing microbes (pathogens) in these wastes can cause diarrhea, cramps, nausea, headaches, etc., and may pose risk of severe illness for infants, the young, and those with weak immune systems.

[3] For Giardia and viruses, the TT is a combination of (1) inactivation by disinfection and (2) filtration. Filtration (or boiling at home) is the technique for Cryptosporidium, which has cysts very resistant to chlorine-based disinfect. Legionella is also controlled through inactivation/filtration. Turbidity a measure of filtration effectiveness and must be < 0.3 NTU in 95% samples/month. (Treatment Technique is a required process intended to reduce level of a contaminant in drinking water.)

[4] None established.

[5] Nde or nde, natural deposit erosion

[6] Million fibers per liter (<10 um in size)

[7] Lead and copper TT controls the corrosiveness of water. When >10% of tap water samples exceed "action level" (in parens.) then additional steps needed.

[8] MRDLG, Maximum Residual Disinfectant Level Goal (residual from water treatment), MRDL, Maximum Residual Disinfectant Level.

PART A. SIGNIFICANCE OF SELECTED MAJOR INORGANIC CONSTITUENTS

Hardness

Hard water usually has significant amounts of calcium and magnesium. Hardness is usually associated with the effects that take place when using soap. Hard water requires the use of large amounts of soap and leaves insoluble residues in bathtubs and sinks. In addition, hard water causes scale to form in water heaters, boilers, and pipes.

Hardness depends mainly on the concentration of calcium (Ca) and magnesium (Mg), but other substances such as sulfate (SO_4) also form insoluble residues from soap. Water having a hardness of less than 60 ppm is considered soft, 61 to 120 ppm is

moderately hard, 121 to 180 ppm is hard, and more than 180 ppm is very hard.

CALCIUM (Ca). Calcium may be leached from most rocks, but limestone and dolomite provide the largest amount. Calcium is a major cause of hardness and forms scale on utensils, boilers, and pipes. Calcium is not considered detrimental to health.

SODIUM (Na). Sodium is readily leached from rocks and tends to remain in solution. Any salt, clay or silt-rich deposits may provide high sodium concentrations to circulating water. In addition, sodium may be taken into solution if the transporting water comes in contact with sewage or industrial wastes. The concentration of sodium is not especially important in water for domestic uses. Persons having an abnormal sodium metabolism should consult their physicians concerning the planning of a sodium-free diet if the supply of drinking or culinary water has a high sodium content. A concentration of sodium in excess of 500 mg/L, when combined with chloride, results in a salty taste. Concentrations in excess of 1,000 mg/L are unsuitable for many purposes. On the other hand, some communities use water containing more than 4,000 mg/L of dissolved solids because other supplies are not available. Generally the more highly mineralized the water, the more distinctive its taste.

Hydrogen-Ion Concentration (pH)

The pH of water is a measure of alkalinity or acidity. A pH of 7 indicates a neutral solution; a pH greater than 7 indicates an alkaline solution and a pH less than 7 indicates an acidic solution. The pH is related to the corrosive properties of water. Low pH water is most corrosive. Acidic water may have a sour taste. Most water has a pH between 5.5 and 8. The recommended range is 6.5–8.5. A low pH may be related to the discharge of acid water from coal mining regions or disposal of spent acids by certain industries.

Other Inorganic Chemicals

Other inorganic chemicals, often present in trace quantities, are also important water quality factors and some have implications for human health. Limits on several inorganic chemicals are given in Table 13.1

SULFATE (SO$_4$). Sulfate is dissolved from rocks containing sulfur compounds, such as gypsum and pyrite (Hem, 1985). Sulfate, when combined with other elements, may produce a bitter taste. Large amounts of sulfate may produce a laxative effect in some people; therefore the U.S. Public Health Service (1962) recommended that the sulfate concentration in public supplies should not exceed 250 mg/L. Although high sulfate concentrations may reflect natural background conditions, various industrial activities may cause severe contamination. In the extensive coal-mining region of Appalachia, much of the surface water contains high levels of sulfate. This is the result of weathering of abundant iron sulfides associated with the coal deposits and the leaching of the water-soluble products into streams. This activity is largely responsible not only for the high sulfate, but the acid condition of the streams as well, since it forms sulfuric acid. Several industries, particularly steel mills, use large amounts of sulfuric acid, which in the past in North America was dumped into streams after it was used. High sulfate content may also indicate sewage pollution.

CHLORIDE (Cl). Although chloride is dissolved from rocks and soil, its presence may indicate contamination by human and animal sewage as well as industrial

effluents. Chloride may combine with sodium to produce a salty taste; the chloride content of public water supplies should not exceed 250 mg/L (Table 13.2). This recommended limit is based solely on taste. Throughout nearly all the land masses, fresh water is underlain by salt water. The depth to the freshwater-salt-water interface may range from a few to several hundred feet. Highly mineralized water from a deep well could reflect the underlying salt water. In addition to sewage, major sources of contamination by wastes of high chloride content include disposal of hydrochloric acid, leaching of oilfield brines, and road salting.

FLUORIDE (F). Most fluoride compounds have a low solubility; hence, fluoride occurs only in small amounts in natural water. Fluoride in drinking water has been shown to reduce the formation of dental caries if the water is consumed during the period of enamel calcification; it may also cause mottling of the teeth under certain conditions. Former recommended fluoride concentrations for public water supplies

TABLE 13.2 Secondary Maximum Contaminant Levels[1]

Contaminant	Level[2]
Aluminum	0.05–0.2 mg/L
Chloride	250 mg/L
Copper	1.0 mg/L
Fluoride	2.0 mg/L
Foaming Agents	0.5 mg/L
Iron	0.3 mg/L
Manganese	0.05 mg/L
Silver	0.10 mg/L
Sulfate	250 mg/L
Total Dissolved Solids (TDS)	500 mg/L
Zinc	5.0 mg/L
Color	15 color units
Corrosivity	noncorrosive
Odor	3 threshold odor number
pH	6.5–8.5

1. Secondary Maximum Contaminant Levels (SMCLs) are federally nonenforceable and establish limits for contaminants in drinking water that may affect the aesthetic qualities (e.g., taste and odor) and the public's acceptance of drinking water.

2. These levels represent reasonable goals for drinking water quality. The states may establish higher or lower levels, which may be appropriate dependent upon adverse effect on public health and welfare and upon local conditions such as unavailability of alternate source waters or other compelling factors. Some contaminants appear in both primary and secondary standards.

varied with the annual average temperature (Table 13.3). An excess of fluoride is associated with excessive bone formation and calcification of ligaments. Fluoride is now included in both the primary and secondary drinking water standards (Tables 13.1 and 13.2).

NITRATE (NO_3). Nitrate may be leached from some rocks by water, but certain plants, plant debris, animal excrement, sewage wastes, and inorganic nitrate fertilizers are probably the major contributors of nitrate. Nitrate concentration (measured as N) in water in excess of 10 mg/L, the maximum limit set by the USEPA (Table 13.1), causes infantile methemoglobinemia (blue babies).

IRON (Fe). Iron compounds are very common in rocks and they are easily leached by water, particularly water with a low pH. Concentrations of iron in excess of 0.3 mg/L will cause staining of laundry and utensils; it is usually objectionable for food processing, beverages, ice manufacturing, and many other processes. It may cause a metallic taste. Streams draining coal or other mining areas commonly contain excessive iron concentrations. Various other industrial processes, such as steel production, may also contribute to excessive iron concentrations.

Minor and trace inorganic substances are included in Tables 13.1 and 13.2 and microbiological and organic chemical data are in Table 13.1. Consult the footnotes for these tables for additional information on the nature, occurrence, and health effects of these substances in drinking water.

QUESTIONS 13, PART A

1. Which of the major inorganic substances might indicate contamination by sewage?

2. What are the major inorganic substances discussed in Part A that may be detrimental to human health?

TABLE 13.3 Former Recommended Fluoride Limits (USPHS, 1962)

Annual Average Of Maximum Daily Air Temperatures (°F)	Recommended Fluoride Limit (mg/L)
50.0–53.7	1.7
53.8–58.3	1.5
58.4–63.8	1.3
63.9–70.6	1.2
70.7–79.2	1.0
79.3–90.5	0.8

3. Iron and chloride are major inorganic substances that adversely affect drinking water. Are they health hazards according to the USEPA tables in this exercise?

4. a. What are the health hazards of excessive fluoride?

b. Why is fluoride listed in Tables 13.1 and 13.2?

5. For each of the following contaminants in drinking water, list the potential health effects and their sources.

Arsenic

Asbestos

Atrizine

Bromate

Carbofuran

Chlordane

Chlorine (as Cl_2)

Chromium

Cryptosporidium

Dioxin

Diquat

Lead

Nitrate

PCBs

Selenium

Tetrachloroethylene

Vinyl chloride

6. Use the information in Table 13.1 to help answer the following.
a. What is the purpose or role of adding disinfectants to drinking water?

b. What are the potential health effects (if any) of the disinfectant byproducts (listed in Table 13.1) that are produced during disinfection of drinking water?

Use information in Tables 13.4 and 13.5 to answer the questions below.

7. a. What water quality factors suggest that the North Dakota farm well is contaminated?

c. For control of *Giardia lamblia* (and most other biologicals in the water), what two techniques are employed?

b. What are the most likely sources of contamination in this well?

d. A treatment technique that controls the corrosiveness of water is used to limit copper and lead contaminants in drinking water. Why is corrosion a factor in copper and lead contamination?

8. a. What is the most likely cause of contamination of the domestic well in Ohio?

b. Could the water be used for cooking? Explain.

TABLE 13.4 Dissolved Solids (mg/L), Inorganic Chemicals (mg/L), and pH of Four Surface-Water Samples in the United States

Water Quality Factor	Source			
	Public Supply Buffalo, NY (Lake Erie)	Public Supply Seattle, WA (Cedar River)	Mahoning River Northeastern Ohio	Big Four Hollow Creek East Central Ohio
Dissolved solids	177	40	890	1620
Sulfate	23	2.4	470	1100
Chloride	23	0.5	100	15
Nitrate	0.2	0.2	4.6	2.1
Iron	0.01	0.05	52	52
pH	8.0	7.4	5.5	2.9

TABLE 13.5 Dissolved Solids (mg/L), Inorganic Chemicals (mg/L), and pH of Four Groundwater Samples in the United States

Water Quality Factor	Source			
	Public Supply Wichita, KS (well field)	Public Supply Shreveport, LA (well field)	North Dakota farm well (30 ft deep)	Ohio domestic well (47 ft deep)
Dissolved solids	844	142	2,400	15,390
Sulfate	128	12	1,750	24
Chloride	221	40	893	7,730
Nitrate	0.4	0.3	222	1.0
Iron	0.05	0.02	5.1	4.1
pH	7.3	6.6	7.3	7.4

9. a. What is the most likely source of contamination of Big Four Hollow Creek?

b. Could this water be detrimental to health if consumed? Explain.

c. What taste should it have?

10. Why do you think that the well water in Wichita is more highly mineralized than the well water in Shreveport? (Consider climate as well as possible differences in geology.)

11. Why does the public water supply in Seattle contain fewer dissolved solids than the supply in Buffalo? (Consider geology, climate, and topography.)

12. Lake Erie was once considered to be strongly contaminated and even "dead" by some environmentalists. Do the data in Table 13.4 support this idea or are other data needed (e.g., water quality factors for fish? Explain.

13. Recall the relationship between dissolved solids (TDS) and specific conductance (see Introduction to Exercise 13) and determine the probable specific conductance for water from the
   a. Mahoning River?

   b. Big Four Hollow Creek?

14. From the data in these tables, would you expect most water supplies to be acidic or alkaline?

15. What is the source of water where you are now living? If possible, check with your local water treatment facility by phone or online, determine its dissolved solids content and record it here.

PART B. SOURCES AND AMOUNTS OF CONTAMINATION

Contamination of surface-water sources, such as lakes and water courses, commonly results from (1) point-source discharge of effluent (wastewater) directly into the water, (2) non-point-source discharge such as runoff from fields, and (3) inflow of contaminated groundwater. Although it may be relatively easy to determine the location of a surface outfall and to measure both the rate of the effluent discharge and its composition, it is much more difficult to evaluate non-point-source pollution and to detect areas where contaminated groundwater affects the quality of surface water.

The principle that "dilution is the solution to pollution" has been used for centuries. During the 1970s in an effort to clean up the streams and rivers of North America, reduction and/or pretreatment of industrial effluent was implemented.

However, dilution is still an important concept. The reasoning behind this is that generally effluent discharge is many times smaller than the discharge of the receiving stream, and consequently the waste ultimately will be diluted to an acceptable concentration for most uses by humans and for aquatic life. For example, contaminated groundwater is constantly being diluted as it slowly moves through the ground.

The quantity of contaminated groundwater seeping into a stream is generally quite small relative to the flow of the stream, and the contaminant may be so diluted that it is undetectable in stream waters even in areas of discharge.

The concentration of a contaminant in a stream is also directly related to the physical properties of the stream. Many organic and inorganic chemicals may become attached to silt or clay particles. These chemicals can be released later and contaminate ground or surface water that otherwise might not be contaminated. Suspended sediment in water (turbidity) is also a contaminant (Table 13.1).

As indicated in the Introduction to Water Resources and Contamination and this exercise, major sources of water pollution come from industrial, municipal (domestic and industrial sources), and agricultural activities (Table 13.6). Physical, chemical, or biological contaminants may adversely affect water quality. Some constituents affect taste (chloride, sulfate, phenols, iron), others may be corrosive (high in dissolved solids, low pH) and a few may be health hazards (heavy metals, nitrate, pathogenic bacteria, and viruses). Increased temperature or increased biological oxygen demand (BOD) from organic wastes causes a reduction in the amount of dissolved oxygen (DO) in water. DO is important for fish survival.

TABLE 13.6 Summary of Sources and Their Contaminants in Ground- and Surface-Water Pollution

Source	Principal Contaminants (groundwater contaminants in italics)
Domestic	Undecomposed organic matter (garbage, grease) that increases BOD (Biochemical Oxygen Demand)
	Partially degraded organics (raw wastes and nitrates from humans) from combined sewers
	Combination of above after limited sewage treatment in municipal sewage plants
	Parasites, bacteria, and viruses (pathogens)
	Grit from washings, eggshells, ground bone
	Miscellaneous organics (paper, rags, plastics, synthetics)
	Detergents
	Inorganics from organic decay (nitrate, sulfate) from septic tank systems
	Salts and ions in public water supply
	Soluble organic compounds
Industrial	Biodegradable organics
	Inorganic solids, mineral residues
	Chemical residues (acids, alkalies, complex molecules)
	Metal ions
	Soluble salts from industrial waste ponds and spills
Agriculture	Increased concentration of salts and ions from animal feedlots
	Fertilizer residues including sewage sludge on land.
	Pesticide residues
	Silt and soil particles
	Concentrated salts in water applied to land from irrigation and sewage sludge
Landfill	*Hardness-producing leachate*
	Soluble chemical and gaseous products

(Modified from McGauhey, 1968)

The pollution load that a stream carries can be calculated if the stream discharge and the concentration of the specific contaminants are known.

$$\text{Load (tons/day)} = Q \times C \times 0.0027$$

where Q = stream discharge (cfs),
C = concentration of specific contaminant (mg/L), 0.0027 = constant to convert seconds and mg/L to days and tons

To determine if a stream is being contaminated, it is necessary to acquire data that were collected before the apparent contamination or to determine the concentrations of selected constituents in uncontaminated reaches of the stream. Commonly, water-quality data reflecting natural conditions prior to contamination, sometimes called background data, are not available. Background information may be obtained, however, by sampling the water course and its tributaries in reaches upstream from the suspected source of pollution. The quality at any sampling station represents all of the upstream input.

QUESTIONS 13, PART B

1. Determine the daily load of iron (tons/day) going over Niagara's Horseshoe Falls (approximate discharge = 200,000 cfs) if the iron concentration is 0.1 mg/L.

2. Biodegradable organic matter promotes oxygen demand in streams. What water quality parameter would be expected to change with increased oxygen demand?

3. Why are silt and clay particles considered to be contaminants?

4. From Table 13.6, what differences in stream-water quality upstream and downstream from a sewage treatment plant should be expected? Explain.

5. a. What is meant by "background" concentrations of chemicals in a watershed?

b. What steps are required to determine the background concentration of chromium in a watershed with a chrome-plating factory?

6. List three microorganisms that indicate contamination by animal wastes (In addition to Tables 13.1 and Table 13.6, see the Introduction to water Resources and Contamination.)

PART C. BOTTLED WATER AND WATER PURIFICATION

Introduction

Bottled water has become a popular drink choice in the United States because it might be better tasting, safer, or more readily available than some tap water or because some people seek an alternative to soft drinks and other beverages (USEPA, 2005). Most bottled water comes from ground water, either from wells or springs; the remainder comes from surface water such as lakes and rivers—in some cases from public water systems. The taste of both bottled and tap water depends on *the source* (e.g., calcium and magnesium minerals add flavor and body) and the *treatment*(s), particularly the disinfection process, which inactivates disease-causing organisms but might leave a taste.

Bottled water from a ground water aquifer is obtained from a well or spring (possibly tapped by a well) and may be labeled as artesian if it comes from a confined aquifer. Distilled water from a surface or groundwater source is condensed after boiling to steam. The process kills microbes (including *Cryptosporidium*) and removes natural minerals (and much of the taste) and some organics.

Tap water is usually disinfected by chlorine or chloramine, because of the effectiveness, cost, and the continuing protection provided while in pipes of municipal water systems and in the home. Some tap water also uses ultraviolet (UV) light or ozone. Ozone is the preferred disinfectant for bottled water because it does not leave a taste and the sealed bottle does not need continuing disinfection. Other bottled water is disinfected using UV light or chlorine dioxide. Ozone (and UV light) kills most microbes depending on the dosage.

Sterilized water or **sterile water** is water that is free from all microbes. **Purified water** is essentially free of chemicals (TDS < 10 ppm) and might be produced by distillation, deionization, or by reverse osmosis (RO). In the latter process, water is forced through a membrane leaving behind all contaminants (minerals, organic and inorganic chemicals, color, turbidity, and microbes). Purified water may also be referred to as demineralized water; or distilled water if purified by distillation, deionized water if purified by deionization, or reverse osmosis water if purified by RO. For

bottled water, micron filtration (screens with microscopic holes) remove most chemicals and microbes. To remove *Cryptosporidium* cysts, the largest hole in the filter must be one micron.

Although the EPA sets the standards for public drinking water supplies, the FDA (Food and Drug Administration) sets those for bottled water (based on EPA standards). The FDA regulates bottled water as a packaged food and sets the quality and identity standards and requirements for processing and bottling. *Carbonated water, soda water, seltzer water, sparkling water, and tonic water are not regulated as bottled water but as* **soft drinks.** Bottled water sold in the United States, including those from overseas, must meet the FDA standards for physical, chemical, microbial, and radiological contaminants.

There are many types of bottled water according to the Code of Federal Regulations (21 CFR 165.110 (a)(2)). See also **sterilized** and **purified** waters above.

Bottled water or drinking water is water intended for human consumption, sealed in bottles, with no added ingredients (except safe antimicrobial agents and fluoride).

Other types of bottled water are:

artesian water or artesian well water—water from a well tapping a confined aquifer in which water stands above the height of the top of the aquifer,

mineral water—water containing not less than 250 ppm TDS that originates from a geologically and physically protected underground water source, with no added minerals, and constant levels and proportions of minerals and trace elements. Mineral water is labeled as *"low mineral content"* if it is < 500 ppm TDS or as *"high mineral content"* if TDS > 1,500 ppm.

sparkling bottled water—water that after treatment and possible replacement of CO_2 contains the same CO_2 as it had when it emerged from the source (compare with *sparkling water,* above, that is regulated as a soft drink).

spring water—water derived from underground formation from which water flows naturally to surface (may also be collected through a borehole).

well water—name of water derived from a well (dug or drilled).

Also, if bottled water comes from a community water system, it will be labeled *"from a community water system"* or *"from a municipal source."*

Note that bottled water meets the EPA's Secondary Drinking Water Standards (e.g., chloride—250 mg/L and TDS 500 mg/L) *except for* mineral water. Bottled water that is certified by NSF International indicates that the bottler complies with all FDA requirements. Members of the International Bottled Water Association (IBWA) meet their trade association's "model code." To learn more about a bottled water and specific contaminants, contact the bottler through the contact information provided on the bottle.

To determine its quality, compare that information with the FDA standards (www.cfsan.fda.gov) and/or the EPA Groundwater and Drinking Water page (www.epa.gov/safewater) and EPA standards for drinking water (http://www.epa.gov/safewater/standards.html). See also EPA's Drinking Water and Health http://www.epa.gov/safewater/dwh/index.html. Additional information is available through the Code of Federal Regulations and NSF International (www.nsf.org) and the International Bottled Water Association (www.bottledwater.org).

Small public water systems are allowed to meet the National Primary Drinking Water Regulations by installing point-of-use (POU), for direct consumption, and point-of-entry (POE) treatment devices. Some consumers on municipal systems install at-the-tap filters or purchase "pitcher filters" for small amounts of drinking water. Filters are designed for removing one or more contaminants.

For information on ground water resources for the home, see the USGS publication on *Ground Water and the Rural Homeowner* (Waller, 2005).

QUESTIONS 13, PART C

With the aid of information in the Introduction to Part C (and possibly other sources) answer the questions below.

1. a. What are the sources of bottled water?

b. What factors control the taste of bottled water?

c. *Cryptosporidium* is a microscopic parasite that lives in human/animal intestines. Cysts of the organism in the public water supply of Milwaukee, Wisconsin, were eventually responsible for 400,000 cases of illness and several deaths. Why did the standard chemical disinfectant used in public water systems not kill this microorganism?

d. Since that time what changes in treatment for public water supplies now remove *Cryptosporidium?*

e. Who sets the standards and regulates bottled water?

f. Explain by defining each, the difference between purified and sterilized water.

2. What is the difference between sparkling water and sparkling bottled water?

3. Define what is meant by artesian water.

4. a. List all the requirements for bottled water to be labeled mineral water.

b. Does mineral water meet the national Secondary Drinking Water Regulations? Explain.

5. Examine a bottled water label, your own or one available in class or online, and record the following information on this water. You might need to use an online source, often listed on the bottle, for more information.

Name of the bottled water: _____ Company: _____

Size of bottle: mL _____ oz _____ Type of water: _____

Listed geologic source of water (check one or more):

    Groundwater aquifer _____ Spring_____ River _____ Lake _____

    Public water system _____ Other _____

Geographic source of water:

Methods of treatment (if any):

    None: _____ None listed: _____

    Distillation _____

    Micron filtration: yes _____ no _____

        Hole size: Nominal (average) = _____ or

            Absolute (maximum) = _____

    Reverse osmosis _____

    Disinfection

        Ozone _____   Chlorine dioxide _____

        UV light _____

        Chloramine _____   Chlorine _____

    Other Treatments

Other Label Information:

6. Check the definitions in the Introduction to this part of the exercise to help answer the following.
    a. What is the definition for purified water?

Is the bottle of water described in the question above purified?

What process was used in the purification?

What is the limit for total dissolved solids (TDS) for purified water?

Has anything been added to the water after purification?

b. What is the definition for sterilized water?

Has the bottle of water described above been sterilized _____

What was the process used for sterilization?

7. a. What is the cost per liter for bottled water (give the brand, size, and where purchased (vending machine, grocery, etc.)? Convert if the bottle is not liter-sized)?

b. How does this compare in price (per liter) with tap water in your city?

(Check with the city or the instructor for tap water cost; 1 U.S. gallon = 3.78 liters)

Bibliography

Feth, J. H., 1973, *Water facts and figures for planners and managers:* U.S. Geologial Survey Circular 601-1, 30 p.

Hem, J. D., 1985, *Study and interpretation of the chemical characteristics of natural water:* U.S. Geological Survey Water-Supply Paper 2254, 264 p.

Mazor, E., 1990, *Applied chemical and isotopic groundwater hydrology:* Somerset, NJ, John Wiley, 274 p.

McGauhey, P. H., 1968, *Engineering management of water quality:* New York, McGraw-Hill, 295 p.

Pettyjohn, W. A. ed., 1972, *Water quality in a stressed environment:* Minneapolis, MN, Burgess, 309 p.

Todd, D. K., 1980, *Ground water hydrology,* 2nd ed.: Somerset, NJ, John Wiley, 535 p.

Waller, R. M., 1994, *Ground water and the rural homeowner:* USGS. http://pubs.usgs.gov/gip/gw_ruralhomeowner/index.html

USEPA National Primary Drinking Water Standards—Contaminants. Retrieved August 24, 2007, from http://www.epa.gov/safewater/contaminants/index.html#listmcl

USEPA 2005, *Bottled water basics:* Water Health Series.

USEPA, 1991, Fact sheet, *Drinking water regulations under the Safe Drinking Water Act, June 1991:* Washington, DC, U.S. Environmental Protection Agency.

USPHS, 1962, *Drinking water standards:* Washington, DC, U.S. Public Health Service, Publication 956, 61 p.

Lake and River Contamination from Industrial Waste

INTRODUCTION

During the 1970s the public observed pollution and the effects of pollution from many sources in the rivers and streams of North America. Detergent foam covered streams and fish floated on the surface, dead from toxins or lack of oxygen; beaches were closed; drinking-water sources were lost; at least one river burned; and fisheries closed due to mercury or organic chemical pollution. In this exercise we explore two areas where industrial wastes resulted in a decrease in water quality. Although the practice of dumping industrial wastes directly into surface waters has decreased markedly in North America since the 1970s there are many places in the world where such point-source pollution is common. Also, past errors by industry and the acts of unscrupulous industrial-waste handling companies have left us with problems that are still in need of being addressed. In 2007 there were more than 1,600 Superfund sites, the most serious toxic waste sites, according to the federal program that is charged with cleaning up these sites. This exercise shows the impact of poor waste disposal practices on a river and one of the Great Lakes and progress in reducing concentrations of contaminants.

PART A. TRACE ELEMENT (MERCURY AND CHROMIUM) POLLUTION IN LAKE ERIE

Lakes Erie and St. Clair and the Detroit and St. Clair rivers were among several areas impacted by mercury pollution in the 1960s and 1970s. In fact, mercury pollution shut down the fisheries industry in many areas, and algal blooms caused reduced levels of dissolved oxygen in other areas. These and other factors caused some environmentalists to proclaim Lake Erie "dead." Although the lake is far from dead, organic and inorganic chemicals (some from agricultural sources) and trace elements are still a concern in Lake Erie. Water quality and the health of the fish population in the lake have shown significant improvement since the 1970s, but excessive consumption of fish from the lake is still a hazard. Although detailed fish consumption advisories (e.g., listing type, size, and location of fish, number of meals per month, and toxins [mercury and PCBs]) are readily available on the Web and elsewhere, some individuals downplay the importance of health hazards. In addition to the trace element and organic chemical contamination of the water, sediments, and biota of the lake, invasive non-native species (fish, invertebrates, and plants) cause economic and environmental damage to the Great Lakes. The sea lamprey invasion occurred with the opening of the St. Lawrence Seaway and more recently zebra mussels have disrupted the ecosystem. The mussel, which was inadvertently introduced from Europe in the 1980s, grows rapidly and blocks water intakes for cities, power plants, and boat motors. Although the zebra mussels actually improved water quality by filtering, the mussels may eventually cause other problems because they concentrate pollutants from the water and remove food needed by other organisms. The lake will continue to change as other non-native species enter the system. The round goby has arrived; the Asian carp is nearby in the Mississippi River System.

In this part of the exercise we investigate contamination by trace elements. Although this exercise covers surface-water quality, the focus is actually on sediments contaminated by trace elements. Trace elements are defined as those substances that normally occur in water at concentrations less than 1 milligram per liter (mg/L or 1 part per million (ppm)). A few trace elements and their recommended maximum concentrations in drinking water are listed in Table 14.1 (see also Table 13.1.). Little is known of the potential adverse or beneficial physiological effects on humans,

TABLE 14.1 Common Trace Elements and Their Maximum Limit (MCL or SMCL) in Drinking Water in the United States

Trace Element	MCL[1] (mg/L)	Trace Element	SMCL[2] (mg/L)
Arsenic	0.01	Copper	1.0
Barium	2.0	Fluoride	2.0
Cadmium	0.005	Manganese	0.05
Chromium	0.1	Silver	0.1
Copper	1.3[3]		
Fluoride	4.0	Zinc	5.0
Lead	0.015[3]		
Selenium	0.05		
Mercury (inorganic)	0.002		

1. MCL: Maximum contaminant level, National Primary Drinking Water Standards. These are maximum permissible levels set for contaminants that have an adverse effect on human health.

2. SMCL: Secondary maximum contaminant level, National Secondary Drinking Water standards. These are federally nonenforceable limits for contaminants that affect aesthetic qualities such as taste and odor.

3. Action level for treatment

other animals, and plants of some trace elements that are present in commonly used or consumed products. Humans require many trace elements, but in some cases there is only a small range between requirement and toxicity. For example, the lack of minute amounts of copper in the diet causes nutritional anemia in infants, but large concentrations may cause liver damage. Another complicating factor is the chemical form of the element. In one chemical state a trace element may pass through the body with little or no harm, but in another it may be absorbed to the point of toxicity. An example of this phenomenon is mercury; metallic mercury is generally harmless but methyl mercury is highly toxic.

In the past three decades there have been numerous trace element studies. In general, these show that although most of the trace elements occur in barely detectable concentrations in water, they may appear in streamside or riverbed deposits in concentrations that are two or three orders of magnitude (100 to 1,000 times) greater. Plants receiving their nutrients from the sediment may contain trace element concentrations that are several hundred or even several thousand parts per million (ppm) more than the sediment. Apparently, many trace elements are barely detectable in water because they become attached to fine-grained sediments, which are subsequently deposited. These elements may then be removed and concentrated by plants and mud- or bottom-feeding organisms.

Anomalous concentrations of trace elements may be due to many natural or artificial causes. Contaminated sediments may arise from industrial and municipal point sources, run-off from farming and construction, dumping on shore or at sea, accidental spills onshore or at sea, leaching from waste disposal sites, and atmospheric loading. *Arsenic* poisoning leading to sickness and death among cattle in New Zealand was attributed to drinking arsenic-rich water of natural origin. The Kansas River once contained local arsenic concentrations that approached the recommended limit set by the U.S. Public Health Service. The source of arsenic in this case was municipal treatment plant and septic tank effluent, which contained arsenic originating as an impurity in presoaks and household detergents.

Near major highways in British Columbia, southern England, Finland, and the United States, cereals and vegetables once contained 4 to 20 times the normal amount of *lead*. These high concentrations were mainly the result of air and soil pollution by the emission of lead-containing automobile exhaust. Lead was added to gasoline to improve the octane rating but its use for this purpose is now generally prohibited.

Mining and metal-producing activities are also major causes of trace-element contamination. Along the upper reaches of the Jintsu River Basin in Japan, milling wastes from a mine producing lead, zinc, and *cadmium* were dumped untreated into the river. Downstream the contaminated water was used by farmers for cooking, drinking, and irrigation of rice fields. Eventually, many people in the basin began to suffer from an unknown but very painful, sometimes fatal, disease. Some years later, samples of river water, soil, and rice were examined; water generally contained less than 1 ppm of cadmium and less than 50 ppm of *zinc*. Soil samples from the irrigated rice fields contained as much as 620 ppm of cadmium and 62,000 ppm of zinc. These metals accumulated to even higher concentrations in the contaminated rice. Apparently, ingestion of the cadmium-contaminated rice by the local farmers had caused this debilitating bone disease known as itai-itai or osteomalacia.

The contamination from trace elements in the Great Lakes is not so severe as these examples; however, the distribution and concentration of several trace elements in Lake Erie illustrate the impact of industrial activity on the lake.

The discovery of *mercury* in Lake St. Clair, the Detroit River, and Lake Erie in the late 1960s led to the banning of sport and commercial fishing in these waterways because of the experience in Minimata Bay, Japan, where dozens of individuals were poisoned after consuming mercury-rich sea-food from the bay. The bay had been contaminated by the disposal of mercury compounds originating in the waste from a plastics factory.

The major mercury contributors to Lake Erie were identified as chemical plants in Wyandotte, Michigan, and Sarnia, Ontario, Canada. These plants manufacture chlorine gas and caustic soda by using mercury cells in the electrolytic process. The rate of accidental loss of mercury from these two plants was estimated to be about 50 lbs/day from 1950 to 1970, but greatly decreased thereafter. In the early 1970s, the normal concentration of mercury in Lake Erie water was less than 0.0002 ppm, but in bottom sediments it ranged from 0.026 to more than 3.0 ppm, and the background averaged less than 0.07 ppm.

The major use of *chromium* is in the electroplating industry. This element also appears in only minute concentrations in water. Background chromium concentrations in water and in the sediment on the bottom of Lake Erie are less than 0.004 and 32 ppm, respectively.

QUESTIONS 14, PART A

Two sets of analyses, from 1971 and 1997/1998 sediment cores, provide the opportunity to determine change in the Lake Erie system over time. We begin with the earlier study.

1. Using the 1971 data in Table 14.2 and the maps in Figures 14.1 and 14.2, construct isolines (contours) showing the concentration of mercury and chromium in the bottom sediment in Lake Erie. For mercury, use isolines representing concentrations of 0.25, 0.5, 1, and 2 ppm. For chromium, use isolines representing concentrations of 20, 60, 80, and 140 ppm. Label the isolines, showing the ppm value of each. Remember, some isolines may terminate at the shore of the lake. Mercury and chromium values for core sites 11-20 have been entered on Figures 14.1 and 14.2, respectively and one or two isolines drawn.

2. From your maps of contaminated sediment, what three areas show major contamination by both mercury and chromium?

3. a. On Figure 14.1 draw several arrows to indicate what you believe are the major directions of water flow from the Detroit River to Buffalo.

b. Explain your reasoning.

4. If concentrations of both mercury and chromium are near background levels at point A, describe the expected land use on the north shore of the lake.

5. a. Where would you expect bottom-dwelling organisms to contain the greatest concentration of mercury?

b. The least?

6. Should commercial fishing have been banned throughout the lake, along the southern shore, or only in the western part? Explain.

7. Is the chromium contamination in Lake Erie reflected in any of the major lakeside municipal water supplies? Explain.

DATA FROM 1997/98 SAMPLING

In 1997/98, another set of samples was obtained to determine trends in contamination and improvements in environmental quality as a result of measures to control mercury and other pollutants. In addition to fewer additions of contaminants to Lake Erie, contaminants in the upper layers of bottom sediment may have changed because of erosion and deposition of sediment. Mercury concentrations in sediment for 1997/98 are reported in Table 14.2 along with the earlier (1971) values.

8. Table 14.2 contains mercury (Hg) concentrations for Lake Erie sediments acquired in 1997–1998. Most of these data have been plotted on Figure 14.3, and some of the isolines have been drawn to show the spatial distribution of mercury. Complete the 0.25 isoline for the Western Basin and the 0.1 isolines for the Central and Eastern Basins. Begin by plotting values from Table 14.2 for stations in the Middle Basin (stations 47, 51, 61, 63, 99, 102, and 105) in Figure 14.3.

TABLE 14.2 Concentration of Mercury and Chromium in the Top 2 Centimeters (1971) and 3 Centimeters (1997/98) of Bottom Sediment in Lake Erie and in Local Municipal Drinking Water

1971 Core Sites*	Chromium 1971 (ppm)	Mercury 1971 (ppm)	1997/98 Core Sites (approx. 1971 locations)*	Mercury 1997/98 (ppm)	1997/98** Core Sites (nearby locations)	Mercury 1997/98 (ppm)
1	143	2.1	97	0.94	96	0.22
2	87	2.0	—	—	—	—
3	88	1.4	83	0.69		
4	47	1.0	81	0.46	82	0.37
5	141	2.8	88	0.08		
6	80	1.4	86	0.69		
7	75	0.71	85	0.64		
8	48	0.82	84	0.23		
9	37	0.44	79	0.24	80	0.21
10	60	0.3	73	0.13	110	0.03
11	18	0.12	74	0.24		
12 Cleve Harbor	143	0.93	92	0.02	68	0.25
13	53	0.46	111	0.30	66	0.18
14	47	0.18	98	0.16	69	0.20
					63***	0.17***
15	5	0.01	51	0.01		
16	24	0.45	99	0.07	47	0.22
17	44	0.19	43	0.09	42	0.07
18	—	0.12	107	0.05		
19	103	2.3	35	0.13	109	0.01
20 Buffalo Harbor	103	2.3	—	—		—
"A"	—	—	61	0.12	102	0.16
					105***	0.05***
Municipal water						
Detroit, MI	0.0015					
Toledo, OH	0.0037					
Cleveland, OH	0.0035					
Buffalo, NY	0.0028					

* The approximate match of 1971 and 1997/98 sites is given by location on the same line.

** Nearby locations, 1997/98: locations near the 1971 sites and their mercury concentration.

*** Secondary nearby locations and their mercury concentration.

Concentrations for some sites have been plotted on Figures 14.1, 14.2, and 14.3.

9. With reference to Table 14.2 and Figures 14.1 and 14.3, briefly describe the changes between 1971 and 1997/98:

a. in the areas of high and low mercury distribution

high mercury values in the West, Middle, and East Basins

low mercury values in the Middle Basin

b. at the sites of the four highest 1971 values of mercury in the West Basin

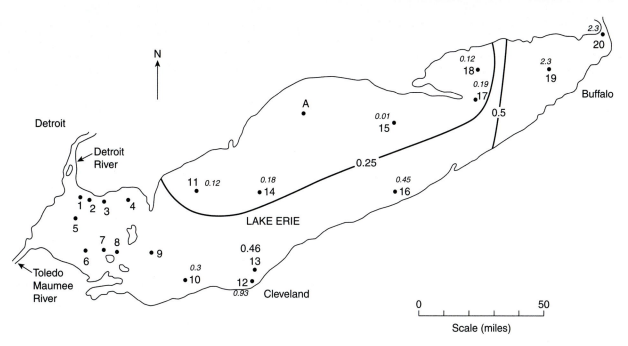

FIGURE 14.1 Concentration of mercury in sediment on the bottom of Lake Erie in 1971. (Numbered dots are sample sites; italicized numbers are selected mercury values in ppm. The 0.25 and 0.5 ppm isolines are given.)

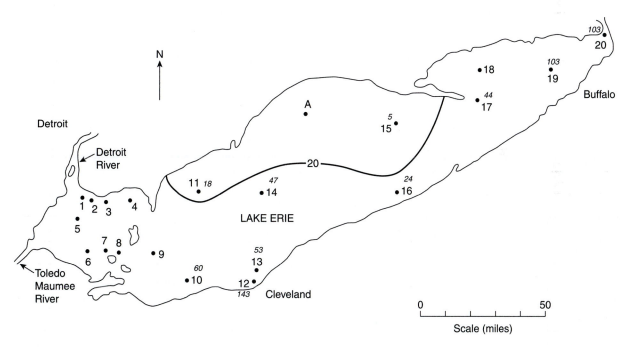

FIGURE 14.2 Concentration of chromium in sediment on the bottom of Lake Erie in 1971. (Numbered dots are sample sites; italicized numbers are selected chromium values in ppm. The 20 ppm isoline is given.)

c. in general for mercury at the same or similar sampling sites

10. Are the regulations against dumping or release of heavy metals working to improve the sediment in the lake? Explain.

11. Did the concentration of mercury in the harbor at Cleveland improve between 1971 and 1997/98?

12. At station 11 (approximately equal to station 74) the value for *chromium* went from 18 to 56 ppm over the period. (There are no chromium values shown in Table 14.2 for 1997/98.)

    a. What was the change in *mercury* values for this period at this site?

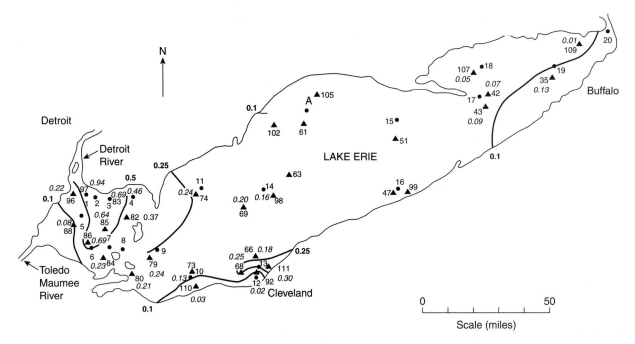

FIGURE 14.3   Concentration of mercury in sediment on the bottom of Lake Erie in 1997/98. Numbered dots [1–20] are the 1971 sampling sites (no mercury values plotted); triangles, the 1997/98 sites. Italicized numbers beside the triangles are mercury concentrations in 1997/98. Some contours given; values in bold. Other 1997/98 values to be added. Sites 1 and 97, 3 and 83, and 4 and 81 are represented by a dot only.

b. What might explain such increases that do not represent the trend for contaminant concentrations?

13. Lakewide, almost all contaminant concentrations have decreased since 1971. In addition to the trend for metals (e.g., Hg), polychlorinated biphenyl (PCB) sediment concentrations also decreased (from 136 ng/g to 43 ng/g). (Note: n is nano or 10^{-9}.)

a. Knowing that organic chemicals such as PCBs, DDT, and Mirex, etc., can bioaccumulate in fish and other organisms, would it now be safe to eat any type and size of fish caught in the lake several times a week? Explain your reasoning.

b. Where would you seek information on eating fish from Lake Erie (i.e., a fish advisory)? Give specific Web (URL optional) or other sources you would check.

PART B. INDUSTRIAL WASTES IN THE MAHONING RIVER, NORTHEASTERN OHIO

The Mahoning River drains the densely populated and once highly industrialized Warren–Youngstown area in northeastern Ohio. In reaches above Pricetown (Figure 14.4) acid-mine drainage from coal mines enters the river, resulting in high concentrations of sulfate, iron, and dissolved solids, and increased hardness and low pH.

Downstream from Leavittsburg, particularly in Warren and Youngstown, was a huge industrial complex with an abundance of steel mills. Steel-mill wastewaters, which were once dumped into the river, contained high concentrations of chloride and sulfate (from spent hydrochloric and sulfuric acid) and of iron and other metals, had low pH (from the acid), and were heated (having served as cooling water).

As temperatures increase, the ability of water to hold dissolved oxygen (DO) decreases. According to State of Ohio regulations for protection of aquatic life in designated warmwater-habitat streams the minimum DO is 4 mg/L.

Conditions along the Mahoning River have changed with the introduction of EPA regulations and the closing of steel mills for economic reasons since these data (1963–65) were obtained. We begin Part B by looking at the river prior to improvements in water quality; this is the type of contamination that may exist today in other regions of the world where industrial pollution is minimally regulated. Then we look at the current conditions that result in today's human health advisory for the Mahoning River and explore options for improving the environmental quality (and economic potential) of the river. In 2005, the Mahoning River was one of the top five contaminated rivers in the United States.

QUESTIONS 14, PART B

1. Use the following four maps, Table 14.4, and a color code to distinguish the four water types given in Table 14.5 (type 1, blue; type 2, yellow; type 3, green; type 4, red). Color the outlined area along the Mahoning River to indicate the range of sulfate, chloride, pH, and temperature in the reach represented by each sample collection site. *The data for each site in Table 14.3*

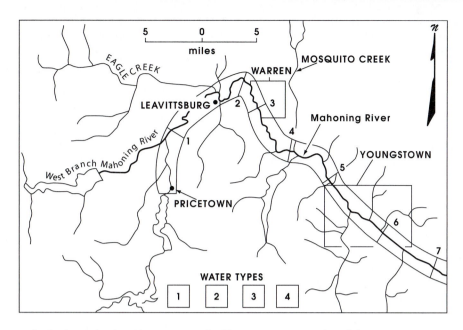

FIGURE 14.4   Water quality in the Mahoning River. Ranges of sulfate concentrations (mg/L) for water types are: 1 = 50–150; 2 = 151–250; 3 = 251–350; and 4 = 351–450.

TABLE 14.3 Range in Water Quality Factors for the Mahoning River, Sites 1–7

Site Number	Sulfate (mg/L)	Chloride (mg/L)	pH	Temperature (°F)
1	151–250	36–50	6.6–7.0	73–79
2	50–150	20–35	7.1–7.5	73–79
3	351–450	36–50	6.6–7.0	80–86
4	351–450	36–50	6.1–6.5	80–86
5	251–350	51–65	5.5–6.0	80–86
6	251–350	51–65	6.1–6.5	87–93
7	351–450	66–80	6.1–6.5	94–100

TABLE 14.4 Numerical Classification of Water (water types)

Water-Quality Factor	Type 1 (blue)	Type 2 (yellow)	Type 3 (green)	Type 4 (red)
Sulfate	50–150 mg/L	151–250 mg/L	251–350 mg/L	351–450 mg/L
Chloride	20–35 mg/L	36–50 mg/L	51–65 mg/L	66–80 mg/L
pH	7.1–7.5	6.6–7.0	6.1–6.5	5.5–6.0
Temperature (°F)	73–79	80–86	87–93	94–100

refer to the reach of the river extending upstream to the next sampling site. Make a separate map for each water-quality factor. You may use distinct patterns if color is not available. Note: water types are defined by concentration range in each figure caption and in Table 14.4.

2. What is the most likely source of the relatively high sulfate concentration at site 1?

3. Why did the sulfate concentration decrease from site 1 to site 2?

4. Why did the pH decrease significantly between sites 2 and 5?

5. Do the data suggest that the greatest quantity of hydrochloric acid is used in the vicinity of site 3 (Warren area) or site 5 (Youngstown area)? Why?

6. Considering the temperature data, would you expect the dissolved oxygen content to be higher upstream from Warren (sites 1, 2) or in and downstream from Youngstown (sites 5,6,7)? Why?

7. The mean discharge at sites 1, 5, and 7 was 408, 673, and 809 cubic feet per second (cfs), respectively. On a date characterized by the mean flow, the sulfate concentration at these sites was 175, 280, and 420 mg/L, respectively.

a. What was the increase in sulfate load, in tons per day, between sites 1 and 5, and 5 and 7? Use the equation: hood (tons/day) = Q × C × 0.0027, where, Q = stream discharge (cfs), C = concentration of contaminant (mg/L), and 0.0027 = constant to convert seconds and mg/L to days and tons.

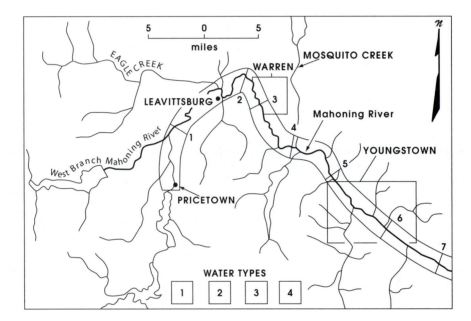

FIGURE 14.5 Water quality in the Mahoning River. Ranges of chloride concentrations (mg/L) for water types are: 1 = 20–35; 2 = 36–50; 3 = 51–65; 4 = 66–80.

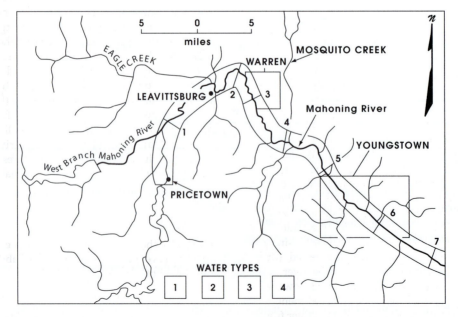

FIGURE 14.6 Water quality in the Mahoning River. pH ranges for water types are: 1 = 7.1–7.5; 2 = 6.6–7.0; 3 = 6.1–6.5; 4 = 5.5–6.0.

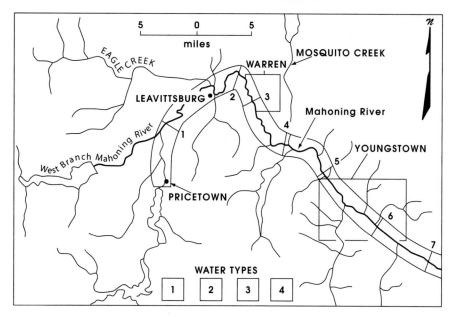

FIGURE 14.7   Water quality in the Mahoning River. Temperature (°F) ranges for water types are:
1 = 73–79; 2 = 80–86; 3 = 87–93; 4 = 94–100.

b. What is the major source of this sulfate increase?

8. What would you expect the chloride concentration to be in Eagle Creek? Explain.

9. Do any constituents in the Mahoning River in the Youngstown area exceed any drinking water quality limits? If so, which ones?

10. Since the 1960s when there was uncontrolled waste disposal from seven steel mills operating in the area, water quality in the Mahoning River has improved significantly due to environmental regulations that require wastewater treatment of effluent from all industries and municipalities. Worldwide competition from modern steel mills resulted in the closing of some steel mills here, further improving water quality. In 2005, the river was still one of the five most contaminated rivers in the United States. Recent measurements of water quality (Stoeckel and Covert, 2002) show that the limits for three physical/chemical parameters for healthy desirable aquatic organisms were only occasionally exceeded (dissolved oxygen [DO], pH, and temperature). The river still does not meet standards for swimming and wading, however. Two additional factors, sewer effluent and toxic sediments in the river, result in a human health advisory for the river from Warren, Ohio, to the Pennsylvania/Ohio line. Overflows of combined storm and sanitary sewers during wet weather contribute *Escherichia coli (E. coli)* and nutrients and the sediments contain toxic organic chemicals from disposal practices prior to regulation (get additional information at: http://oh.water.usgs.gov/reports/Abstracts/wrir02-4122.html).

a. How should the region address the *E. coli* problem in this river system? (*E. coli* indicates recent fecal contamination from gastrointestinal tracts of humans or animals and possible presence of pathogens that may adversely affect human health.)

b. A preliminary study of river sediment by the U.S. Army Corps of Engineers (USACE) recommended removal of 750,000 cubic yards of sediment with organic toxins to a secure waste disposal site to remove the human health advisory. The proposed *Mahoning River Environmental Dredging Project* would be the second largest USACE restoration project (the Everglades is first) and would require about $35 million in local money (Eastgate Council of Governments, n.d.). You will find useful information to help answer parts of this question online (e.g., http://www.eastgatecog.org/env-mahoningriver.asp, http://www.ysu.edu/mahoning_river/river_restoration.htm, or http://www.lrp.usace.army.mil/pm/mahonoh.htm).

c. Should the region and the nation expend funds to remove this environmental hazard, a liability to economic development? Explain.

d. What are the options for dealing with the contamination of the Mahoning River?

e. According to on-line information, how much will the federal share be?

f. What are the options for disposal of any sediment that is removed?

11. On the EPA website, http://www.epa.gov/owow/watershed/, go to your own watershed somewhere in the United States, or select one in an area of interest to you by clicking on the following: Surf Your Watershed, Locate Your Watershed, Search by Map or Zip Code. When you find Your Watershed, click on Stream Flow.
  a. Select and click on one of the USGS Station Numbers and complete the following:

_____ _____ AT _____ _____
8-digit USGS   Creek or   Town or City   State
Number   River

  b. What is the elevation of the gage?

c. What is the gage height today?

d. What is the stream flow or discharge (cubic feet per second) at this station?

12. Go to the EPA website to learn about projects to improve watersheds (http://www.epa.gov/owow/watershed/). Begin by clicking on *Targeted Watersheds* and then from *Project Summaries* select one of the targeted watersheds of interest.

  a. What is the name of the watershed?

  b. In what state(s) is it located?

  c. List the three most interesting projects for this watershed.

  d. Return to the previous page, and select *Project Executive Summary*. From the *Executive Summary* page, describe the two most interesting projects from the Innovative Ideas list.

Bibliography

Bedbar, G. A., Collier, C. R., and Cross, W. P., 1968, *Analysis of water quality of the Mahoning River in Ohio*: U.S. Geological Survey Water-Supply Paper 1859-C, 32 p.

Eastgate Council of Governments, n.d., *Mahoning River Ecological Restoration Project*. Retrieved August 15, 2007, from http://www.eastgatecog.org/env-mahoningriver.asp

Great Lakes Information Network, n.d., *Toxic Contamination in the Great Lakes Region*. Retrieved August 14, 2007, from http://www.great-lakes.net/envt/pollution/toxic.html

Mazor, E., 1990, *Applied chemical and isotopic groundwater hydrology*: Somerset, NJ, John Wiley and Sons, 274 p.

McGauhey, P. H., 1968, *Engineering management of water quality:* New York, McGraw-Hill, 295 p.

OhioEPA, 2007, *Ohio Sport Fish Consumption advisory*. Retrieved August 14, 2007, from http://www.epa.state.oh.us/dsw/fishadvisory/index.html

Ontario Ministry of the Environment, n.d., *The 2007–2008 Guide to Eating Ontario Sport Fish*. Retrieved August 14, 2007, from http://www.ene.gov.on.ca/envision/guide/index.htm

Painter, S., et al., 2001, Sediment contamination in Lake Erie: A 25-year retrospective analysis: *Journal of Great Lakes Research*, v. 27, no. 4, 434–448.

Pettyjohn, W. A., ed., 1972, *Water quality in a stressed environment:* Minneapolis, MN, Burgess, 309 p.

Stoeckel, D. M., and S. A. Covert, 2002, Water quality in the Mahoning River and selected tributaries in Youngstown, Ohio (Prepared in cooperation with the City of Youngstown, Ohio.), USGS WRI 02-4122, 45 p. Retrieved August 16, 2007, from http://oh.water.usgs.gov/reports/wrir/wrir02-4122.pdf Abstract retrieved January 25, 2005, http://oh.water.usgs.gov/reports/Abstracts/wrir02-4122.html

U.S. Army Corps of Engineers, 2006, *Mahoning River, Ohio, Environmental Dredging*. Retrieved August 15, 2007, from http://www.lrp.usace.army.mil/pm/mahonoh.htm

U.S. *EPA Surf Your Watershed*. Retrieved August 15, 2007, from http://www.epa.gov/surf/

U.S. *EPA Watersheds*. Retrieved August 15, 2007, from http://www.epa.gov/owow/watershed/

Youngstown State University, n.d., *Mahoning River Watershed—OVERVIEW: A Brief Description of the Mahoning River and the Restoration Project*. Retrieved August 15, 2007, from http://www.ysu.edu/mahoning_river/river_restoration.htm

Groundwater and Surface Water Contamination from Resource Extraction

INTRODUCTION

Groundwater and surface water may be contaminated by accident or by improper storage or disposal of wastes at the surface. Improper storage or disposal has occurred in many areas due to our ignorance about groundwater flow and potential health effects, the lack of concern for water supplies, and a short-term view of the behavior of groundwater and our future needs for water.

In this exercise we look at cases in which pits and holding ponds were used to dispose of or store liquid wastes. In the past it was expedient to create waste ponds, where the wastes decreased in volume through evaporation or infiltration. In this exercise we explore the cause and extent of contamination from oil field brines and runoff from a lead mine.

PART A. GROUND WATER CONTAMINATION FROM OIL FIELD BRINES IN CENTRAL OHIO

In many oil-producing areas, severe problems of groundwater contamination were common. These were caused primarily by the infiltration of saltwater into the ground. Saltwater, or brine, is produced with the oil and, since the brine often is a by-product of little or no economic value, when unregulated it was commonly disposed of in the most economical manner possible. In most areas this is done by reinjection into the oil-producing zone by means of a well. In others it is accomplished by pumping the brine into holding ponds or pits, where a small percentage evaporates but most of it infiltrates. Infiltration can lead to severe groundwater pollution since the chloride concentrations of the brines may exceed 35,000 mg/L. In contrast many areas have groundwater with background or naturally occurring chloride concentrations of less than 25 mg/L. Sea water is less salty than the brines, with a chlorinity of 19,000 mg/L, which makes up 55 percent of the total salt content of sea water.

Once the oil wells and pits are closed, the chemical quality of the groundwater tends to improve, usually very slowly, as the concentrated solutions migrate to areas of discharge such as springs, streams, or wells. The natural flushing of the groundwater system depends on the hydraulic conductivity of the rocks, the hydraulic gradient, the effective porosity, and the amount and rate of infiltration of rain and snowmelt. It may require decades for the groundwater system to return to its natural chemical state. The rate of flushing and the amount of time that the groundwater reservoir remains contaminated are of profound interest in legal cases.

The brines sterilize the soil, kill vegetation, and create an undesirable taste in drinking water. The concentration at which a brine becomes harmful to vegetation depends on the type of plant, the depth of the root system, the season, and the depth of the water table, to mention only a few factors. Dead trees and other vegetation, however, commonly mark areas where brine-contaminated groundwater discharges into streams or where it flows from springs. The USEPA recommends that drinking water contain no more than 250 mg/L of chloride, since higher concentrations cause a salty taste. Higher concentrations are not likely to cause illness in humans because the water is too salty for consumption.

Most of the problems developed prior to 1980; however, research on the fate of contaminant plumes continues to the present.

Groundwater Contamination near Delaware, Ohio

In this part of the exercise we study the extent, movement, and changes in concentration of oil field brines that contaminated a site on the nearly flat floodplain of the Olentangy River in Ohio. Three oil wells were drilled in this area in June 1964. The brine-to-oil ratio was about 10:1, and nearly 236,000 barrels of salt water were pumped into three ponds from June 1964 to July 1965. Dissolved solids in the brine averaged 60,000 mg/L, and of this about 35,000 mg/L consisted of the chloride ion (Pettyjohn, 1971).

The accompanying figures (Figures 15.1, 15.2, 15.3) show the location of four brine-disposal pits, three oil wells, 25 observation wells, and a water well. The observation wells averaged 25 feet in depth and were installed in late 1965, following cessation of brine disposal, to monitor the movement of the contaminated groundwater. Shale bedrock is over-lain by up to 30 feet of alluvial material consisting of a mixture of sand, silt, and clay. The average hydraulic conductivity (K) of the alluvial material, which contains the contaminated water, is about 25 ft/day, and the average effective porosity (n_e) is 0.15. The water table gradient (I) can be determined from a water-table map.

The objectives of the exercise are to determine the direction and rate of flow of the contaminants in the ground and to evaluate the possible contamination of a nearby water well.

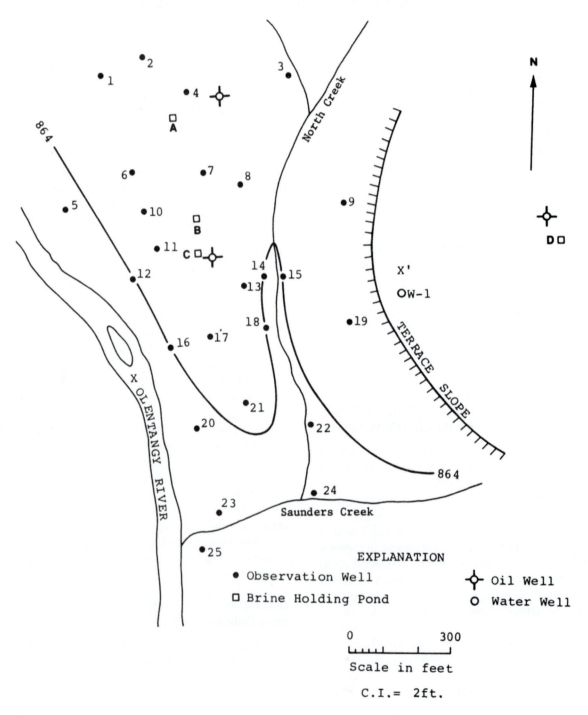

FIGURE 15.1 Map showing configuration of the water table in March 1969.

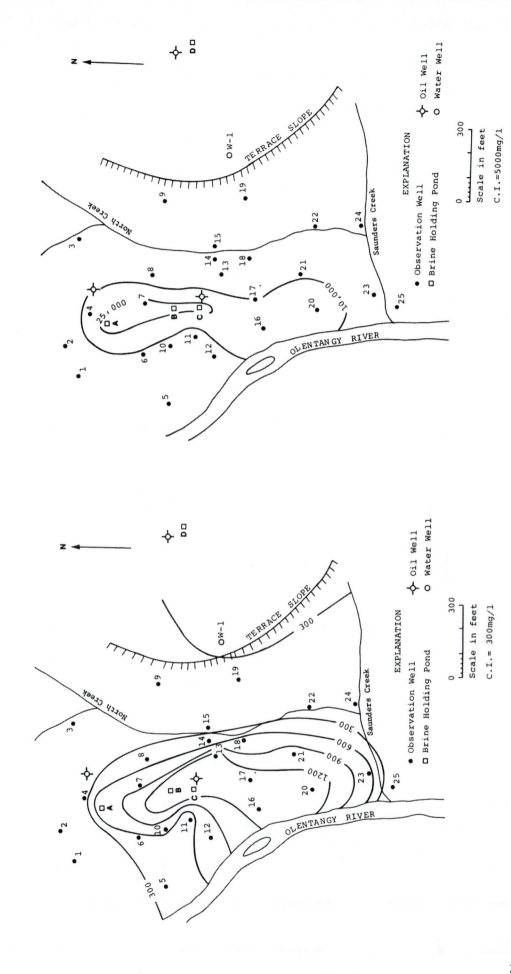

FIGURE 15.3 Map showing groundwater isochlors in March 1969.

FIGURE 15.2 Map showing groundwater isochlors in October 1966.

QUESTIONS 15, PART A

1. Using the data in Table 15.1, construct a water-table map (Figure 15.1). Begin by transferring the water-table elevations from Table 15.1 to the appropriate test hole locations in Figure 15.1. Then contour the water-surface elevations using a contour interval of 2 feet. The contours should roughly parallel the 864-feet contour already drawn.

2. Draw several flow lines originating at the brine holding ponds to the most likely area of groundwater discharge. Remember that during dry weather streams flow only because groundwater discharges into them.

3. What is the gradient from pond C to the Olentangy River? ft/ft.

4. What is the gradient from pond C to Saunders Creek? ft/ft.

5. Calculate the velocity of groundwater moving from pond C to the Olentangy River and from pond C to Saunders Creek using the following formula and the data given earlier in this exercise on the hydraulic characteristics of the unconsolidated material.

$$v = (KI)/(n_e)$$

where v = velocity (ft/day), K = hydraulic conductivity (ft/day) I = gradient, (ft/ft), and n_e = effective porosity (% as a decimal)

a. The velocity of groundwater from pond C to the Olentangy River is about _____ ft/day.

TABLE 15.1 Chloride Content of Wells and Water-Table Elevation in the Delaware Area

Well No.	Water-Table Elevation (March 1969)	Chloride Content (mg/L)		
		Nov. 1965	Oct. 1966	March 1969
1	867	4,500	288	24
2	871	—	875	36
3	866	12	12	12
4	869	—	12,000	200
5	862	—	8,000	400
6	868	18,000	8,875	407
7	868	—	26,250	662
8	865	—	1,000	490
9	870	—	14	16
10	868	25,500	9,850	917
11	868	31,000	7,500	550
12	864	—	8,750	740
13	865	—	6,875	1,355
14	864	—	3,125	292
15	864	—	1,725	302
16	864	22,750	15,500	1,230
17	868	—	10,000	1,300
18	864	5,600	1,500	600
19	869	27	25	300
20	862	—	15,000	1,400
21	865	—	9,000	1,100
22	862	4,800	4,800	117
23	859	5,250	6,625	779
24	861	33	235	27
25	860	95	—	40
W-1	880	—	20	320

b. The velocity of groundwater from pond C to Saunders Creek is about _____ ft/day.

6. If we divide the distance of travel (measured along a flow line) by the rate of flow of groundwater, we obtain the travel time. What are the travel times for water from pond C to
    a. Olentangy River:

    b. Saunders Creek:

7. On another map (Figure 15.2) construct contours representing lines of equal chloride concentrations (isochlors). Use the data for October 1966 (Table 15.1) and a contour interval of 5,000 mg/L. Consider the direction of groundwater flow when drawing these contours. Thus isochlors are given in Figure 15.2.

8. Should the Olentangy River and Saunders and North Creeks contain higher than normal concentrations of chloride in the vicinity of the contaminated area? Why?

9. Wells 23 and 24 (Table 15.1) contain higher concentrations of chloride in October 1966 than in November 1965, while the other wells contained less. Consider your answer to question 6b in your explanation of why this is happened.

10. What do you think the chloride concentration of the groundwater was before brine-pit disposal began (i.e., what was the background concentration)?

11. What techniques might be used to increase the rate of flushing of the high-chloride water in the areas of contaminated soil?

12. A second isochlor map, based on the March 1969 data, is shown in Figure 15.3. A contour interval of 300 mg/L was used. This map is useful in determining the change in contamination with time. Compare Figures 15.2 and 15.3 and describe the changes that have occurred.

13. The shallow farm well (12 ft deep) at W-1 increased in chloride concentration between 1966 and 1969 (Table 15.1). Has this contamination resulted from brine disposal into ponds A, B, C, or D? Explain your answer with the aid of the cross-section sketch (Figure 15.4), which goes from points X to X' in Figure 15.1. Complete the water table and indicate groundwater flow by arrows in Figure 15.4.

PART B. SURFACE WATER CONTAMINATION FROM OIL FIELD BRINES IN CENTRAL OHIO

In this part of the exercise river pollution can be traced to techniques for disposing of saltwater (brine) that is pumped with oil. Regulations in most areas now require subsurface disposal of oil field brines and have significantly reduced groundwater and surface-water pollution.

    Surface waters can be contaminated directly through effluent discharge and surface runoff or indirectly through discharge of contaminated groundwater. This exercise illustrates the effects on a drainage basin of poor waste-disposal practices of oil field

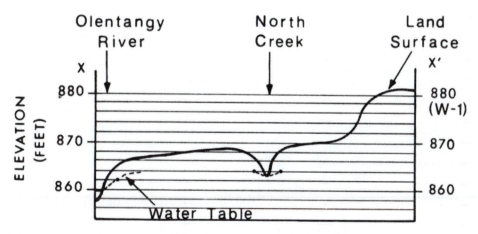

FIGURE 15.4 Cross section from points X to X' in Figure 15.1.

brines; specifically we will look at a case where groundwater contamination has led to the deterioration of surface-water quality. A generalized diagram showing movement of groundwater into a stream is shown in Figure 15.5.

The water table usually lies at a depth of a few feet and follows the general topography of the land surface; that is, the water table is at a higher elevation under hills than it is in nearby low-lying areas. Groundwater moves in the direction of the water-table gradient, from higher pressure to lower pressure, which often may mean from higher elevation to lower elevation. Where the land surface is lower than the water table, such as at a swamp, lake, or stream, groundwater will flow onto the land surface.

Rainwater has a low mineral content, but as it slowly infiltrates soils and bedrock and perhaps flows great distances, its mineral content increases. The types and concentration of constituents in groundwater reflect the composition of the soils and rocks through which the water has moved. The naturally occurring concentrations of elements in groundwater are called *background concentrations*. If a water-soluble contaminant is allowed to infiltrate the ground, it will increase the concentrations of elements and compounds in the groundwater and may contaminate it so it is unusable.

In some parts of the world, an indirect but significant cause of surface-water contamination is disposal of oil-field brines. The brines, which are highly concentrated solutions consisting largely of sodium chloride (NaCl), are pumped from the ground with the oil. The mixture is routed through a separator which removes the oil from the brine. The oil flows to storage tanks while the brine most commonly is discharged to an unlined holding or evaporation pond (see Part A of this exercise) or is pumped back into the bedrock. In some climates where ponds are used, only a very small part of the brine evaporates; most of it infiltrates.

Brines sterilize soil, kill vegetation, and create an undesirable taste in water. The U.S. Environmental Protection Agency has recommended that drinking water should contain no more than 250 mg/L of chloride because higher concentrations cause a salty taste.

Even with careful regulation of the extractive industries, higher background readings of some components can be expected in some areas due to natural weathering and erosion. Before there was any human exploitation of oil in North America, natural brine seeps and springs degraded water quality in some areas.

Alum Creek Basin, Ohio

Low relief and a relatively high water table characterize this watershed in central Ohio. Much of the agricultural land is artificially drained. The region is underlain by a clayey glacial till that in many places contains thin layers of gravel. The till ranges from a few inches to several tens of feet in thickness. Oil-bearing reservoirs underlie much of the region at an average depth of about 3,500 feet.

Sixty-five water samples were collected in late autumn from streams in the upper part of the basin. Because there was no rain or surface runoff for several days prior to collection, and because groundwater discharges to the streams, the stream data represent the quality of the groundwater. The samples were analyzed to determine chloride ion concentrations. Chloride is a common constituent in oil field brines (Table 15.2). Stream sampling sites and locations of existing or abandoned gas or oil wells, including dry holes, are shown in Figure 15.6.

QUESTIONS 15, PART B

1. Using the data in Table 15.2 and Figure 15.6, construct a surface-water quality map. Assume that the chloride content at a station reflects the quality between that site and the next upstream station. Mark in blue the stream reaches that contain 25 mg/L or less of chloride. Use brown for reaches that contain more than 25 but less than 50 mg/L, and red for reaches that exceed 50 mg/L. A map of this type not only presents an obvious picture of the change in water quality from one area to the next, but it also can be used to determine the major source areas of contamination. Use only those data representing sample sites 1–41. Sites 42–65 on the other streams can be examined if you are interested or if the instructor assigns them. You may substitute other colors or patterns for the colors suggested above.

2. Briefly describe the quality of the water in Alum Creek basin using the map you completed in Question 1.

3. What areas are the major sources of chloride contamination in the drainage basin?

4. Why did the chloride concentration decrease between sites 21 and 25?

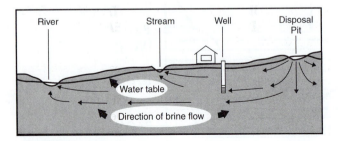

FIGURE 15.5 Generalized diagram illustrating contamination of groundwater and surface water by pit disposal of oil field brines (Pettyjohn, 1972, p. 168).

TABLE 15.2 Sample Site Numbers and Concentration of Chloride Ion, in mg/L, in Alum Creek Basin

Site	CI (mg/L)	Site	CI (mg/L)	Site	CI (mg/L)
1	6	23	23	45	79
2	64	24	24	46	101
3	79	25	84	47	16
4	41	26	37	48	16
5	210	27	33	49	62
6	15	28	39	50	62
7	195	29	13	51	58
8	5	30	8	52	53
9	10	31	41	53	47
10	7	32	629	54	42
11	159	33	77	55	22
12	33	34	103	56	27
13	62	35	21	57	48
14	6	36	113	58	8
15	66	37	38	59	25
16	62	38	11	60	16
17	122	39	23	61	24
18	11	40	21	62	22
19	55	41	109	63	26
20	53	42	22	64	30
21	119	43	11	65	29
22	10	44	116		

5. Would you expect the stream's chloride concentration to be greater or less during the spring? Why?

6. Obviously, much of the groundwater in the basin is contaminated, at least locally. Do you think that all of the groundwater is contaminated? Explain.

7. Outline two areas in Figure 15.6 that should show background concentrations of chloride.

8. What are possible sources of chloride contamination in Alum Creek other than oil field brines?

9. In the upper part of the basin, many of the agricultural fields are underlain by drainage tile. The tiles intercept groundwater and divert it away from the fields. This causes the water table to remain at a lower elevation in fields with tile than in fields without tile. Ultimately, drainage from these tiles flows into a stream. How could you use data from water-quality samples taken from field tiles to aid in determining the areas of groundwater contamination by oil field drilling and extraction activities?

PART C. SURFACE WATER CONTAMINATION FROM LEAD MINES IN SOUTHEASTERN MISSOURI

In this part we look at the impact of heavy metals from a lead-mining district on water and aquatic organisms. Similar impacts can be documented in other areas of North America; however, new mine operation and closing practices have lessened the environmental impact of mining.

In 1955 new deposits of copper, lead, silver, and zinc were discovered in southeastern Missouri. This

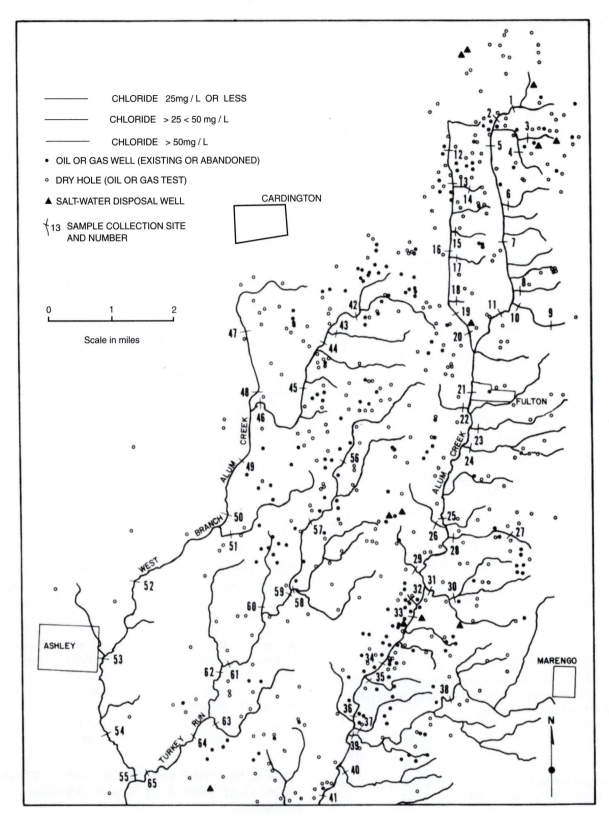

FIGURE 15.6 Chloride ion concentration in Alum Creek basin in the autumn.

region, known as the New Lead Belt, became one of the largest lead-producing areas in the world. It continues to be a primary producer of lead with secondary production of zinc, and small quantities of copper and silver (Brumbaugh et al., 2007). In order to evaluate the impact of mining on this sparsely populated,

heavily forested, and hilly region, a number of surface-water and aquatic life forms were examined. Samples were collected from streams free of mining and milling operations (control samples) and from contaminated streams.

Changes in water quality and aquatic life in these streams since 1955 are related to the discharge of milling and mining wastes which include excess water pumped from mines, finely crushed rock, chemical reagents, and waste oils and fuel. These wastes are allowed to settle in holding ponds, and the effluent is either reused or allowed to discharge into streams. The greatest share of the noxious substances is retained in the settling basins.

Most of the heavy metals in the streams are in very fine particles; a minute amount is dissolved but most heavy metals travel on suspended sediment in the water. Where large excesses of groundwater have been pumped from mines and discharged into streams, significant algal blooms have occurred. They may be a result of the nutrients in the groundwater. The dense algal communities act as filters and remove many of the fine particles that escape the settling basins.

In this exercise we begin by investigating metal contamination of water and aquatic organisms on the West Fork Black River and tributaries Strother Creek, Neal Creek, and Bee Fork Creek (Figure 15.7) using data published in 1973 (Gale et al., 1973). We conclude with an assessment of conditions in this same area based on sampling of water and sediments between 2002 and 2005 (Brumbaugh et al., 2007).

QUESTIONS 15, PART C

1. Examine Figure 15.7 and Table 15.3. Sample sites 6, 9, 10, and 14 are in uncontaminated areas. What are the background concentrations of the following elements in surface water?

Lead:

Zinc:

Copper:

Manganese:

2. In Figure 15.7 mark in red (or use a pattern [e.g., dots] that you identify in the explanation) the stream reaches that exceed background concentrations of lead. Consider the most likely source of the lead as a guide in marking the stream reaches.

3. In Figure 15.7 mark in green (or with a dash pattern) the stream reaches that exceed 0.011 ppm of zinc.

4. What relationship appears to exist between settling basin location and the quality of water in the stream?

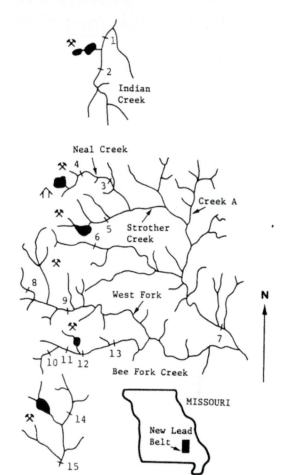

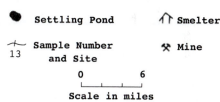

FIGURE 15.7 Trace element sampling sites and lead and zinc concentrations in water in the New Lead Belt, southeastern Missouri (Gale et al., 1973).

5. On the basis of available data, would you expect background concentrations of the contamination in Creek A? Explain.

6. Examine the Mn concentrations in aquatic organisms in Strother Creek as shown in Table 15.4.
   a. What general relationship is evident?

TABLE 15.3 Mean Concentrations of Lead, Zinc, Copper, and Manganese in Water in the New Lead Belt

Site	Flow (cfs)	Mean Concentrations			
		Pb (ppm)	Zn (ppm)	Cu (ppm)	Mn (ppm)
1	11.22	.021	.034	.010	.077
2	11.22	.011	.017	.010	.049
3	–	.010	.011	.011	.012
4	5.80	.014	.069	.010	.035
5	8.70	.090	.034	.011	.127
6	4.10	.011	.010	.010	.011
7	382.6	.014	.011	.010	.021
8	33.8	.044	.011	.010	.013
9	33.8	.011	.010	.010	.013
10	9.22	.011	.010	.010	.013
11	—	.035	.140	.017	1.637
12	—	.033	.134	.012	1.691
13	9.22	.030	.038	.010	.488
14	6.75	.012	.010	.010	.025
15	6.75	.019	.020	.018	.156

TABLE 15.4 Mean Concentrations of Lead, Zinc, Copper, and Manganese in Aquatic Organisms of Strother Creek

Organism	Miles Below Pond	Mean Concentrations (mg/g)			
		Pb	Zn	Cu	Mn
Snails	0.2	16	54	57	1,770
	1.3	75	27	21	710
	2.8	44	18	16	59
Crayfish	0.2	69	97	142	195
	1.3	24	92	130	750
	2.8	38	86	86	410
	4.2	28	94	97	485
Tadpoles	0.2	36	210	26	5,650
	1.3	780	265	44	4,560
	2.8	310	210	26	690
	4.2	1,590	160	17	500

b. Similar but less obvious trends exist for Pb, Zn, and Cu; however, Pb in tadpoles exhibits the opposite trend. What might account for this different trend?

7. Examine Figure 15.8.
  a. What is the background concentration for lead in aquatic vegetation in the Bee Fork drainage basin?

b. How many times greater is the concentration of lead at the mine than 6.5 miles downstream from the mine?

c. How many times greater is the value for lead at 6.5 miles downstream than the apparent background level?

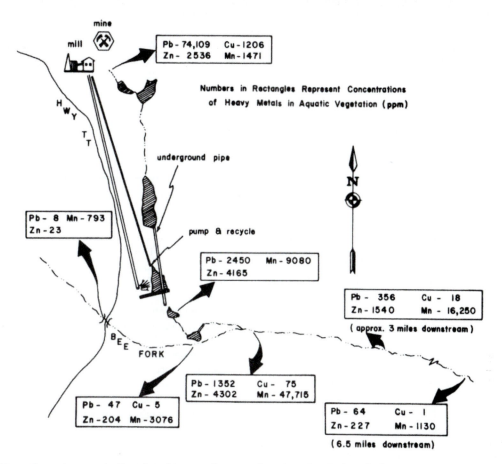

FIGURE 15.8   Trace element concentrations (ppm) in aquatic vegetation along Bee Fork and tributary, New Lead Belt (Gale et al., 1973).

8. The pollution load that a stream carries can be calculated if the stream discharge and the concentration of the specific contaminants are known.

$$\text{Load (tons/day)} = Q \times C \times 0.0027$$

where Q = stream discharge (cfs), C = concentration of specific contaminant (mg/L), and 0.0027 = constant to convert seconds and mg/L to days and tons

Using the available data and the formula given above, calculate how many pounds per day of lead and manganese were being transported past site 7 (Figure 15.7; refer to Appendix A for conversion of tons to pounds).

9. What techniques might be used to reduce the contamination of these streams in the New Lead Belt?

10. Using the above information from the USGS studies that began in 2000, briefly describe, in a bullet statement for each, the key findings or conclusions that you might make from this study of part of the New Lead Belt.

Concern about potential degradation of water quality and aquatic biota of nearby federally protected streams prompted a multidisciplinary study of the area that began in 2000. The results of the sediment and surface water analyses on samples collected between 2002 and 2005 have been published by the USGS (Brumbaugh et al., 2007). That publication is the source of much of the following environmental information.

The greatest concentrations in sediment collected in 2002 were from sites downstream from mines on Strother Creek and West Fork, with noticeable enrichment in lead in sediments from Bee Fork. Compared to reference sites, sediments downstream from mine areas were enriched "by factors as large as 75 for cadmium, 62 for cobalt, 171 for nickel, 95 for lead, and 150 for zinc."

The impact of mining was recorded at least 75 kilometers downstream in Clearwater Lake where metal concentrations were 1.5–2.1 times greater than in sediments in an area of the lake with no upstream mining. Sediment samples collected in 2004 on West Fork showed "dramatically lower" concentrations of metals, which was attributed to the closing of a mill on West Fork.

Concentrations of metals in surface water generally tracked those in sediments. Water samples from July 2005 on Strother Creek showed a "considerable increase in metal loadings" for a few days in which there was a moderate increase in stream discharge.

11. Using the online resources (or the actual publications) from the USGS study on the New Lead Belt that began in 2000, prepare two or three lab questions based on those reports that would be suitable for use by your fellow students. Include the full reference, any diagrams or tables, and the questions. Also include the answers that you expect for each of the questions. You can focus on either the geological or the biological data or use both. (The instructor might make this a take-home assignment, depending on available resources and time, and possibly a group assignment. Select from suggested references below or others that you find that might be useful. Full references are in the Bibliography section of this manual.)

Besser et al. 2006 http://www.springerlink.com/content/r4t8q457354847t0/; Brumbaugh et al., 2007 http://pubs.usgs.gov/sir/2007/5057/, Schmitt et al., 2007a, 2007b.

Bibliography

Besser, J.M., Brumbaugh, W.G., May, T.W., and Schmitt, C.J., 2006, Biomonitoring of lead, zinc, and cadmium by streams draining lead-mining and non-mining areas, southeast Missouri, USA: *Environmental Monitoring and Assessment*. Retrieved October 2006 from http://www.springerlink.com/content/r4t8q457354847t0/

Boster, R. S., 1967, A *study of ground-water contamination due to oil-field brines in Morrow and Delaware Counties, Ohio, with emphasis on detection utilizing electrical resistivity techniques*, M.Sc. thesis: Columbus, OH, The Ohio State University, 191 p.

Brumbaugh, W.G., May, T.W., Besser, J.M., Allert, A.L., Schmitt, C.J., 2007, *Assessment of elemental concentrations in streams of the New Lead Belt in Southeastern Missouri, 2002–05*: U.S. Geological Survey Scientific Investigations Report 2007–5057, 57 p.

Feth, J. H., 1973, *Water facts and figures for planners and managers*: U.S. Geological Survey Circular 601–1, 30 p.

Gale, N. L., Wixson, B. G., Hardie, M. G., and Jennett, J. C., 1973, Aquatic organisms and heavy metals in Missouri's New Lead Belt: *Water Resources Bulletin*, v. 9, no. 4, p. 673–688.

Pettyjohn, W. A., 1971, Water pollution by oil-field brines and related industrial wastes in Ohio: *Ohio Journal of Science*, v. 71, no. 5, p. 257–269.

Pettyjohn, W. A., 1972, *Water Quality in a Stressed Environment*: Minneapolis, MN, Burgess, 309 p.

Pettyjohn, W. A., 1973, *Sources of chloride contamination in Alum Creek, Central Ohio*: Columbus, OH, Ohio Department of Natural Resources, 54 p.

Reiten, Jon C., 2006, Oil-field brine plumes in shallow ground water, Sheridan County, Montana: Sixteen years later: AAPG Rocky Mountain Section Annual Meeting, June 11–13, 2006, Billings, Montana.

Sassen, Douglas S., 2004, Oil-field-brine-induced colloidal dispersion: A case study from southeast Texas: Geological Society of America *Abstracts with Programs*, vol. 36, no. 1, p. 24.

Schmitt, C.J., Brumbaugh, W.G., and May. T.W., 2007a, Accumulation of metals in fish from lead-zinc mining areas of southeastern Missouri, USA: *Ecotoxicology and Environmental Safety*, v. 67, p. 14–30.

Schmitt, C.J., Whyte, J.J., Roberts, A.P., Annis, M.L., and Tillitt, D.E., 2007b, Biomarkers of metals exposure in fish from lead-zinc mining in southeastern Missouri, USA: *Ecotoxicology and Environmental Safety*, v. 67, p. 31–47.

Shaw, J. E., 1966, *An investigation of ground-water contamination by oil-field brine disposal in Morrow and Delaware Counties, Ohio*, M.Sc. thesis: Columbus, OH, The Ohio State University, 127 p.

Groundwater Overdraft and Saltwater Intrusion

INTRODUCTION

Groundwater accounts for about 25 percent of the total fresh water used in the United States, and untapped resources have great potential to meet future needs. Even though vast amounts of groundwater are available, in some areas pumping rates are such that water levels have declined hundreds of feet, wells have gone dry, the cost of pumping the water has increased substantially, and water of poor quality has been induced to flow into aquifers. Areas of large irrigation systems, such as that part of the Great Plains states underlain by the High Plains or Ogallala aquifer, exemplify the problems of groundwater overuse. Irrigation is the largest user of groundwater, although areas of dense population and heavy industry also consume large quantities.

Techniques have been devised to halt or reduce the water-level decline in some water-short areas. The most obvious method is to conserve water and reduce pumping. In other instances it may be possible to divert surface water, including treated wastewater, to infiltration basins, pits, or wells, which will allow the water to percolate into the ground at a rate that is considerably greater than that permitted by natural conditions. These techniques are collectively known as artificial recharge and have been used successfully throughout the world.

In this exercise we examine an agricultural area along the Mississippi River that has seen the groundwater level decline for over 100 years and a southeastern urban coastal region with an even longer period of declining water levels. In the rice-growing Grand Prairie region (east of Little Rock, Arkansas), the Mississippi River Alluvial Aquifer is predicted to soon become useless and a deeper aquifer there is declining as well. In the Savannah, Georgia, area, declining water levels in the coastal Floridan Aquifer are impacting the water supplies of that city and nearby communities including those in South Carolina.

The general objectives of this exercise are to explore changes in water levels and flow directions in overpumped aquifers, the environmental and economic impacts of such changes, and responses to groundwater overdraft in these two settings.

PART A. OVERUSE OF A GROUNDWATER RESOURCE: GRAND PRAIRIE REGION, ARKANSAS

The Grand Prairie region is in east-central Arkansas. This region is characterized by low relief, which, in conjunction with an extensive aquifer and warm climate, provides an ideal setting for rice irrigation. In 1904 the Mississippi River Alluvial Aquifer became the irrigation source for rice farming in the district and since 1915 the water table has declined about 1 ft/yr in some areas. The configuration of the water level in the Alluvial Aquifer in 1915 is shown in Figure 16.1. Concentrated pumping of irrigation water from this sandy aquifer has caused a substantial overdraft in the groundwater supply and a decline in water levels of several tens of feet. The objective of Part A of this exercise is to examine water-level and flowline changes due to overpumping and the responses of the Grand Prairie region.

Groundwater plays a key role in rice production here. As the water level declines, the cost of pumping water increases. In the long run, pumping costs and other farm operating costs could be greater than the value of the crop. The economic impact of a declining water level is obvious.

In order to evaluate the rate and areal extent of water-level decline and to determine remedial measures, maps of conditions in 1915 and 1954 were prepared and evaluated.

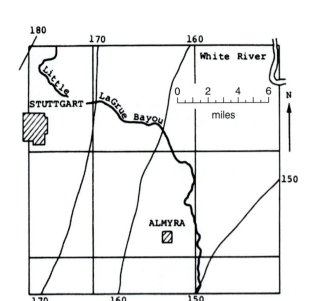

FIGURE 16.1 Altitude (in feet) of water level in Grand Prairie region in 1915.

(Modified from Sniegocki, 1964)

Note: Subsequent studies of the Alluvial Aquifer predict that it will become completely useless by 2015.

QUESTIONS 16, PART A

1. a. On Figure 16.2 construct a map that shows the configuration of the water-level surface in March 1954, using a contour interval of 10 feet. Compare this map with Figure 16.1. What are the major differences?

b. On Figure 16.3 construct a map that shows the net decline of water levels from 1915 to 1954. Use a contour interval of 10 feet. What area has had the greatest decrease in water levels?

c. Starting at the east edge of Stuttgart, draw a flowline across the 1915 map (Figure 16.1). Draw a similar flowline passing through Almyra. What was the general direction of groundwater movement in 1915 at

Almyra?

Stuttgart?

d. On Figure 16.2 draw the flowlines passing through Stuttgart and Almyra. What was the general direction of groundwater movement in 1954 near

Almyra?

Stuttgart?

2. Calculate the gradients that existed in 1915 and 1954 using the flowlines that pass through Almyra.

a. Gradient in 1915:

b. Gradient in 1954:

3. The hydraulic conductivity of the water-bearing deposits averages 260 ft/day, and the effective porosity averages 17 percent. What was the groundwater velocity in the vicinity of Almyra (refer to Exercise 12 for formula details; $v = KI/n_e$)

a. in 1915

b. in 1954

4. Assume that the saturated sand in the northeastern part of Figure 16.2 (along line A–A') is 40 feet thick. How much groundwater, in ft³/day, flowed across A–A' during a single day in March 1954? ($Q = KAI$)

5. Figure 16.3 indicates that there has been a significant lowering of water level. This means that more water is pumped from the aquifer than is flowing into it. This negative change in groundwater storage is termed overdraft. What could be done to decrease the rate of decline, maintain the existing level, or cause the water level to rise?

6. "Ever think you'd run out of water where you live? Neither did the people in the Grand Prairie area of eastern Arkansas. That's why this web site has been created .... so you can understand what the Grand Prairie Area Demonstration Project (GPADP) is all about." At the Army Corps of Engineers website you can see what the area's response was to a study that predicted the loss of the upper Alluvial Aquifer and the decline of the deeper Sparta Aquifer. It was estimated that without the project, rice production would drop to 23 percent of current value, partially replaced by less productive dryland farming, and waterfowl recreation income would decrease. The Sparta Aquifer is the source of high-quality drinking water and, because it was being used

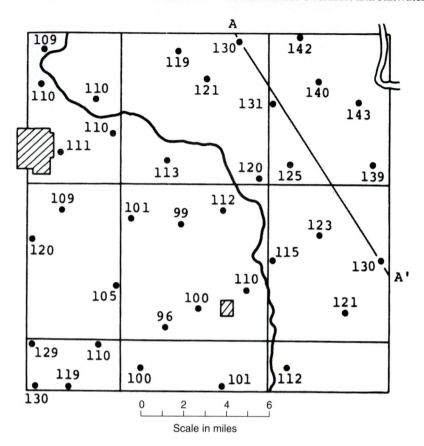

FIGURE 16.2 Altitude (in feet) of water level in Grand Prairie region in March 1954.

(Data from Sniegocki, 1964)

for irrigation too, it also was declining at > 1 foot per year. Farmers did not expect much increase from the Sparta Aquifer because of a low specific yield (~ effective porosity) of only 0.01 (cf., Alluvial Aquifer 0.30) so the capacity was not there. Also the recharge was slow and the cost of pumping was higher and increasing.

Check out the following websites (and others) to determine the expected economic and environmental impacts from the loss of irrigation water in this area of Arkansas known as the rice and duck capital of the world (Stuttgart is the headquarters of Riceland Foods). The brochure (website below) for the project will be particularly useful in answering this question. The site for the Grand Prairie Area Demonstration Project (also known as the Grand Prairie Irrigation Project) is http://www.mvm.usace.army.mil/grandprairie/maps/pdf/GPADProj.pdf. See also USGS Fact Sheet 111-02 at http://pubs.usgs.gov/fs/fs-111-02/, and the news item of July 21, 2006, by L. Satter in the *Arkansas Democrat Gazette* (http://www.nwanews.com/adg/News/161090/) for more information on the project and its status.

a. List two or three types of structures that are to be built in the Grand Prairie Area Irrigation Project.

b. What is the purpose of the project?

c. From where will much of the water come for this project?

d. List several economic and environmental benefits of the project.

e. List several objections that environmental groups have had to the project.

f. Although there have been many starts and stops to the project over decades, what did Judge Bill Wilson do on July 20, 2006?

What was the name of the bird that was a factor in the decision?

g. Why would area farmers not plan to continue expanding their use of the lower aquifer (Sparta Aquifer) but instead would like to use water from the GP Irrigation Project?

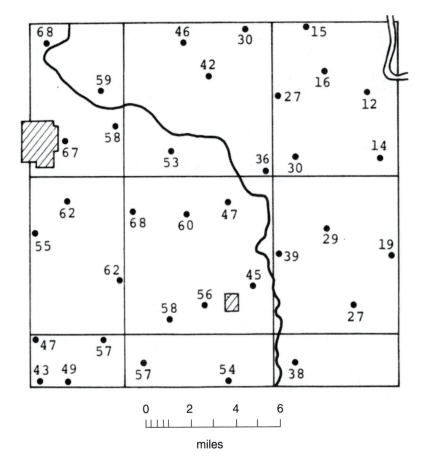

FIGURE 16.3 Net decline of water levels in the Grand Prairie region from 1915 to 1954.

(Data from Sniegocki, 1964)

h. Search online to determine the current status of the irrigation project and any economic and environmental changes that have occurred since 2006 in this area that has been aware of groundwater overdraft for more than 100 years.

PART B. OVERUSE OF A GROUNDWATER RESOURCE: SALTWATER INTRUSION IN A COASTAL AQUIFER

The intrusion of saltwater into a fresh groundwater reservoir is surprisingly common throughout the world. Saltwater intrusion is generally caused by excessive pumping of groundwater and leads to deterioration of water quality. In inland areas it is caused by the upward movement of the fresh- and saltwater interface; in coastal regions it is commonly caused by both vertical and horizontal migration of sea water into a coastal aquifer. For example, during the Dec 26, 2004, tsunami in the Indian Ocean, coastal areas were flooded by seawater. Some wells were filled with saltwater; some freshwater aquifers were contaminated by saltwater infiltrating the land surface.

Although geologic and hydrologic conditions may be exceedingly complex, the mechanics of saltwater intrusion may be visualized in the following manner. Assume that an aquifer crops out on the continental shelf and is hydrologically connected to the ocean. Normally, fresh water discharges from a coastal aquifer into the ocean along seepage faces or through springs (Figure 16.4A). As fresh water is pumped from the aquifer, the water pressure in the aquifer is lowered, reversing the hydraulic gradient (Figure 16.4B). With a reversal in gradient, sea water migrates inland and may eventually contaminate pumping wells.

Occurrences of saltwater intrusion are relatively common in coastal areas. Problem areas include much of the Atlantic coast from Florida to New England and many regions of the west coast where there are large withdrawals of groundwater.

Several techniques have been used to control saltwater intrusion. These include reducing the amount of pumping; constructing physical barriers, such as pumping cement into the rocks (Figure 16.4C); pumping wells nearer to the coast and allowing the water to flow back into the ocean (Figure 16.4D); and artificial recharge (Figure 16.4E). The simplest solution is to reduce pumping, but commonly this is not feasible because of existing water demands. The most promising approaches are artificial recharge and water conservation. In the artificial recharge method, water is injected into the ground through pits and wells. This forms a hydraulic barrier (injection ridge) due to the higher water or water-pressure surface in the vicinity of the recharge sites, which lie between the well field and the coast (Figure 16.4E). The hydraulic barrier tends to reverse the water gradient and forces the saline water out of the aquifer.

In many coastal areas, saltwater intrusion has not yet occurred, but an examination of existing groundwater levels and pumping data indicates that there is a strong potential for future intrusion. If potential saltwater intrusion sites are analyzed before contamination actually occurs, it may be possible to develop adequate solutions before the supply situation becomes critical.

The objectives of Part B of this exercise are to examine water-level declines in a coastal region, areas

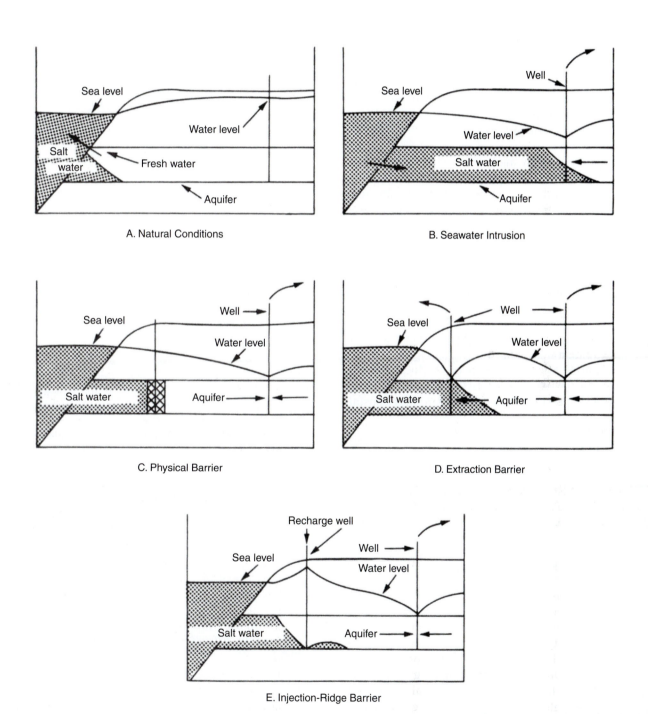

FIGURE 16.4 Saltwater intrusion of a confined coastal aquifer and the use of barriers to prevent contamination of water supplies. Water level is the water-pressure surface (See text for explanation of A through E.)

of potential saltwater intrusion, methods that could be used to halt the intrusion, and community plans for future water needs.

Potential Saltwater Intrusion in the Savannah Area

Large quantities of groundwater are used in the Savannah, Georgia, area for industrial, municipal, and domestic purposes. The first water well in the Upper Floridan Aquifer was drilled in 1885; between 1900 and 2000, the pumping rate increased from 10 million gal/d to 300 million gal/d. Over the years, water levels in wells have declined to more than 120 feet below sea level. This has caused concern that the water supply might become seriously depleted or contaminated. Most of the groundwater used in the Savannah area is pumped from a confined, limestone aquifer that lies about 100 feet (northeast) to 350 feet (southwest) below land surface.

Although groundwater in the Savannah area has not yet become salty, the supply at Parris Island, about 25 miles northeast, has deteriorated due to saltwater intrusion. The pumping of groundwater in the Savannah area will no doubt increase and, as a result, intruding saltwater may eventually reach the pumping center and contaminate the water supply.

The Savannah River, which flows through the area, has been used as a partial source of water, but locally it is contaminated by industrial and municipal wastes. Furthermore, water can only be withdrawn from it at certain times because the river is influenced by tidal waters of high salinity.

In addition to these impacts on the local ground water supplies, rising sea levels will further increase the salinity and decrease storage space in some coastal aquifers.

QUESTIONS (16, PART B)

1. A water-level map of the Savannah area representing conditions that existed in 1880 is shown in Figure 16.5. Construct four equally spaced flowlines showing the direction of groundwater movement in 1880. Remember that flowlines cross the water-pressure contours at right angles. What was the general direction of flow? Was groundwater at Parris Island likely to have been salty in 1880? Why or why not?

2. Using Figure 16.6, construct a water-level map showing the conditions that existed in 1961. Use a contour interval of 10 feet down to the −40 feet level, then use an interval of 20 feet.

3. Starting at the southwest and northwest corners of Figure 16.6, and at Parris Island, construct flowlines showing the general direction of groundwater movement in 1961. In what general direction was the water moving

a. in the southwest corner?

b. in the northwest corner?

c. at Parris Island?

4. Study Figure 16.7 and describe the changes from 1961 to 1984 at Hilton Head Island, Savannah, Georgia, and 25 miles up the Savannah River from the center of the cone of depression.

5. How much has the water-pressure surface been lowered at Savannah between 1880 and 1984?

6. From what area do you expect the fresh water/saltwater interface to first reach the Savannah area? SW or NE (circle one) Why? (Hint: Examine the water-level contours, and the variation in depth of the aquifer described in the introduction.)

7. Figure 16.8 is a cross section extending into Port Royal Sound from the NE end of Hilton Head Island showing simulated changes in the brackish and salt water zones for the years 2000, 2016, and 2032. The model assumes no change in the rates of groundwater withdrawals on Hilton Head Island or inland near Savannah.

a. What value is used as the transition between freshwater and brackish water?

b. In the period between 2000 and 2032, how many meters will the brackish/freshwater interface have moved?

c. What is the average annual rate of projected advance of this interface between 2000 and 2032?

d. About when will the interface reach the edge of the island?

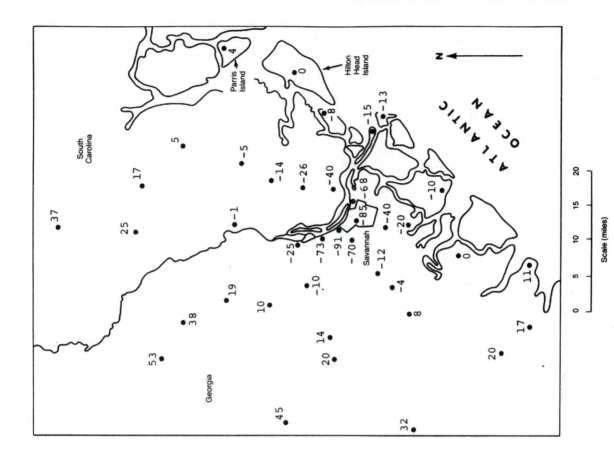

FIGURE 16.6 Altitude (in feet) of water level in the Savannah area in 1961. (Data from Wait and Callahan, 1965)

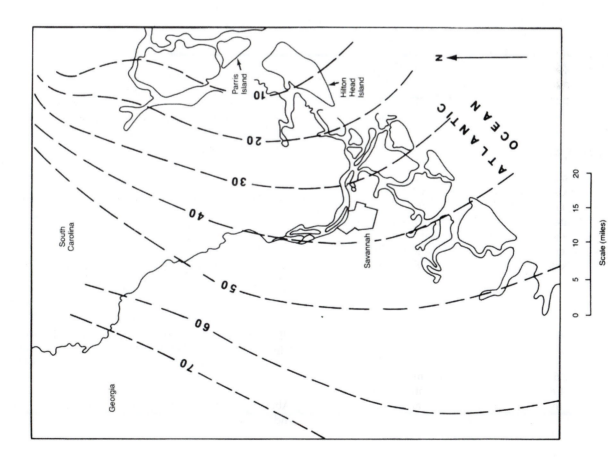

FIGURE 16.5 Altitude (in feet) of water level in the Savannah area in 1880. (Modified from Wait and Callahan, 1965)

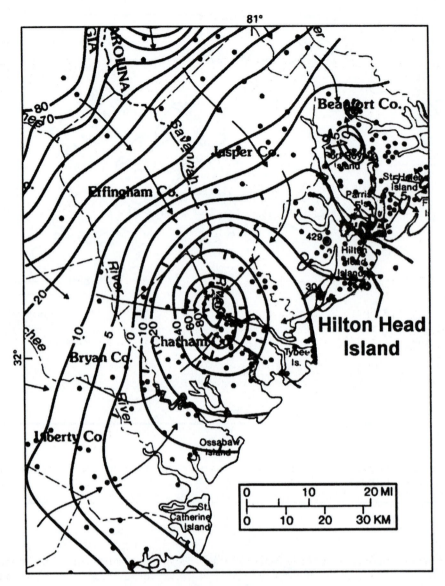

FIGURE 16.7 Altitude (in feet) of water level in the Savannah area in 1984. Dots are wells; arrows indicate general direction of groundwater flow. Parris Island is north across Port Royal Sound from Hilton Head Island.

(Modified from Smith, 1988)

8. a. What techniques might be used to halt or slow saltwater intrusion into the area of Hilton Head Island? (Consider engineering and management techniques [Figure 16.4] to stop the advance of the brackish water in the aquifer beneath the sound and the island).

b. Explain Figure 16.4E or one other provided by your instructor.

Read the following information and then answer **Questions 9 and 10** below. Groundwater overdraft in the Floridan Aquifer System of the southeastern coastal region has produced cones of depression in SE Georgia and neighboring South Carolina. The decline in artesian pressure in this porous limestone/dolostone aquifer has resulted in saltwater intrusion in the coastal areas. Some of the intrusion is lateral from the ocean or downward from eroded river channels (as near Hilton Head, SC) or upward through fractured limestone units in the Lower Floridan Aquifer as at Brunswick, Georgia. Georgia recognized that the rate of decline in the water pressure surface (water table) and the increasing salinity posed a threat to sustainable water resources in the region. A 1997 report described the conditions and trends in the Floridan Aquifer, a primary water supply source for many in the region. An *interim strategy* (1997–2005) was developed to reduce further overdraft of the aquifer, pending the Sound Science Report (released in 2005).

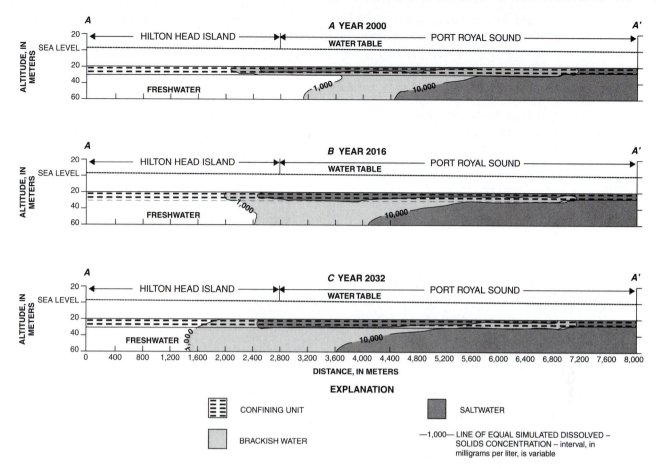

FIGURE 16.8 Simulated dissolved-solids concentrations in the aquifer beneath Hilton Head Island and Port Royal Sound. Modeling shows changes in the confining bed and the aquifer for the years 2000, 2016, and 2032, assuming no changes in withdrawal rates for Hilton Head and Savannah area. The clayey-silt confining unit contains slowly moving fresh, brackish, and salt waters.

(From Smith, 1994)

The interim regulations (until release of the 2005 report) included the following: (1) reduced the quantity of water available for withdrawal under current permits (essentially capping future withdrawal at 1997 levels), (2) allowed no new permits in worst areas, instead suggested surface sources, (3) made 36 million gallons per day additional extraction available outside the worst areas, (4) required 24 coastal communities to prepare plans for future water needs [this focused the need for land-use planning and consideration of possible population growth], and (5) required additional monitoring wells and modeling of groundwater flow [Florida and South Carolina also increased monitoring of their *Upper* Floridan Aquifer]. Additional use of the *Lower* Floridan Aquifer was discouraged until connections between the two aquifers was understood.

In addition to what you learned about the scope of the declining water pressure surface and possible engineered barriers to saltwater intrusion from other questions in this exercise, use the interim regulations and description above to help answer **Question 9.**

9. As a community leader of one of the 24 small communities in the region, you have been asked to join a team to develop your community's plan for future water needs. Before the first meeting of this new committee, all the members of the committee are asked to prepare a list of short statements or items that should be part of the discussion for planning future water needs. You organize and complete your response under the following headings:

a. List of data needed to make the plans for future water needs.

b. List of possible objectives for a sustainable community (for next 50 years) and your preferred objective.

c. List of possible options that could be considered for meeting your water needs for the community.

d. List of factors that might make an unsustainable community with respect to water resources (even with a plan for sustainable water resources).

10. Now use information from the Sound Science Report (2005) that is available from several sources online and indicated below to prepare a water supply plan for your community. (Your instructor may assign this as an in-class or homework activity. Check the reference section of this manual for additional references. You may choose to submit your responses on a separate sheet of paper.)

The key questions to be answered now are:

a. Will engineered barriers to prevent saltwater intrusion be suitable for your community? (See http://www.gadnr.org/cws/Documents/What_Have_We_Learned.pdf)

b. A plume of salt water in the vicinity of Hilton Head Island has advanced 6 miles since 1965. By 2005, what was the average rate of advance in feet per year over that period?

c. From the information available in the 2005 Sound Science Initiative report, how long would it be before there is a problem with the aquifer in the Savannah area of Georgia?

d. What are the alternatives for meeting water needs according to what you have found in your searches? (http://www.gadnr.org/cws/Documents/Determining_Future_Water_Demand_Needs_Water_Systems.pdf)

e. What are the recommendations that you (or your group) would make to provide (possibly through several sources) sustainable water supplies for your coastal community. Consider changing populations and climates in your recommendations. (Optional additional information: http://www.gadnr.org/cws/Documents/saltwater_management_plan_june2006.pdf)

Bibliography

Bruington, A. E., 1972, Saltwater intrusion into aquifers: *Water Resources Bulletin,* American Water Resources Assn., v. 8, no. 1, p. 150–160.

Carr, J. E., et al. (comp.), 1990, *National water summary 1987–hydrologic events and water supply and use:* U.S. Geological Survey Water-Supply Paper 2350, 553 p.

Coastal Georgia Water and Wastewater Permitting Plan for Managing Salt Water Intrusion by Sue Grunwald, EPD Planning & Policy Advisor (PowerPoint presentation to the 2007 Georgia Water Resources Conference), http://www.gadnr.org/cws/Documents/Coastal_Permitting_2007.pdf

Counts, H. B., and Donsky, E., 1963, *Saltwater encroachment, geology and ground-water resources of Savannah South Carolina:* U.S. Geological Survey Water-Supply Paper 1611, 100 p.

Foxworthy, G. L., 1978, Nassau County, Long Island, New York—water problems in humid country. In *Nature to be Commanded* ...: eds. G. D. Robinson and A. M. Spieker, U. S. Geological Survey Professional Paper 950, p. 55–68.

Garza, R., and Krause, R.E., 1996, *Water-supply potential of major streams and the Upper Floridan aquifer in the vicinity of Savannah, Georgia:* U.S. Geological Survey Water-Supply Paper 2411, 36 p.

Georgia Department of Natural Resources, 2005, What have we learned from the Sound Science Initiative, http://www.gadnr.org/cws/Documents/What_Have_We_Learned.pdf

Georgia Department of Natural Resources, 2006, Coastal Georgia water and wastewater permitting plan for managing salt water intrusion, http://www.gadnr.org/cws/Documents/saltwater_management_plan_june2006.pdf

Georgia Department of Natural Resources, 2007, Coastal Permitting Plan Guidance Document: Method for determining future water demand needs for public/private water systems, http://www.gadnr.org/cws/Documents/Determining_Future_Water_Demand_Needs_Water_Systems.pdf

Georgia Department of Natural Resources, Environmental Protection Division home page, http://www.gaepd.org

Georgia Department of Natural Resources, n.d., *Coastal Sound Science Initiative—A scientific study of groundwater use in coastal Georgia:* Retrieved August 19, 2007, from http://www.gadnr.org/cws/

Grand Prairie Irrigation Project by Natural Resources Conservation Service, http://www.ar.nrcs.usda.gov/programs/grand_prairie.html

Hays, P. D., and Fugitt, D. T., 1999, *The Sparta Aquifer in Arkansas' critical ground-water areas—Response of the aquifer to supplying future water needs:* U.S. Geological Survey Water-Resources Investigations Report 99–4075, 6 p.

Hosman, J., and Weiss, J. S., 1991, *Geohydrologic units of the Mississippi embayment and Texas coastal uplands aquifer systems, south-central United States:* U. S. Geological Survey Professional Paper 1416-B, 19 p.

Krause, R. E., and Clarke, J. S., 2001, *Saltwater contamination of ground water at Brunswick, Georgia and Hilton Head Island, South Carolina:* Retrieved August 19, 2007, from http://ga.water.usgs.gov/publications/gwrc2001krause.html.

Krause, R. E., and Randolph, R. B., 1989, *Hydrology of the Floridan aquifer system in southeast Georgia and adjacent parts of Florida and South Carolina:* U.S. Geological Survey Professional Paper 1403-D, 65 p.

Landmeyer, J. E., and Belval, D. L., 1996, Water-chemistry and chloride fluctuations in the Upper Floridan aquifer in the Port Royal Sound area, South Carolina, 1917–93: U. S. Geological Survey Water-Resources Investigations Report 96–4102, 106 p.

McGuiness, C. L., 1963, *The role of ground water in the national water situation:* U.S. Geological Survey Water-Supply Paper 1800, 1121 p.

McKee, P. W., and Hays, O. D., 2004, *The Sparta Aquifer: A sustainable water resource:* U.S. Geological Survey, Fact Sheet 111-02, http://water.usgs.gov/pubs/fs/fs-111-02/

Peck, M. F., Clarke, J. S., Ransom III, C., and Richards, C. J., 1999, Potentiometric surface of the Upper Floridan aquifer in Georgia and adjacent parts of Alabama, Florida, and South Carolina, May 1998, and water-level trends in Georgia, 1990–98: *Georgia Geologic Survey Hydrologic Atlas* 22, 1 plate.

Reed, T. B., 2004, *Status of water levels and selected water-quality conditions in the Mississippi River Valley Alluvial Aquifer in eastern Arkansas, 2002:* USGS Scientific Investigations Report, 2004–5129, http://water.usgs.gov/pubs/sir/2004/5129/, 60 p.

Satter, L., 2006, Ivory-billed deflates river-tapping project: *Arkansas Democrat Gazette.* Retrieved Friday, July 21, 2006, from http://www.nwanews.com/adg/News/161090/

Smith, B., 1988, Ground-water flow and saltwater encroachment in the Upper Floridan aquifer, Beaufort and Jasper Counties, South Carolina: U. S. Geological Survey Water-Resources Investigations Report 87–4285, 61 p.

Smith, B., 1994, Saltwater movement in the Upper Floridan aquifer beneath Port Royal Sound, South Carolina: U. S. Geological Survey Water-Supply Paper 2421, 40 p.

Sniegocki, R. T., 1964, *Hydrogeology of a part of the Grand Prairie Region, Arkansas:* U. S. Geological Survey Water-Supply Paper 1615-B, 72 p.

Sniegocki, R. T., Bayley, F.H., Engler, K, and Stephens, J. W., 1965, *Testing procedures and results of studies of artificial recharge in the Grand Prairie region, Arkansas:* USGS Water Supply Paper, No. 1915-G, 56 p.

Solley, W. B., Merk, C. F., and Pierce, R. R., 1988, *Estimated use of water in the United States in 1985:* U.S. Geological Survey Circular 1004, 82 p.

U.S. Army Corps of Engineers Grand Prairie Area Demonstration Project with links to agencies, http://www.mvm.usace.army.mil/grandprairie/overview/agencies.asp

U.S. Army Corps of Engineers, 1999, Grand Prairie Area Demonstration Project Environmental Impact Study, http://www.mvm.usace.army.mil/grandprairie/maps/study_synopsis.asp

U.S. Army Corps of Engineers, 2000, Grand Prairie Area Demonstration Project Record of Decision, http://www.mvm.usace.army.mil/grandprairie/pdf/RecordOfDecision.pdf

U.S. Army Corps of Engineers, n.d., Grand Prairie Area Demonstration Project, Project Brochure. Retrieved January 30, 2005, from http://www.mvm.usace.army.mil/grandprairie/maps/pdf/GPADProj.pdf

USGS in Arkansas, http://ar.water.usgs.gov/

USGS News Release 2004, http://ar.water.usgs.gov/NEWS/May-NB.html

Wait, R. L., and Callahan, J. T., 1965, Relations of fresh and salty ground water along the Southeastern U. S. Atlantic Coast: *Ground Water,* v. 3, no. 4, p. 5–17.

IV. Introduction to Sustainability: Resource Planning and Global Change

INTRODUCTION

"Our entire society rests upon—and is dependent upon—our water, our land, our forests, and our minerals. How we use these resources influences our health, security, economy and well being."

—J. F. KENNEDY, 1961.

"It is difficult for people living now, who have become accustomed to the steady exponential growth in the consumption of energy from fossil fuels, to realize how transitory the fossil-fuel epoch will eventually prove to be when it is viewed over a longer span of human history."

—M. K. HUBBERT, 1971.

"If we don't grow or harvest it, we must mine it."

—ANONYMOUS

SUSTAINABILITY

Sustainability for the human colony on Earth may be described as balancing natural resources against human resources and the rights of the present generation against the rights of future generations. This balance must attempt to account for expected natural and human-induced changes in the Earth system—the lithosphere, atmosphere, hydrosphere, and biosphere. This is achieved partly by not fully using a variable renewable resource such as water. Thus, when a drought occurs, hardship in the region is minimized.

CLIMATE CHANGE, RESOURCES, AND GEOLOGIC HAZARDS

As for the unexpected changes in the Earth system, we seek improved understanding of how the Earth system works in order to minimize surprises. Significant long-term climate change produces major differences in weather, water resources, the biosphere, and landscapes. With those changes, some cultures are no longer sustainable and they abandon their communities, as in the American Southwest, Peru, and Greenland. Some climate change in the Earth system over the last 15,000 years has been very rapid. In the past, where possible, cultures migrated to the resources (favorable climate, water, etc.). Today, the migration option has been reduced by human population growth. Most niches are filled.

Future rapid climate change could also occur by asteroid impact (it also might produce a huge tsunami). In some impact events the human colony would be sustainable, but much reduced. In other cases ... well you know the story of the dinosaurs. There is a limit to planning for sustainability in the face of natural events.

REGIONAL PLANNING: POPULATION GROWTH, FINITE RESOURCES, AND GEOLOGIC HAZARDS

Energy, in addition to water, is a major factor in the sustainability of a species. Humans have been intelligent enough to exploit buried sunlight—those fossil fuels that produced the industrial revolution, power the technological age, and make it possible to accommodate >6.6 billion people on Earth (at least for a while). For long-term sustainability, replacements must be found for fossil fuels, which are finite.

Population increase also provides a challenge to resource use and sustainability; without it we would not need most regional planning. As population centers grow, converting surrounding rural and suburban land use, they run the risk of inappropriate land use. In the simplest terms, we need to find building materials, water, energy, stable nonflooding building sites, and waste disposal areas for an expanding human colony. Some areas are not only short of suitable materials and sites, they are also hazardous to humans and their structures. To avoid loss of resources, geoscientists and most citizens support the concept of regional planning. Such planning requires basic data, acquired by geologic and soil mapping. Essentially we need to determine where in a region we can accommodate people and how many. The guiding concepts are to design with nature in mind in order to avoid loss of life and waste of resources, including those infrastructure resources lost in disasters.

LONG-RANGE PLANNING

For sustainability, we must take the long-term view of the Earth system, which is not difficult for geologists who think well beyond the next election or budget cycle. Long-range planning begins with determining how to accommodate the human resources (number of humans) with the known and expected natural resources (our geological and biological support system, which includes biodiversity and energy factors). As the Earth is finite, there is a limit to the number of people it can support. Approximating this limit has been attempted in various ways: water and land per person, food calories per person, and renewable energy per person, etc. In long-range planning for sustainability we move from the regional to the global system.

Also in this carrying-capacity approach to planning for human impact on Earth, we must consider the quality of life we can have or will accept, and the changes brought about in the system by humans. Humans are now geologic agents; we capture about 40% of the biosphere and move more material on the lithosphere than the natural geomorphic systems of wind, water, and ice. We are changing the chemistry and temperature of the atmosphere. With climate change, geomorphic processes, and therefore, landscapes change. We are part of and a factor in our changing Earth system—our life support system.

How do we achieve sustainability? How do we balance natural resource availability (assisted by our ingenuity) and human resource use (number of humans × average consumption) in this changing system, which has unknown or poorly defined limits? Many scientists understand that without long-range planning we have actually planned to discover the limits for the human colony by overshoot and collapse.

Are humans capable, intellectually and socially, of a humane approach to long-range sustainability? If we accept this objective as a component of sustainability, we should determine the global resources available to us and adjust to them. There will be technological advances that increase our sustainability; there will be surprises that decrease it. We know that oil production is nearing its peak and "natural" biologic hazards are increasing. We have models of societal failure (e.g., Easter Island); satellite views of current conditions on the land and in the sea; historical perspectives of change in the natural system from marine/lake sediments, ice cores, and human records; and immediate communication (e-mail, phone, TV) between humans (private, governmental, and nongovernmental organizations [NGOs]) about natural and societal change anywhere in the world. We have many tools to assist us in our quest for sustainability. In the end, the laws of nature may well govern sustainability, as individuals and societies in the human colony work to mitigate harmful natural and human-induced changes at the same time that they seek to adapt to those changes. The sooner we understand the options facing Spaceship Earth, the better the chance for a sustainable outcome.

In Section IV the two exercises, Geology and Regional Planning (17) and Global Change and Sustainability (18), address sustainability on both local and global levels. Exercise 17 begins with the environmental factors in solid waste disposal, followed by application of geologic data to selection of land uses in Waco, Texas. In Exercise 18, real-time ozone, temperature, and glacier retreat data are presented to demonstrate change and the human component of some of that change. In this exercise we also examine the ice-core record of longer-term temperature change and the projections for future climate changes. The last activity is an exploration of possible futures, using group discussion, projection of trends, and construction of a brief scenario.

In preparation for the scenario activity and improved understanding of the sustainability issue, the last part of this Introduction examines the finite nature of some key geologic resources.

GEOLOGIC RESOURCES

A *resource* is something that has economic value. Natural resources such as oil, iron ore, fish, or trees are often defined by the monetary value placed on them. Unfortunately, this definition leaves out some natural resources such as scenic views, favorable topography for exploiting solar and wind energy, and even some gravel deposits and favorable building sites. A simple working definition for a *natural resource* is a "useful material or environment"; a broader definition is "biological or physical phenomenon that exists in nature and can be adapted for use by humans." *Geologic resources* are those natural geological phenomena that can be adapted for human benefit. Natural resources and geological resources may be described as renewable or nonrenewable and exhaustible or inexhaustible.

Renewable often is interpreted to mean derived from living matter and replenished with the growing season or at least over a short time period. Some nonliving geological resources, such as water, are renewable. In fact, oil is produced in sediments on a scale of hundreds of thousands of years; gold and other metals are slowly produced in volcanic settings such as the rift at the bottom of the Red Sea, and gravel is renewed annually at the base of an alluvial fan. Most of these rates of "renewal" of geologic resources are slow compared to the rates of human use, and the resources are therefore considered nonrenewable.

Nonrenewable is a "one-crop" resource. Nonrenewable is the usual category for mineral resources. When a mining district or an oil field is exhausted, there is no more of that resource to be had. Geothermal energy, when extracted at a high rate, may be a nonrenewable resource. The same geothermal resource, if extracted at a rate lower than the heat and water are recharged, is a renewable resource.

Natural resources are also characterized as exhaustible or inexhaustible. *Exhaustible* resources are (1) of limited quantity compared to rate of use and, (2) those renewable resources whose productivity may be destroyed (whale oil). *Inexhaustible* resources include those available in exceptionally great quantities (magnesium in seawater), those easily renewed, and those with a low rate of use. With unlimited energy, we would be able to mine low-grade deposits for specific mineral resources and to avoid or repair environmental damage. Energy might be the ultimate geologic resource.

If a natural resource is exhaustible, we would like to know when it will be used up. If we are using the resource at a certain rate per year and there is a known quantity, then we should be able to estimate its lifetime, by:

Estimated lifetime (yrs) = Original amount

(tons)/rate of use (tons/year)

However, it is not as simple as that. The rate of use may increase or decrease, usually increasing with population and decreasing with substitution and improved technology (we are much more efficient now at generating electric energy per pound of coal than we were in the early 1900s). Also we do not know the full extent of the resources that might eventually become reserves and be used. For this reason, a classification of mineral resources is particularly useful in looking at the availability of all resources in the future.

MINERAL RESOURCES: CLASSIFICATION AND ORIGINS

Mineral resources are concentrations of naturally occurring material (solid, liquid, gas) in or on the crust of the Earth in a form such that economic extraction is currently or potentially feasible. The U.S. Geological

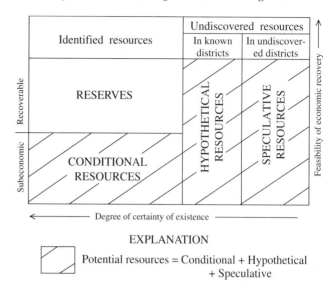

FIGURE IV.1 Classification of mineral resources by feasibility of economic recovery and certainty of geologic existence (Brobst and Pratt, 1973).

Survey and the U.S. Bureau of Mines have classified mineral resources using a scheme developed by Brobst and Pratt (1973). One of the key components of this scheme (Figure IV.1) is the difference between mineral *reserves* and other mineral resources. A *reserve* is a mineral resource whose extent and grade are known and for which extraction is now economically, technologically, and politically feasible. It is a resource that can readily be withdrawn from the "bank."

The processes of formation of mineral resources are basically the geologic processes of the Earth system. Mineral deposits form by igneous, metamorphic, and sedimentary processes. The superficial process of weathering also concentrates elements and materials that are useful to humans. Biological processes are largely responsible for phosphorus deposits and sedimentological, biological, and chemical processes play a role in formation of fossil fuels.

IMPORTANCE OF GEOLOGIC RESOURCES

Mineral resources are an essential part of society and improvement of the quality of life. Use of mineral and energy resources has been positively correlated with gross domestic product, although with conservation measures, renewable energy use, technological advances, and the growing concern about environmental degradation, these relationships could change. In addition to energy resources, both metallic and industrial (nonmetallic) rocks and minerals are essential to a quality life. According to V. McKelvey (1973), a former director of the U.S. Geological Survey, the level of living (L) depends on the following relationship:

$$L = (R \times E \times I)/P$$

where R = natural resources, E = energy, I = Ingenuity, and P = Population sharing those resources. A resident of the U. S. will use more than 2 million pounds of minerals in his or her lifetime, almost half of which will be aggregates (gravel and crushed stone).

We are very near the peak in oil production in the Fossil Fuel Age, living on buried sunlight. The degree of dependence on fossil fuels by the U.S. is readily seen with the mix of energy we now use. For example the average of July 2003 and January 2004 U.S. energy consumption amounted to 8.85×10^{15} Btus with petroleum accounting for 38%, natural gas for 24%, coal for 24%, hydroelectric and other renewables for 5.7%, and nuclear power for 8.3%.

Hubbert (1969) and others have documented the change over time of our fossil fuel use. Against a scale of 10,000 years of human history, our use of fossil fuel shows as a "blip." In the 1950s Hubbert forecast the decline in U.S. oil production, which began in 1970. We now import more than 60% of our crude oil. The global peak and decline in production of conventional oil is occurring now (Simmons, 2005; Strahan, 2007; Zwicker and others, 2004).

Alternative, particularly renewable, energy resources are the key to sustainability. Fossil fuel use will decline over the next 30 to 40 years due to availability and its impact on local and global environments. Solar energy will grow in importance. It will be used directly as heat or as electricity derived by heat or photovoltaics, wind energy, ocean currents, ocean thermal gradients, or biomass. Nuclear fusion is not expected to be a factor in the near future but could play a very important role if attained; nuclear fission might be one option to counteract global warming by CO_2 from fossil fuels, and this industry is showing signs of recovery. Deep geothermal energy (not groundwater heat pump systems heated mainly by solar energy) is another alternative energy source that could become more important. For many remote localities, photovoltaic cells are providing viable alternatives to connections to an electric grid. In developing countries, photovoltaics will provide major improvements in the quality of life and sustainability without construction of centralized power plants and distribution systems.

GEOLOGIC RESOURCE AVAILABILITY

Given the importance of mineral and energy resources, it is understandable why so many commissions in the 20th century explored their availability and expected lifetimes. Two philosophical views about mineral resources are common. One view is that the supply of mineral resources is not a problem. Minerals will either be replaced by something else if they run out or an increase in price will make so many resources economic that there will not be a limit to supply. Another view is that many resources are exhaustable, and it is not wise to rely on expected new discoveries. Ultimately, as vividly illustrated by Apollo mission photographs of Earth as a sphere in space, most resources are limited.

There was a $10,000 wager between opposing camps in the concern for mineral resources. In the "cornucopian" camp (mineral resources will be available) the late Julian Simon, an economist at the University of Maryland, declared in the 1980s that in ten years there would be no shortages or major increases in costs for mineral commodities. Paul Ehrlich, a population biologist at Stanford in the "sustainability" camp, said there would be shortages or price increases. Ehrlich lost the bet. However, will this condition continue? In 1968, Preston Cloud suggested that the cornucopians had five premises on which their lack of concern for mineral resource availability was based: (1) cheap inexhaustible nuclear energy, (2) economics as the major factor in resource availability, (3) uninterrupted variation from ore to natural crustal abundance, (4) assumed population control, and (5) a technological fix.

Resource availability is impacted by environmental and social concerns. In the 1990s an environmental conscience began to spread globally in the mining world and formal guidelines were proposed for the mineral production sector. At least one example of much delayed mine development by a major mining firm can be traced to inappropriate assessment of environmental and social impacts and poor public relations. Environmental concerns must include impacts of exploration, removal or mining, processing, manufacturing and transportation, resource use, and post-resource-use waste management. Poor waste management results in direct (waste) or indirect (pollution) loss of resources.

The modern viewpoint is that wastes are resources out of place. Urban ore, the mineral resource values concentrated by humans in their landfills, can be recycled and reused. In some cases the "deposits" in landfills are richer than the mineable virgin mineral deposits.

For any finite nonrenewable mineral resource, we should understand the nature of extraction of that resource over time. "Hubbert's Bubble" (Figure IV.2) shows the actual and forecast production curves for world oil. Depletion of any nonrenewable mineral resource, from a mining district or from a country, follows a similar trend.

QUESTIONS IV, INTRODUCTION

1. What is a geologic resource, according to the discussion in Geologic Resources?

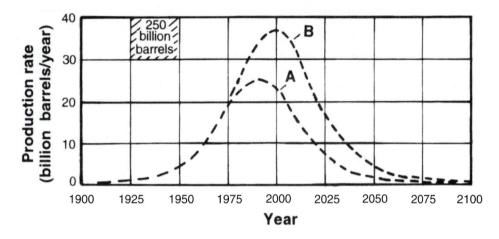

FIGURE IV.2   Cycle of world crude-oil production at assumed total resources of: A, 1,350 billion barrels, and B, 2,100 billion barrels (modified from Hubbert, 1969). Total world crude-oil production in 2001: 77 million bpd; forecast for 2020: 112 million bpd. In 2000, the U.S.A used 19.7 million bpd; forecast for 2020 is 26.7 million bpd.

2. Name one renewable and one nonrenewable geologic resource.

3. If you knew the current rate of production of a mineral resource from a mine, what additional information would you need to determine the remaining lifetime of the mine (number of years until it is exhausted)?

4. According to Brobst and Pratt (Figure IV.1), what is the difference(s) between hypothetical and speculative mineral resources?

5. Would a mining company be likely to extract conditional resources? Explain.

6. If you owned a gold mine that could produce gold at a profit only if the price were equal to or greater than $800/ounce, what would happen to your reserves if the price went to $600 per ounce?

7. What are the five Cornucopian Premises, according to Preston Cloud?

8. Hubbert, in 1956, forecast the peak in U.S. oil production would occur in 1970 and it did. Assuming the total world oil resource is 2,100 billion barrels, when should we peak in world oil production, according to Hubbert (Figure IV.2)?

9. In Figure IV.2, the production rate is given in billion barrels per year. If we were producing a steady 5 billion barrels per year, how much would we use in 25 years?

In Figure IV.2, outline (and shade in with pencil) that amount in one of the rectangles.

10. Using the literature or other sources, determine the current daily and annual rates of oil production for the world.

11. What impact will "peak oil" have on the U.S.A.?

Bibliography

Brobst, D., and Pratt, W., 1973, Introduction, in, D. Brobst, and W. Pratt, eds., *United States Mineral Resources*, U.S. Geological Survey Professional Paper 820, p. 1–8.

Cloud, P. E., 1968, Realities of mineral distribution: *Texas Quarterly*, v. 11, no. 2.

Drew, L. J., 1997, *Undiscovered petroleum and mineral resources—assessment and controversy*: New York, Plenum Press, p. 210

Hubbert, M. K., 1969, Energy resources, in Committee on Resources and Man, NAC-NRC, *Resources and Man*, San Francisco, W.H. Freeman, p. 157–242.

McKelvey, V., 1973, Mineral resource estimates and public policy, in D. Brobst and W. Pratt, eds., *United States Mineral Resources*, U. S. Geological Survey Professional Paper 820, p. 9–19.

See *Geotimes* magazine for monthly energy updates; U.S. Geological Survey Circular 831 for a detailed mineral resource classification. Websites: www.api.org and www.geotimes.org/Res_index.html:.

Simmons, M. R., 2005, *Twilight in the Desert:* Hoboken, NJ, J. Wiley, p. 422

Strahan, D., 2007, *The Last Oil Shock: A Survival Guide to the Imminent Extinction of Petroleum Man:* London, UK John Murray General Publishing Division, p. 304

Zwicker, B., Greene, G., and Silverthorn, B., 2004, The End of Suburbia: Oil Depletion and the Collapse of the American Dream (video): Toronto, The Electric Wallpaper Company. www.endofsuburbia.com.

Geology and Regional Planning

INTRODUCTION

"The social value of geology increasingly derives from the environmental tensions created by the resource and land-use needs of an expanding population. Quality of life for the eight or ten billion people who will inhabit the planet by the end of the coming century will depend on how well these unavoidable tensions are managed. Thus, our scientific agenda is inextricably bound not just to geological phenomena, but also the relation between those phenomena and the behavioral patterns of human beings."

—SAREWITZ, 1996

"Obedience to nature could well be the motto of every planning agency."

—LEGGET, 1973

"Sooner or later we all pay, directly or indirectly, for unintelligent use of land."

—MCGILL, 1964

In this exercise the objective is to understand the geologic controls on land use, and the steps in and geologic information for regional planning. In Part A we look at the environmental factors in disposal of municipal solid waste; in Part B, we select suitable sites for several land uses in Waco, Texas, using properties of geologic formations or units, basic data maps, and resource capability maps. We begin with an overview of regional planning.

GEOLOGY AND LAND USE

Human use of the physical environment must consider our need to avoid hazards and maximize resource availability. Hazards, as explored in Section II of this book, topography, soil strength, depth to bedrock, and resources (gravel, rock, water, and agricultural soils) all impact land-use decisions.

An acre or hectare of land may have the potential to provide multiple resources for humans—both simultaneously and sequentially. Potential land uses include mineral extraction, forestry, groundwater supply, water reservoirs, waste disposal, and cemeteries in addition to those resource uses listed in Table 17.1. In some cases there is a definite best use of the land because the resource is not found elsewhere in the region.

COMPREHENSIVE REGIONAL PLANNING

In regional assessments for resource use, planning agencies must consider more than geologic hazards and resources; they must include biological, social, political, and historical data, too. By inventorying, analyzing, and displaying these data, knowledgeable land-use options can be made.

There are many approaches to regional planning. These are the basic steps in one common approach:

1. Analysis of goals and objectives of the region (public input)
2. Collection of basic data
   a. inventory of geologic and biologic factors in the environment
   b. inventory of economic, social, cultural, and political factors
3. Analysis of resource capability (using 2a)
4. Analysis of resource suitability (using 3 and 2b)
5. Synthesis of data to develop goals and objectives, in the form of alternate plans for the region
6. Selection of the best regional plan (public input)
7. Determination of techniques for implementing the plan.

Capability usually is defined as the ability of the land to provide a resource; maps depicting capability are called resource capability maps. An area that is

TABLE 17.1 Basic Geologic Data Needed for Evaluating Resource Uses in a Region

Factor (Basic Data)	Function (Resource Use)									
	Golf	Camp-ground	Trails	Picnic	Agri-cuture	Septic Tanks	Resi-dential	Comm. Indust.	Transport.	Utilities
Slope	×	×	×	×	×	×	×	×	×	×
Bedrock depth					×	×	×	×	×	×
Permeability	×				×	×				
Flooding	×	×	×	×	×	×	×	×	×	×
Water-table depth	×				×	×	×	×		×
Stoniness					×	×	×	×		
Shrink-swell							×	×	×	×
Frost action						×	×		×	×
Water supply	×	×		×	×	×	×	×		
Corrosion potential					×			×		×

physically and biologically capable of providing an identified resource may not be acceptable for other reasons (such as those listed in 2b above). Some planners make a distinction between capability and suitability, using the term suitability for regions that are capable (geologically) and also acceptable in terms of economic, social, political, and cultural factors.

Maps such as geologic maps and topographic maps are known as basic data maps. Examples of how geologic variables affect various resource uses are given in Table 17.1. Note that if septic tanks are to be used for home sewage treatment, bedrock depth, soil permeability, slope, flooding, water-table depth, and stoniness are considered to be important. However, simply knowing that these are important factors in the use of the land for septic tanks is not enough if we want to make a map showing which areas are actually capable of this resource use. We also need limiting values for the factors that have an impact on septic tank

use. For example, what bedrock depth would be deleterious for septic tank use? Usually such values for any factor are quantified into a threefold classification as most capable, capable, and not capable. For example, in Table 17.2 water-table depth as it affects residential development is classified as most capable (>5 ft), capable (2.5–5 ft), and not capable (<2.5 ft). Sometimes these capabilities are depicted by a color system on a map, with red = not capable, yellow = capable with some restrictions, and green = most capable. This color system is known as the stoplight code.

Each factor important in residential development must be evaluated for each small unit of the map area. If a unit on the map contains a "not capable" value for a factor for residential development, then that area has to be excluded from the capable category of the resource capability map. Other criteria for residential development are evaluated on a weighted basis to determine the overall capability of a site.

TABLE 17.2 Defining Factor Limits for Determining Resource Capability for Residential Development

Factor (Basic data class)	Most Capable	Capable	Not Capable
Water-table depth	>5 ft	2.5–5 ft	<2.5 ft
Bedrock depth	>5 ft	3.3–5 ft	<3.3 ft
Flooding	none	rare	frequent/occasional
Shrink-swell	low	moderate	high
Frost action	low	moderate	high
Groundwater availability	>10 gpm	5–10 gpm	<2–5 gpm
Slope	4–9%	0–3%, 10–15%	>15%

The geologist's primary role in planning is to provide basic data and resource capability maps to the planner. The geologist may also assist the planner in integrating them with other data and maps to produce resource suitability maps which will be the basis for a long-range regional plan. Such a plan is subject to change but does provide guidance in making wise decisions about land use.

PART A. SITING A SANITARY LANDFILL

Management of municipal solid waste presents a growing problem for urban regions. The sanitary landfill is now the most common way to "dispose" of this waste, although incineration, composting, and traditional recycling are also used. Municipal solid waste includes food waste, beverage containers, yard and garden waste, automobiles and parts, appliances, furniture, newspapers, and disposable diapers, etc. As this waste decomposes it produces leachate, a nasty liquid that contains components of the soluble materials in the landfill—from heavy metals to organics. Leachate in poorly constructed landfills can travel beyond the sanitary landfill causing pollution of soils, groundwater, and surface water. Methane and other landfill gases may cause problems when they migrate underground into surrounding buildings.

The number of geologically suitable sites for sanitary landfills is limited. As a result, landfill sites are expensive to acquire and to engineer for environmental protection. In 1970 in North America, 85 percent of municipal solid waste was disposed of in a dump, many located in old pits and quarries. In these dumps waste was not covered regularly; open burning and leachate runoff were common. About this time, the environmental damages and hazards from these dumps were recognized and legislation was enacted for sanitary landfills. A **sanitary landfill** is a method of disposing of refuse on land without creating nuisances or hazards to public health or safety. Engineering techniques are used to confine the waste to the smallest practical volume, to cover it daily, and to prevent leachate, gas, blowing debris, and rodent and odor problems.

The sanitary landfill system is expensive. Disposal fees and limited space in some areas make it economical to ship solid waste many miles by truck or train for disposal in a landfill. Waste from New York City goes to Ohio and other states; Toronto, Ontario, has sent some of its waste to Michigan. In addition to the geologic factors that control the siting of landfills, social factors that must be considered have also increased costs for disposal sites. The terms NIMBY (Not In My Back Yard), LULU (Locally Unwanted Land Uses), and NIMTO (Not In My Term of Office) exemplify the social response to proposed landfills.

Several factors of the physical environment must be considered in selecting a site for a sanitary landfill. These include the following:

Climate. In cold climates freezing of the soil restricts excavation and availability of cover material. A low-lying site may be undesirable in areas of heavy rainfall because of flooding and muddy working conditions. Windy sites need special consideration because of dust problems and blowing paper.

Bedrock geology. The type of bedrock is important; sandstone, conglomerate, and limestone could rapidly transmit water containing pollutants, while shale and igneous and metamorphic rocks would not. Rock structures (e.g., faults and joints), the dip of rock strata, and underground mines must be considered.

Hydrology. Possibilities of groundwater pollution from leachate and surface-water pollution from leachate and runoff, the seasonal fluctuation of the water table, and the direction of groundwater movement are all important factors. Well-head areas, recharge zones for aquifers, and floodplains should also be considered in siting landfills.

Topography. Surface slopes before and after development must be considered, since erosion of cover materials may expose trash. Floodplain proximity and flood level must be taken into account, as well as the effects of topography on surface and underground drainage.

Regolith and soil conditions. A sufficient quantity of easily workable compactible cover material must be available. The type of regolith (i.e., alluvium, glacial drift, clay soils, etc.) must be considered.

Other factors in site selection include economics, transportation routes, engineering techniques to reduce offsite impact, adjacent historical, cultural, and environmental values, and politics.

QUESTIONS 17, PART A

Trashmore is a hypothetical city of about 40,000 in the Great Lakes region of North America. The gently rolling landscape has been glaciated and the regolith contains silty-clay till, outwash gravels, glacial lake deposits of clay, and modern alluvium over bedrock at depths of 0 to 100 feet. The bedrock consists of limestone and shale formations. Trashmore has a sanitary landfill that will run out of space in two years. As a member of Trashmore's City Council you have been asked to assist the regional planners in selecting a suitable site for a new sanitary landfill. You consult a portion of a geologic map and cross section of the region (Figure 17.1) and recognize three excavated sites. If one of these pits and quarries could be used, you think you could save on excavation costs.

1. What geologic resource (rock or sediment) was extracted at each of the three sites?

State Aggregate:

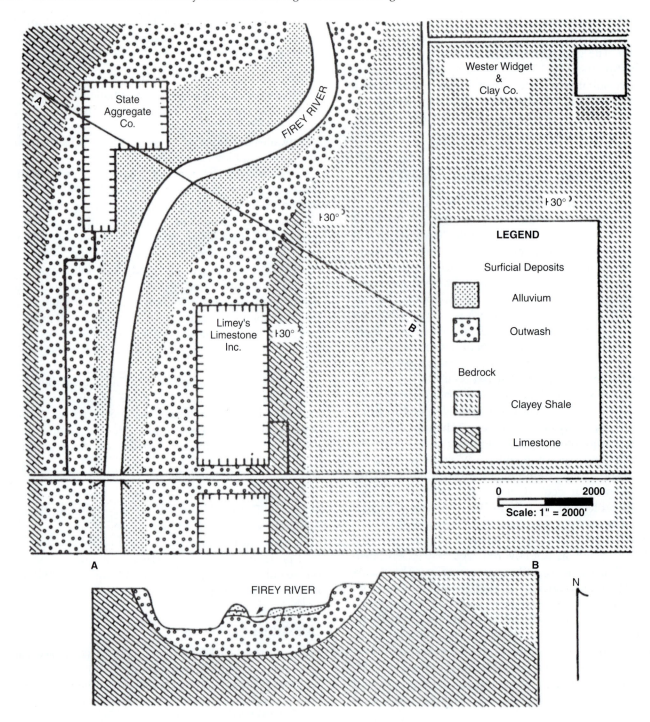

FIGURE 17.1 Geologic map and cross section of western Trashmore.

Western Widget:

Limey's Limestone:

2. a. Which of the three excavated sites was your first choice for the new landfill? (Consider the physical environmental factors, described in the Introduction to this exercise, that are important in selecting sanitary landfill sites.)

b. What physical environmental factors were important in making your selection of a landfill site?

c. Explain why your number one factor is important.

3. List the problems that are associated with each of the *other two* sites not selected in Question 2a.

a. Site: _____; Problems:

b. Site: _____; Problems:

4. If you were to excavate to a deeper level at the State Aggregate pit, what additional geologic material could you obtain?

5. Sanitary landfills must be covered daily (and after they are closed) with a low permeability material to keep out precipitation and runoff. (Otherwise, they do not meet the regulations for sanitary landfills.)

a. Which of the four (4) materials shown in the legend of Figure 17.1 might be suitable for covering the waste daily? Why?

b. Trashmore's regolith has been deposited on bedrock by glacial processes. Thus the unconsolidated material at the surface might include outwash gravel, lake silt and clay, clayey till, and wind-blown silt. What low permeability material from the region might be suitable for landfill cover?

6. Which way is the bedrock dipping at Limey's Limestone and what is the approximate angle of dip? (Hint: Use the cross section and a protractor or use the symbols on the map.)

7. In Figure 17.2, a landfill is in an outwash valley on a flood plain. Many such sites were selected in glaciated environments because a gravel pit was available.

a. Complete the water table (dashed line) on both sides of Firey River.

b. Place arrows to show the flow of groundwater in your profile. (See Exercise 12 for assistance, if needed.)

c. Knowing that, to be classified as a sanitary landfill, trash cannot be dumped into standing water, would this site be a suitable place for an unlined sanitary landfill?

d. List the problems that this site by the river would present if it were selected as a modern sanitary landfill. Refer to the guidelines given in the Introduction.

PART B. PLANNING AND GEOLOGY IN WACO, TEXAS

Waco, Texas, is situated in central Texas, between Austin and Dallas, on the Bosque escarpment or fault-line scarp and is underlain by eastward-dipping bedrock of the Gulf Coastal Plain. The Upper Cretaceous formations outcropping in Waco include Taylor Formation (Kta), Austin Chalk (Kau), South Bosque Shale (Ksb), Lake Waco Formation (Klw), and Pepper Shale (Kpe). Lower Cretaceous formations include Del Rio Clay (Kdr) (which outcrops) and the Georgetown Limestone and Edwards Limestone, which are in the subsurface, outcropping to the west of the city. Tertiary rocks outcrop to the East; Quaternary sediments occur in the Bosque and Brazos rivers and terraces and in the alluvium of these rivers. The cities of Austin, Dallas, and San Antonio are also underlain by the same Upper Cretaceous rock formations as found in Waco.

Surface bedrock formations in the Waco area affect city growth because of their effect on topography, engineering properties, soils, drainage, and construction; subsurface formations may be important for water and mineral resource extraction. The properties of these geologic units are given in Table 17.3.

QUESTIONS 17, PART B

Use information in Figure 17.3 (Waco map at back of book) and Table 17.3, to answer the following questions.

1. a. List the geologic symbols of formations or units (Table 17.3) that might be used for extraction of sand and gravel in the Waco area.

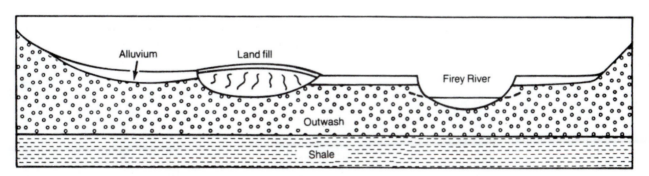

FIGURE 17.2 Cross section through a landfill on an alluvial plain.

TABLE 17.3 Properties of Geologic Units in Waco, Texas (after Burket, 1965; Elder, 1965; and Rupp, 1976)

Geologic Formations (w/symbols)	Max. Thick. (ft)	Rock Unit	Slope Stability	Excavation Difficulty	Bearing Capacity[1]	Infiltration Capacity	Rock/Mineral Resource Potential	Ground-Water Potential	Flood Potential
Alluvium (floodplain) Qal	?	Silt and clay; some gravel	Moderate (>10°)	Slight	Moderate to low, piers for heavy structures	Interm. to High	Minor sand and gravel	Low; irrigation and domestic use	Very high
Bosque/Brazos Flood Plain. = Qbof and Qbrf Terrace = Qbot, Qbr 1, Qbr2, Qbr3 and Qbr4	?	Sand and gravel	Moderate (>10°)	Low except for Qbot and Qbr2 (N. of Lake Waco) Intermediate	Moderate; low in some Qbot sands	High	Major sand and gravel	Intermed./low; irrigation and domestic use	Qbof/Qbrf high; terraces none
Taylor Kta	1170	Marl with some sandstone and limestone	Low (<10°)	Slight	Moderate to low, pier support for heavy structures	Low	None	Low	None
Austin Kau	295	Chalk, Marl	High (>45°)	Severe	High	Moderate	Subgrade for roads	None	None
South Bosque Ksb	140	Salt, Shale	Very low in upper 40 ft, low in lower 120 ft	Slight	Low; flotation or pier support	Low	Expanded aggregate	None	None
Lake Waco Klw	145	Calcareous shales, thin limestone interbeds	Low to moderate	Intermediate	Moderate; some pier support	Low	None	Domestic quantities	None
Pepper Kpe	100	Noncalcareous shales, sand	Very low. fails when wet if slope <10°	Slight	Very low; flotation or deep piers	Low	Possible ceramic materials	Low	None
Del Rio Kdr	85	Blue clay, calcareous siltstone	Moderate to low	Intermediate	Moderate to low (in middle part)	Low	Possible ceramic materials	None	None
Georgetown	210	Nodular limestone, shale	High (>45°)	Intermediate	High	Moderate	Marginal subgrade 5 miles from Waco	None	None
Edwards	45	Limestone, siltstone	Not exposed	Not exposed	Not exposed	Not exposed	Very good dimension stone 15 miles from Waco	Low	None

1. Very low and low require piers and floating foundations for heavy structures; intermediate allows significant part of foundation load of heavy structures on footings or beams; high allows conventional footings for heavy structures, but local faulting and joints could modify these conditions.

b. On Figure 17.3, select and mark (with "S&G") three sites where sand and gravel that could be used as aggregate might occur.

2. a. From what formation could you obtain ceramic grade materials?

b. On Figure 17.3, select and mark ("CM") a potential site for extraction of high-quality (ceramic-grade) clays.

3. a. What geologic formation would provide "dimension" building stone for Waco?

b. Why can you not mark such a site on the geologic map (Figure 17.3)?

4. From the information given in Table 17.3, determine the geologic units that might be investigated as potentially acceptable sites for a sanitary landfill in the West Waco area. Justify your choices by completing the table below.

5. In which geologic units (Table 17.3) in Waco would the greatest potential exist for contamination of potable groundwater by leachate from sanitary landfills or abandoned dumps?

6. a. Refer to your answer to question 4 and select a possible site for a sanitary landfill. Outline this area on the Waco West Quadrangle (Figure 17.3). Place the symbol "SL" in the center of the area. (Or describe the location and geologic material at the site).

b. State the most important factor that resulted in your decision and a possible minor problem for the site.

RESOURCE-CAPABILITY MAPS AND BASIC-DATA MAPS

7. By now you probably have determined that consulting basic data tables and maps and evaluating these data for a specific resource use can be confusing and time consuming. Resource capability maps are often used to ease the problem of determining the best use of the land from data in tables. Figure 17.4 is a resource capability map showing capability for septic-tank tile fields. It has been drawn using the following basic data or basic data maps:

> soil permeability: low permeability is a severe limitation
> depth to bedrock: less than 10 ft is a severe limitation
> slope: over 10 percent is a severe limitation
> flood hazard: soils with any degree of overflow present a severe limitation

Using the information in Figure 17.4, determine the potential for using septic tank systems with drain fields in the following areas (Figure 17.3) that are being considered for subdivision development.

a. X, 1.7 miles NE of the dam of Lake Waco, on Qbr1:

b. Y, immediately SW of Heart O'Texas Fairground, on Kau:

c. C, east of the Brazos River, on Qbr2:

8. Using the resource capability maps (Figures 17.4, 17.5, and 17.6), basic data map (Figure 17.7), and basic data for the geologic units in Table 17.3, evaluate *one* of the following areas in Figure 17.3 for possible residential development: A, D, or Z (Kta). Describe the problems and advantages for low-density housing in the area you select. Begin by reviewing the characteristics of each site. Then select one for your evaluation. Note: "A" is on the west side of the curve in the road on Kpe; "Z" is on Taylor Marl, Kte.

Geologic Unit	Material	Infiltration	Rock and Mineral Resources	Groundwater Potential

9. a. As an engineering geologist you are asked to assist in the selection of possible sites for construction of a large distribution-center warehouse. It will be supported by conventional footings resting below the weathered or soil zone. Considering only geologic factors, which two of the following geologic settings would you suggest to your client for further study? See Table 17.3.

b. Explain the reasons for your two choices, and your rejection of the third option.

Site 1,   on Austin Chalk (Kau)

Site 3,   on South Bosque Shale (Ksb)

Site c,   on Brazos Terrace (Qbr2)

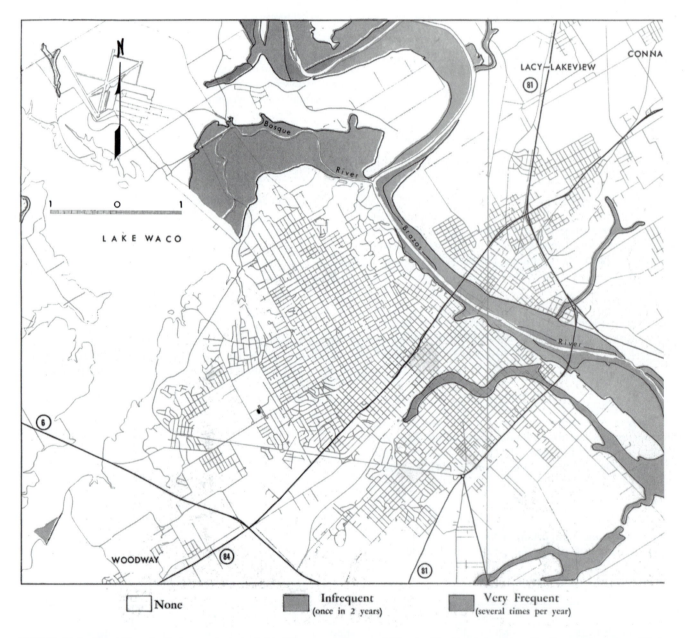

FIGURE 17.4 Capability of soils for septic drain fields, Waco area. The capability classification is based on the three categories of limitations shown in the explanation for the map: *slight, Moderate and Severe. Slight* translates to a <u>capable area,</u> *Moderate* to a *Moderately Capable* area with some limitations, and *Severe* to a <u>Not Capable or Low Capability</u> area. Limitations (capability) are based on normal density of residences and soil permeability (very slowly permeable soil is severe limitation), depth to bedrock (soil less than 10 inches thick is severe limitation), slope (soil with >12 percent slope is a severe limitation), and flood hazard (any degree of overflow is a severe limitation). Much of this map is in the severe category; classification does not apply to Lake Waco.

(Modified from Elder, 1965; used with permission).

HAZARDS DUE TO SHRINK-SWELL

Low
0.5" rise in 6" or to bedrock

Moderate
0.5" to 1.25"

High
1.25" to 2.0"

Very High
2.0"

FIGURE 17.5   Capability of soils for residential foundations based on shrink-swell potential. The <u>most capable</u> soils have a *low shrink-swell,* here defined as less than 0.5 inch rise (in 6 feet of soil or to bedrock) when a fully dry soil is thoroughly wetted. The <u>least capable</u> soils have a *very high shrink-swell* hazard with a rise of more than 2 inches. Soil thickness, clay type and amount of clay influence shrink-swell. Climate, on-site mangement of water, drainage and vegetation also influence the hazard.

(Modified from Elder, 1965; used with permission.)

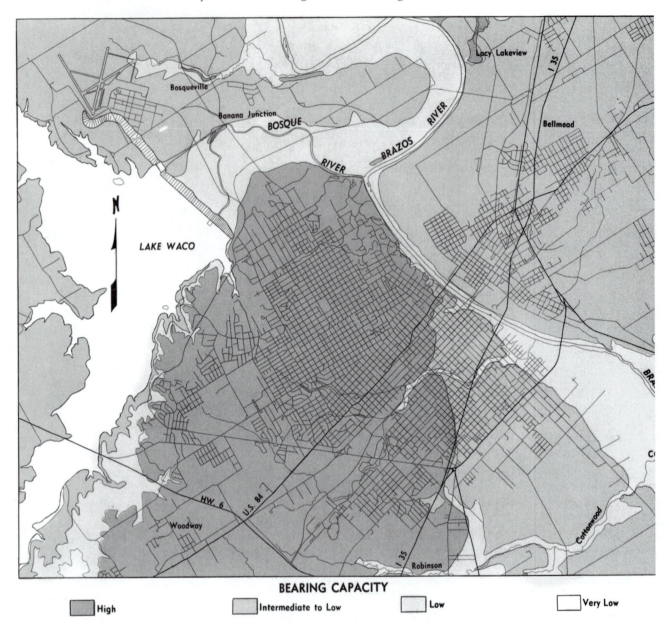

BEARING CAPACITY

High   Intermediate to Low   Low   Very Low

FIGURE 17.6 Bearing capacity of bedrock formations for heavy structures in the Waco area. High bearing capacity means that the formation allows heavy structures to be supported by conventional footings and grade beams and generally no pier or caisson support is required (jointing or faulting could modify these conditions). A *high bearing capacity* area represents a <u>high capability</u> area for this resource use (large structures).The Austin Chalk and limestone units of the Georgetown Formation are in this category. *Low to very low bearing capacity* requires heavy structures be supported by piers (into adjacent more resistant units or floating foundations. The Pepper Shale and some units of the South Bosque Shale, the Lower Taylor Marl, and the Del Rio Clay are low to very low bearing capacity and thus of <u>low capability</u> for large sructures. See Table 17.3 for further explanation.

(Modified from Font and Williamson, 1970; used with permission.)

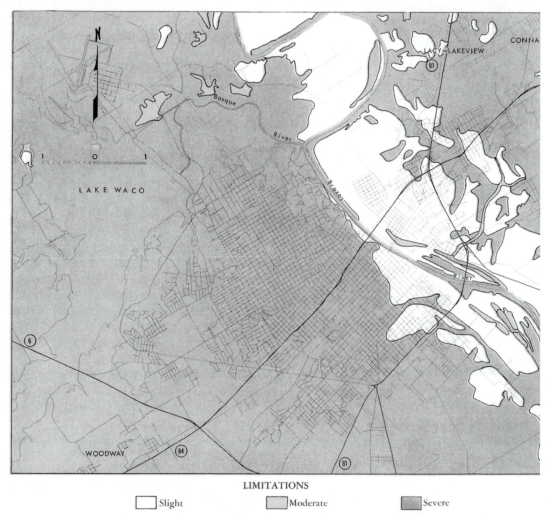

FIGURE 17.7   Flood hazard for soils, Waco area. Three degrees of hazard are shown: None, Infrequent (about once in two yeras), and Very Frequent (several times per year). Changes in flood detentation basins and other water management works could change the hazard regions on the map.

(Modified from Elder, 1965; used with permission.)

Bibliography

Bernknopf, R. L., Brookshire, D. S., Soller, D. R., McKee, M. J., Sutter, J. F., Matti, J. C., and Campbell, R. H., 1993, *Societal value of geologic maps:* U. S. Geological Survey Circular 1111, 53 p.

Breen, B., 1990, Landfills are #1: *Garbage,* v. 2, no. 5, p. 42–47.

Brown, R.D., and Kockelman, W. J., 1983, *Geologic principles for prudent land use, a decisionmaker's guide for the San Francisco Bay region:* U. S. Geological Survey Professional Paper 946, 97 p.

Burket, J.M., 1965, *Geology of Waco:* Baylor Geological Studies, Bulletin 8, p. 9–46.

Church, G.J., Holmes, S., and Taylor, E., 1988, Garbage, garbage, everywhere: *Time,* Sept. 5, p. 81–82.

Elder, W. R., 1965, *Soils and urban development of Waco:* Baylor Geological Studies, Bulletin 9, 66 p.

Font, R.G., and Williamson, E. F., 1970, *Geologic factors affecting construction in Waco:* Baylor Geological Studies, Bulletin 12, 34 p.

Forsyth, J. L., 1968, *Physical features for the Toledo regional area:* Toledo, Ohio, Toledo Regional Area Plan for Action: Regional Report 8.2, 111 p.

Hannon, B. M., 1972, Bottles, cans, energy: *Environment,* v. 14, no. 2, pp. 11–21.

Hughes, G. M., 1972, *Hydrogeologic considerations in the siting and design of landfills:* Illinois Geological Survey Environmental Geology Notes, no. 51, 22 p.

Hughes, G. M., Landon, R. A., and Farvolden, R. N., 1971, *Hydrology of solid waste disposal in northeastern Illinois:* U.S. Environmental Protection Agency, Washington, D.C., U.S. Government Printing Office, 154 p.

National Academy of Engineering, 1970, *Policies for solid waste management:* Public Health Service Publication no. 2018, Washington, D.C., U.S. Government Printing Office.

Rupp, S., 1976, *Subsurface Waters of Waco:* Baylor Geological Studies, Bulletin 11, 68 p.

Sarewitz, D., 1996, Geology as a social science: *GSA Today,* vol. 6, no. 2, p. 22–23.

Spencer, J. M., 1966, *Surface Waters of Waco:* Baylor Geological Studies, Bulletin 10, 49 p.

Young, J. E., 1991, *Reducing waste, saving materials:* in Brown, L. (ed), State of the World 1991: New York, W. W. Norton, p. 35–55.

Global Change and Sustainability

INTRODUCTION

"The public in the industrialized nations, particularly in the United States, must be made more aware of the pervasive trends in environmental degradation and resource depletion, and of the need to modify patterns of life to cope with these trends."

—KAULA AND ANDERSON, PLANET EARTH COMMITTEE, AMERICAN GEOPHYSICAL UNION, 1991

"The time to consider the policy dimensions of climate change is not when the link between greenhouse gases and climate change is conclusively proven ... but when the possibility cannot be discounted and is taken seriously by the society of which we are part. We in BP have reached that point."

—JOHN BROWNE, GROUP CHIEF EXECUTIVE, BP AMERICA, 1997

The quotations above capture the concern of many geologists who understand Earth history and change in the Earth system. In the distant past humans did not play a role in these changes. Now we do. In this exercise we look at changes in climate using ice cores and glacier retreat, changes in the atmosphere through ozone and CO_2, and the human component of change through population increase and the response to projected temperature change.

Other components of global environmental change include deforestation and habitat loss, loss of biodiversity (which may be the most important one), desertification, soil erosion and agriculture, spread of infectious disease by microorganisms, rising sea level, modifications in weather patterns and storminess, and decreasing water resources.

A very few scientists do not accept the evidence of warming, or if they do, its relation to human activity. These few doubters receive "equal time in the press" with those actively working on the science of global change. It has been revealed that many of the doubters are funded by special interest groups and fossil fuel coalitions that do not want research on, or evidence of, global change that could impact their economic condition (Gelbspan, 1997). Island nations, environmentalists, active geoscientists, environmental economists, and insurance organizations were among the first to recognize that global change is occurring. Now many groups are working on adaptations to change and on ways to reduce human impact globally.

We should be able to adapt to the limits of our habitat on Earth without overshooting them. Many organisms undergo increases and collapses in their populations but they do not have our technology, information, and understanding. We have models, examples, and explanations of why societies collapse (Diamond, 2005). We probably will never understand how the total Earth system works; however, we need to write the "Operating Manual for Spaceship Earth." The final activity in this exercise is constructing a scenario—your description of the future viewed from 50 years in the future. This forecasting approach is used by think tanks to understand possible futures (Schwartz, 1991).

PART A. ICE-CORE PALEOTEMPERATURES AND GLACIER RETREAT

Ice cores provide excellent records of past atmospheric conditions, including temperatures and CO_2 concentrations. Interpreting cores from ice caps in mid-latitude and equatorial regions has been the research focus of Lonnie and Ellen Thompson's Ice Core Paleoclimatology Research Group at the Byrd Polar Research Center, The Ohio State University. Material in this part of the exercise has been provided by the Thompson team. Images of the Quelccaya Ice Cap are available at http://bprc.osu.edu/Icecore/

Basic concepts in this section include: oxygen iso-topes in ice provide information on atmospheric tempera-ture fluctuations—lower delta ^{18}O values (written δ^{18}O) indicate lower temperatures; cold glaciers, well below the melting point, are required for extraction of annual records—any melting mixes the annual signal; with increasing altitude atmospheric temperatures decrease—if ice cap melting now occurs where it did not in the past, the atmosphere has warmed; and, retreat of valley or outlet glaciers is due to warmer temperatures or less accumulation of snow and ice to replenish the glacier. Worldwide, almost all mountain glaciers are retreating due to increased atmospheric temperatures. In the recent past, a few glaciers were advancing because the warmer air temperatures have produced more snow.

QUESTIONS 18, PART A

1. Study Figure 18.1a, the oxygen isotope record for the Quelccaya Ice Cap in Peru. This ice cap is fed by moisture from the Amazon Basin.

    a. Were temperatures warmer or colder than the mean for the record during the period from A.D. 1550 to 1900?

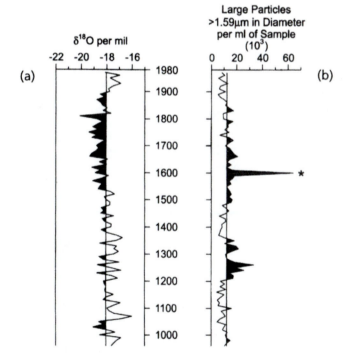

FIGURE 18.1 Oxygen isotope (a) and particle record (b) from the 1983 Quelccaya, Peru, ice core. More negative δ^{18}O values (per mil) indicate cooler temperatures during precipitation. Mean values in a and b given by vertical lines. Wind blown or volcanic dust-particle concentrations are given as number of particles per milliliter of sample. The Huaynaputina, Peru, volcanic eruption is indicated by *.

(Modified from L. Thompson et al., 1986; L. Thompson, pers. comm. 1991, 2008; used with permission)

b. How could the record of the Huaynaputina volcanic eruption be used to date this ice core (Figure 18.1b)?

2. During this period, would you expect mountain glaciers in the region to have advanced (due to less summer melting) or retreated (due to more summer melting)? Explain.

3. From the record in 18.1a, has the last century at Quelccaya been warmer or cooler than the three centuries before 1900?

4. Between 1963 and 1991, significant retreat and lowering of an outlet glacier of the Quelccaya Ice Cap was observed. Although glaciers may not respond immediately to climate change, if increased temperature was the cause of this retreat, is this explanation consistent with temperature data in Figure 18.1a? Why or why not?

5. In 1600 the Rhone Glacier in Switzerland was in an advanced position. A four-story building, erected about 1950, now covers the glacier's former terminal position. Does this evidence agree or disagree with the ice core record in Figure 18.1a?

6. In parts of Glacier Bay, near Juneau, Alaska, glacier ice was near its maximum in 1820. Since that time, the ice has retreated more than 45 miles in this fiord. Does this Little Ice Age (about 1450 to 1850) advance of ice in Glacier Bay, and the retreat that followed, agree with the ice-core record from Peru (18.1)? Explain.

7. Study Figure 18.2 A, B, C, D, and E. This figure presents the oxygen isotope profiles for the tropical Quelccaya Ice Cap drilled in 1991 on the summit dome (A) and cores drilled in 1976 on the summit dome (B), the middle dome (C), and the south dome (D). The north-to-south cross sec-tion of the Quelccaya Ice Cap (E) illustrates the locations from which the oxygen isotope records in A through D were drilled and the position of the percolation line in 1976 and 1991. The percolation line is the altitude on the glacier below which melting of snow and subsequent percolation of water through the snow and firn occurs.

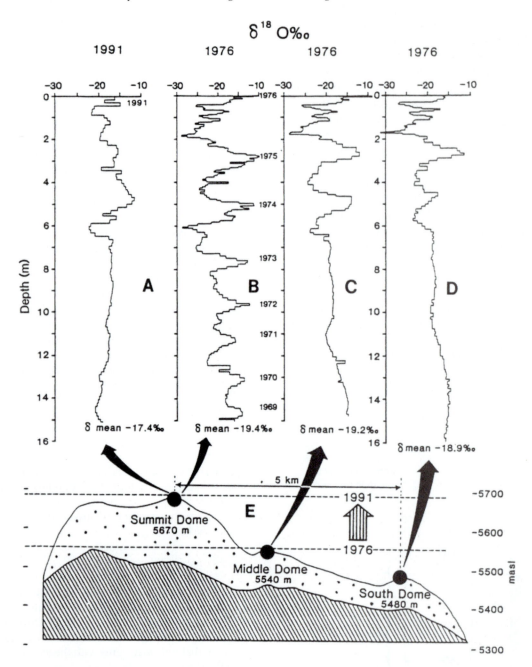

FIGURE 18.2   Oxygen isotope profiles (A to D) and north–south cross section (E) for the Quelccaya Ice Cap, Peru. Records A and B are from Summit Dome; C is from Middle Dome and D is from South Dome. Dashed lines are the altitudes (meters above sea level, masl) for the percolation or melting line.

(Modified from L. Thompson, pers. comm., 1991, 2008; used with permission)

a. How do the mean isotopic values at the summit core drilled in 1991 compare with those drilled in 1976? (See mean values at bottom of each record)

b. What does the change in mean isotopic value between 1976 and 1991 at Summit Dome suggest about atmospheric temperatures in this part of the World?

c. Why are the distinctive annual spikes seen in the 1976 summit core missing or subdued in the 1991 summit core? (Hint: Compare 1991 summit record with the record from South Dome.)

8. In 1991 the percolation line was above the Summit Dome (Figure 18.2E). How many meters has the percolation line risen between 1976 and 1991?

9. Melting now occurs at Summit Dome that did not occur there in 1976, indicating that the air temperature has warmed. We know that the temperature of the atmosphere changes by 0.7° C for each rise in elevation of 100 meters (0.007° C/1 meter). This is the atmospheric lapse rate. Using your answer in Question 8 and this lapse rate, calculate the temperature change that took place at Summit Dome between 1976 and 1991.

10. Study the photos (Figure 18.3) of the Qori Kalis Glacier, the largest outlet glacier from the Quelccaya Ice Cap. What has happened to this glacier between 1978 and 2000?

11. **a.** Study the map (A) in Figure 18.4, which illustrates positions of the Quori Kalis glacier terminus mapped from aerial photos taken in 1963 and terrestrial photos taken in 1978, and later. What is the approximate straight-line distance between the maximum position in 1963 (near the 1 in the date 1963 in the figure) and the center of the *average* position of the terminus in 1991?

b. From your measurement, what is the average annual rate of change for this period (1963–1991) in m/yr?

c. Study the inset graph in Figure 18.4. Has the rate of retreat of the glacier increased or decreased since 1978?

12. **a.** Determine the rate of retreat from 1991 to the most recent position of the end of the glacier as shown in Figure 18.5.

b. Has the rate increased from that during the 1963–1991 period?

13. To help understand the impact of even small changes in temperature, we now look at "the year without a summer." Volcanic events are recorded in ice cores as seen in Figure 18.1b where the eruption of Mt. _____ is shown by an ash layer. On April 5, 1815, Mount Tambora in Indonesia began erupting and injected ash and gases (e.g., H_2O, CO_2, HCl, SO_2, N_2) into the stratosphere. Some of these gases (for instance, HCl) have the potential to affect the ozone (O_3) layer; most act to enhance the greenhouse effect, but sulfate (SO_4) droplets in the stratosphere can increase cooling. In 1816 temperatures declined by 1° C in the northern hemisphere (Bryant, 1991). The following description, modified from the Berlin, Ohio, Sesquicentennial History (1966) captures the conditions in this area of the Midwest in the "year without a summer."

In each of the 12 months of 1816, ice and frost were experienced in Ohio, the Midwest, and New England. The month of January was very mild, and only a few cold days in February. March was moderately cold; April came in warm, but changed to snow and cold by the end of the month.

May was so cold that all buds and fruits were frozen and ice formed an inch thick. Corn that came up was frozen, replanted and frozen again. Other crops were likewise frozen. June was the coldest known with frost, ice and snow. Snow fell to a depth of 10 inches in Vermont, 7 inches in Maine, and 3 inches in Massachusetts and New York.

Frost was common in July. A half-inch of ice was recorded in August; the stunted corn was fed to cattle as fodder. September began cold with snow in New York on September 11. New Berlin was mild in mid-September, then cold, frost and 1/4 inch of ice. Frost and ice were common in October; November was stormy with much snow. December was mild and comfortable.

Switzerland was also experiencing unusual weather in the summer of 1816. The rainy and overcast weather was a factor in

FIGURE 18.3 Terminus of the Qori Kalis outlet glacier of the Quelccaya Ice Cap. Note changes in margin, snow, debris cover, and profile from 1978 (A) to 2000 (B). (From L. Thompson, pers. comm., 2008; used with permission)

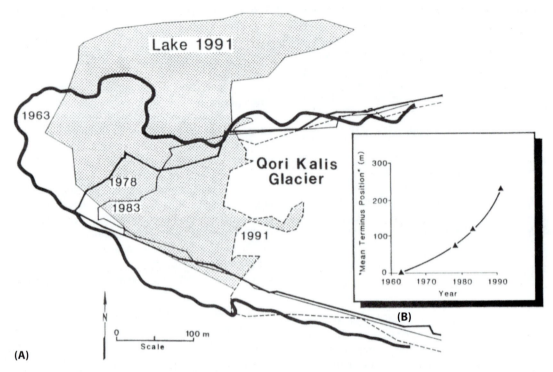

FIGURE 18.4 Map (A) of the Qori Kalis terminus in 1963 (thickest line), 1978 (intermediate line), 1983 (thinnest line), and 1991 (dashed line). Dot pattern is the proglacial lake as seen in 1991. Inset figure (B) is a plot of the retreat of the mean terminus position from 1963 to 1991. (From L. Thompson, pers. comm., 1991, 2008; used with permission)

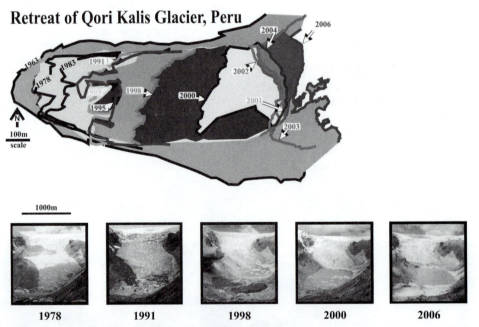

FIGURE 18.5 Map of changes in the Qori Kalis terminus from 1963 to 2006. Other Peruvian glaciers show similar trends.

(Modified from L. Thompson, pers. comm., 2008; used with permission)

Mary Shelley's writing a novel about Frankenstein (Shelley, 1965). Consider the introduction for this question and:

a. Describe the cause of this cooler weather in many places in the world in 1816.

b. Speculate on the impact on society of a larger eruption.

PART B. GLOBAL RESPONSE TO DECREASE IN STRATOSPHERIC OZONE

Stratospheric Ozone and the Ozone Hole

Ozone (O_3) is produced in the stratosphere when O_2 molecules are split by solar radiation and an oxygen atom (O) then bonds with O_2 to form O_3. This trace gas makes up only 0.000007% of the atmosphere (oxygen makes up 20.95%), so why were humans concerned about it in the 1970s when scientists (who later won Nobel prizes for their work on ozone) predicted that it would decrease due to interaction with CFCs? One reason is the fact that ozone blocks UV radiation, too much of which is harmful for humans (skin cancer and cataracts), animals, and some plants. Such a change would present human health problems and impact our life-support system. A second concern was economical. A large industry depended on CFCs—refrigerants for air conditioners and freezers, propellants in aerosol cans, etc. Industry's first response was that the CFC–O_3 Connection was bad science. How could heavy inert CFCs (chlorofluorocarbons) reach the stratosphere and then how would their chlorine interact with ozone molecules? Figure 18.6 contains the model for the reaction. CFCs may remain in the atmosphere for many years. Chlorine and bromine atoms remain in the atmosphere for one or two years and a chlorine atom may destroy 100,000 ozone molecules as it cycles through the O_3–ClO reaction!

Although most ozone is produced in the tropics, it makes its way to the polar regions, too. During the winter over isolated Antarctica, O_3 attaches to ice particles in the cold atmosphere. With sunrise in September/October, reactions with chlorine decrease the concentration of ozone (an ozone "hole" or thinning develops when the concentration of ozone is less than 220 Dobson Units or ppb). It is here in the Antarctic during spring that the lowest ozone readings occur.

Similar reactions occur in the Arctic, however to a lesser degree because it is warmer. The ozone decrease extends to the mid-latitudes and beyond. See Shanklin (2005) and other sources for additional information on Antarctic ozone.

The Antarctic thinning in October was discovered in 1985 at the British Antarctic station, Halley, where ozone had been measured in atmospheric studies since 1955. Swift international action produced the Montreal Protocol in 1987; later agreements hastened further the end of CFC production. The quick response occurred because less harmful substitutes were developed and there were only a limited number of manufacturers. The ozone problem showed that the world could unite to solve a global environmental-change problem. (The CO_2 global warming problem is much more difficult because there are many sources of CO_2, almost everyone depends on fossil fuel energy, and no readily available substitutes for fossil fuels exist.) The concentration of ozone-destroying chemicals in the stratosphere is near its peak. According to Shanklin (2004), the ozone hole over Antarctica is not expected to completely heal until mid-century—barring massive volcanic eruptions or a bolide impact such as the Tunguska event.

Questions on ozone are based primarily on data from Halley, Antarctica (Figure 18.7).

QUESTIONS 18, PART B

1. In Figure 18.6, identify the following in the space below or by marking on the diagram: CFC, ozone, and diatomic oxygen molecules and chlorine and oxygen atoms.

2. In step 5 to step 6 of Figure 18.6, what causes the chlorine atom to separate from the chlorine-oxygen molecule?

3. a. Why does one chlorine atom destroy many O_3 molecules?

b. Where and when in the year are the lowest values for stratospheric ozone?

c. Why is the ozone "hole" over Antarctica?

4. The concentration of ozone is measured in "Dobson units." From Figure 18.7 one Dobson unit is equal to _____

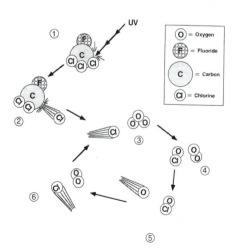

FIGURE 18.6 Chemical reactions for destruction of ozone in the stratosphere.

(Modified from NASA, 1993)

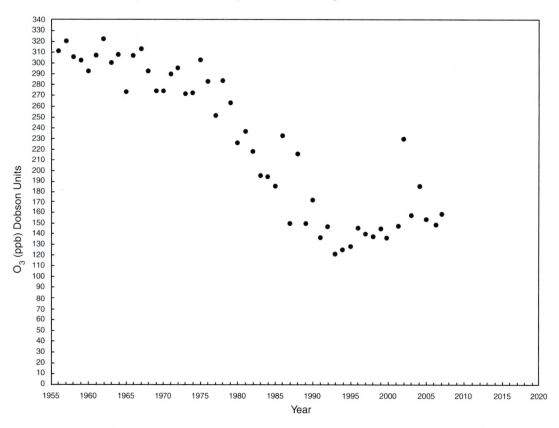

FIGURE 18.7 Provisional mean total column ozone at Halley (76° S), Antarctica, for October. Data from Jon Shanklin (personal communication, 1998) and updated (2008) with permission from National Environment Research Council - British Antarctic Survey site (http://www.antarctica.ac.uk/met/jds/ozone/data/ZOZ5699.DAT). Additional ozone information at http://www.antarctica.ac.uk/met/jds/ozone/.

5. What is the average value for the October concentration of ozone in the stratosphere over Halley Station, Antarctica, in 1956? _____ in 1972 _____ in 1993 _____

6. On Figure 18.7 sketch a smooth best-fit curve through the data. About when did the rate of decline change significantly?

7. What was the average annual decrease or rate of change (ppb/year) from 1972 to 1993?

8. Draw a straight trend line through the 1972 and 1994 data points to the bottom axis. What might you conclude about the ozone problem from dates available at the end of 1994?

9. From data up to 2004, what might you conclude about society's response to the ozone problem?

10. In addition to the October decrease in ozone, there has been a corresponding increase in the area of the ozone hole. Fortunately few people live in Antarctica (mainly scientists), but ozone could impact marine life (productivity of plankton), an important component of the world's life support system. Check the WEB for more recent data on the October average ozone value at Halley Station.

11. Why were the nations able to reach agreement on CFC reduction? (give several factors)

PART C. CO$_2$, POPULATION, AND °C CHANGE

A rise in CO$_2$ concentration and other greenhouse gases "(including water vapor, methane, nitrous oxide, carbon monoxide, and others.) in the atmosphere will produce warmer climates in the future. With a doubling of CO$_2$ in the atmosphere to about 700 ppm it has been estimated that there would be an increase in temperature of about 2 to 4°C. According to models, temperature changes in high latitudes would be greater than in equatorial regions. Portions of the globe will experience some benefit from climate warming (e.g., longer growing and ice-free transportation seasons); others will not. Adjusting to climate change may be difficult for some sectors (e.g., agriculture, urban water and sewage systems, and river and lake transportation). Current responses to rising greenhouse gases (GHG) probably will not prevent CO_2 from exceeding 700 ppm.

QUESTIONS 18, PART C

Study Figure 18.8 and then answer the following questions.

1. Where were the data collected that are shown in the graph?

2. According to Figure 18.8, when did CO_2 levels first begin to increase?

3. Calculate the average annual rate of CO_2 increase (in ppm/yr) between 1960 and 1980?

4. Using Figure 18.8, estimate the CO_2 concentration in the year 2020 by linear projection. Extend the CO_2 line by placing a ruler on the line joining 1980 and the last point (2004). Draw this line through the year 2020. The CO_2 value expected at that time by this technique is _____ ppm.

5. Using the data in Table 18.1, plot the change in world population between 1700 and 2005 on the same figure as CO_2 increase (Figure 18.8). Use the right axis for the population scale. Write a revised caption for Figure 18.8 here.

6. Note the shape of the two curves in Figure 18.8. What is the apparent cause–effect relationship between the two?

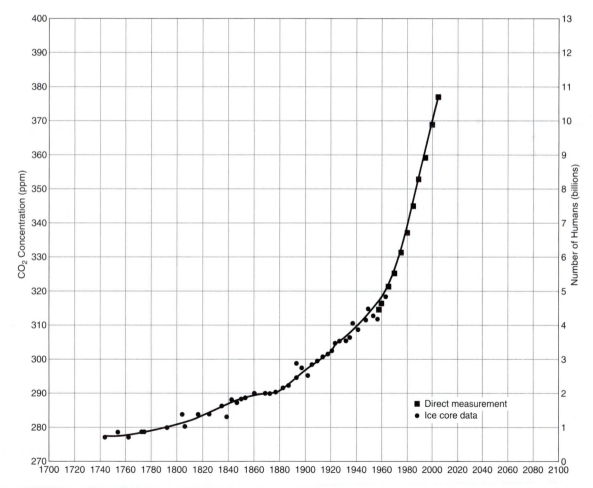

FIGURE 18.8 CO_2 concentration in the atmosphere. Squares are direct atmospheric measurements at Mauna Loa and circles are measurements from ice cores from Siple Station, Antarctica.

(Modified from Siegenthaler and Oeschger, 1987; data also from Keeling, 1976, and Mauna Loa Observatory.)

TABLE 18.1 Approximate World Human Population, 1700–2005

Year	World Population (billions)
1700	0.660
1750	0.800
1800	1.000
1850	1.300
1900	1.600
1930	2.000
1950	2.500
1960	3.000
1975	4.000
1990	5.300
2000	6.100
2005	6.450

7. Given that the world's energy use and population continue to increase, over the next 10 years will the CO_2 content of the atmosphere (check one)

_____ increase at its current rate

_____ increase at a slower rate

_____ increase at a more rapid rate

_____ decrease

Why?

8. In 1997, the complex diagram (Figure 18.9), provided a comprehensive summary of changes in past temperatures and atmospheric CO_2. The records extended from the present, through the last ice age (Wisconsinan) and though the last interglacial to 150,000 years before the present. These data from the Vostok, Antarctica, ice core extend well beyond the directly measured and earlier ice-core derived temperatures of Figure 18.8. Use the information in Figure 18.9 to help answer the following questions.

a. When was the lowest value for atmospheric CO_2? What was the value?

b. In the 150,000 years prior to 1850, about when did the highest atmospheric CO_2 occur? What was the CO_2 value at that time?

c. What was the maximum temperature variation after that peak in CO_2?

d. About how much colder was the temperature during the coldest parts of the glacial periods than during the warmest part of the last interglacial?

9. What might you infer (Figure 18.9) about the general relationship between CO_2 and temperature of the Earth in the past?

10. From Figure 18.9

a. What is the projected temperature variation in 2100?

b. What is the projected CO_2 level with "no controls" in 2100?

c. How does the projected temperature for 2100 compare with the temperature during the height of the last interglacial (125,000 years ago)?

d. If CO_2 levels go beyond 700 ppm, what change, if any, would you expect in temperature?

11. The Third Assessment Report (TAR) in 2001 of the Intergovernmental Panel on Climate Change (IPCC, 2001) provides an update on values in Figure 18.9. Highlights from the Summary for Policy Makers include the following.

Projected concentrations of **atmospheric CO_2** for the year 2100 range from 540 to 970 ppm compared with about 280 ppm for pre-industrial (pre-1750) values and about 368 ppm for the year 2000. These projections are based on population, social, technical, and economic factors; however, additional uncertainty arises when natural carbon removal and climate feedback impacts are considered. The range becomes 490–1250 ppm, which are increases of 75–350 percent over the concentration in the year 1750.

TAR also looks at revised values for temperature. Using the GHG projections and several climate models, **globally averaged surface temperatures** are expected to increase by 1.4–5.8° C in the period 1990–2100.

Furthermore, the same report shows weakening in the **ocean thermohaline circulation** system; however, no models show a shutdown by 2100. That could occur beyond 2100 and would indicate a major climate threshold had been passed. If local warming of 5.5°C occurred in Greenland for 1,000 years, this would add about 3 m to global **sea level.**

a. Given the above summary, what action should the world take now (or in the future) in the way of long-range planning to address these climate changes? Consider several responses in your point-form answer. If you opt for action only in the future, indicate when you would take action.

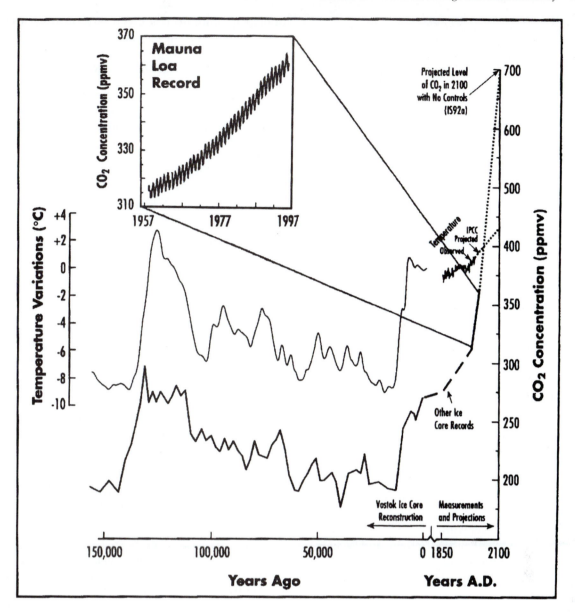

FIGURE 18.9 Past and projected global temperatures and atmospheric CO_2. Note scale change (National Science and Technology Council, 1997).

b. Your instructor may assign specific topics in the TAR for you or your group to research before next class. Topics might include expected temperature increase over 1,100 years (beginning in the year 1000), sea level rise and its causes, scenarios for GHG emissions, options for reducing emissions, and how long temperature will continue to increase after GHG are stabilized? Also, What are the socioeconomic scenarios? What are the 20th century changes in the Earth system? What are the impacts of extreme climate changes?

You may be asked to prepare an oral report (with written notes) on your topic.

c. Additionally, answer the question, What action should the world take now (or in the future) to address these projected climate changes?

The TAR Summary for Policy Makers is online with other IPCC Reports at: http://www.ipcc.ch/

12. The IPCC Fourth Assessment Report (AR4) further refines the certainty of human activity as a major factor in global warming and the possible range in temperature to be expected in 2100. Consult IPCC(2007) online and prepare a paragraph on one of the following:

a. expected temperature in 2100

b. CO_2 content of the atmoshphere in 2100

c. Expected sea level change by 2100.

The AR4 Synthesis Report is online with other Assessment Reports at http://www.ipcc.ch/ipccreports/assessments-reports.htm.

PART D. LONG-RANGE OUTLOOK: A SCENARIO

There are advantages to being aware of possible futures (e.g., for selection of financial investments or planning for sustainability). In either case, we are concerned about quality of life—beginning with our own and that of our children. But we also recognize that the quality of life of others impacts us. Our approaches to forecasting include intuition, trend extrapolation, system analysis with models/games, Delphi forecasts, and scenarios. In this short exercise you construct a scenario, a description of a possible future, viewed from the future looking back on what has happened to reach that future.

In addition to drawing on what you have learned from your course to this point (geologic hazards, pollution, geologic and other resource availability, regional planning, climate change), there are other sources of information you should consider. Although many sources are neutral in their outlook, some are optimistic (cornucopian) and others are pessimistic or maybe geologically realistic. Several useful references include Bartlett, 2004; Beckwith, 1986; Deffeyes, 2001; Diamond, 2005; IPCC, 2001 and 2007; Kahn and Wiener, 1967; Lapp, 1973; Meadows et al., 1972; Schwartz, 1991; Simon, 1990; Smil, 1972; and UN Population Fund, 1999. Other references include the journal of the World Future Society (the Futurist), annual State of the World books from the World Watch Institute, and Eco-Economy Updates and Plan B 3.0 (Brown, 2008) from the Earth Policy Institute (http://www.earth-policy.org).

There are two simple relationships that should be kept in mind when seeking to understand the future. One is a simplified quality-of-life relationship that was suggested by a former head of the USGS (McKelvey, 1972). It is:

$$Q = (REI)/P$$

where Q is the quality of Life; R is for Resources, E is for Energy; I is for ingenuity (our knowledge and technology); and P is the population that must share in the REI.

This is not a perfect expression for the Q of L; for some people "stuff" and/ or "$" are the most important, for others it is friends, family, and making a contribution to society.

The other relationship is the expression for Doubling Time with exponential growth. This describes the time it takes for an investment or other process to double in amount at a given rate of growth (for an investment it is compound interest). It is:

$$DT = 70/GR$$

where DT is Doubling Time in years and GR is annual Growth Rate in percent. For example, DT = 70/10 = 7 years (to double your money at 10% GR).

The exponential growth equation is more complicated than this; however, this simplified version works for small percentages. Concern over our lack of appreciation for exponential growth has prompted some to alert us to potential problems, particularly with respect to finite resources and population (Bartlett, 2004).

QUESTIONS 18, PART D

1. Using at least one of the references in the Introduction to Part D, prepare a *one-page scenario* describing the next 50 years. Write it as if the next 50 years have already passed. Consider starting your scenario with, "The last 50 years have produced a few surprises ... " or "The year is 20XX and it is a fine day in ____" or as a letter to a former classmate, "I am now ____ years old and things have changed since we were in class 50 years ago." You should focus on changes in energy and mineral resources, pollution, and quality of life, plus any unusual natural disaster events. You might want to include a few related social, economic, and political events. Unless your instructor suggests otherwise, you may take any viewpoint, optimistic/pessimistic, sustainable/unsustainable, etc.

2. Using the doubling time equation, the population growth rate of 1 percent (expected in 2015), and your ideas of how the world is progressing, answer the following.

a. Considering the trends in population and the availability of resources (including fossil fuels), what is your best guess for the *maximum human population* on Earth?

b. This maximum human population will occur in the year ____.

c. The major *factor(s) that will end population growth* will be (consider the Introduction to Section IV and any changes that you expect in population, resources, biodiversity, and global climate).

d. How will we achieve sustainability for humans on Earth?

Bibliography

Bartlett, Albert, 2004, Thoughts on long-term energy supplies: Scientists and the silent lie: *Physics Today,* v. 57, (0.7) p. 53. And comments, http://www.aip.org/pt/vol-57/iss-11/p12.html

Beckwith, B., 1986, *The next 1000 years:* New York, Exposition Press, 431 p.

Blong, R. J., 1984, *Volcanic Hazards: A sourcebook on the effects of eruptions:* Sydney, Academic Press.

Bradley, R. S. and Jones, P. D., 1992, *Climate since AD 1500:* New York, Routledge, 679 p.

British Antarctic Survey: Antarctic Ozone. Additional updated ozone information available at http://www.antarctica.ac.uk/met/jds/ozone/.

Brown, Lester R., 2008, Plan B 3.0: Mobilizing to Save Civilization: New York, W.W. Norton, 398 p.

Bryant, E. A., 1991, *Natural hazards:* New York, Cambridge University Press, 294 p.

Deffeyes, K. S., 2001, *Hubbert's Peak: The impending world oil shortage,* Princeton, NJ, Princeton University Press, 208 p.

Development Education Program, World Bank. Retrieved February 21, 2005, from http://www.worldbank.org/depweb/english/modules/index.html

Diamond, Jared, 2005, *Collapse: How societies choose to fail or succeed:* East Rutherford, NJ, Viking, 592 p.

Gelbspan, Ross, 1997, *The heat is on:* Reading, MA, Addison-Wesley, 278 p.

Harington, C. R., ed., 1992, *The year without a summer: World climate in 1816:* Ottawa, Canadian Museum of Nature, 576 p.

Horel, J., and Geisler, J., 1997, *Global environmental change, an atmospheric perspective:* New York, Wiley, 152 p.

Ice Core Paleoclimatology Research Group, Images of Quelccaya Ice Cap, Peru. Retrieved February 21, 2005, from http://www-bprc.mps.ohio-state.edu/Icecore/

IPCC, 2001, *Climate Change 2001: Synthesis Report. Summary for Policymakers, An Assessment of Intergovernmental Policy on Climate Change.* 34 p. Retrieved February 20, 2005, http://www.grida.no/climate/ipcc_tar/vol4/english/pdf/spm.pdf

IPCC, 2007, *Climate Change 2001: Synthesis Report. Summary for Policymakers, An Assessment of Intergovernmental Policy on Climate Change.* 34 p. Retrieved February 20, 2005, http://www.grida.no/climate/ipcc_tar/vol4/english/pdf/spm.pdf

Kahn, H., and Wiener, A. J., 1967, *The year 2000:* New York, Macmillan, 431 p.

Keeling, et al., 1976, Atmospheric carbon dioxide variations at Mauna Loa Observatory, Hawaii: *Tellus,* v. 28, p. 538–551.

Lapp, R. E., 1973, *The logarithmic century:* Englewood Cliffs, NJ, Prentice-Hall, 263 p.

Lorius, C., Jouzel, J., Raynaud, D., Hansen, J., and LeTrent, H., 1990, The ice-core record: climate sensitivity and the future greenhouse warming: *Nature,* v. 347, p. 139–145.

Lorius, C., Raisbeck, G., Jouzel, J., and Raynaud, D., 1989, Long-term environmental records from Antarctic ice cores. In *The environmental record in glaciers and ice sheets,* eds. H. Oeschger and C. C. Langway. New York, Wiley and Sons, p. 343–361.

Math in Daily Life: Population Growth, http://www.learner.org/exhibits/dailymath/getpicture.html

Mauna Loa Observatory, NOAA, 2005, *Mauna Loa Monthly Mean Carbon Dioxide:* Retrieved February 16, 2005, from http://www.mlo.noaa.gov/Links/Flinks.htm p. 26

McKelvey, V. E., 1972, Mineral resource estimates and public policy: *American Scientist,* v. 60, p. 32–40.

Meadows, D. H., Meadows, D. L., Randers, J., and Behrnes, W. W., III, 1972, *The limits to growth:* New York, Potomac Associates-Universe Books, 205 p.

NASA, 1993, *Ozone: What is it, and why do we care about it?* NASA Facts, NF 193, 4 p.

National Science and Technology Council, 1996, *Our Changing Planet: The FY 1997 U.S. Global Change Research Program:* Washington, DC, Office of Science and Technology, 162 p.

National Science and Technology Council, 1997, *Our Changing Planet: The FY 1998 U.S. Global Change Research Program:* Washington, DC, Office of Science and Technology, 118 p.

Population Connection: Retrieved February 16, 2005, from http://www.populationconnection.org/

Population Reference Bureau, *Human population: Fundamentals of growth.* Retrieved February 21, 2005, from http://www.prb.org/Content/NavigationMenu/PRB/Educators/Human_Population/Population_Growth/Population_Growth.htm

Population Reference Bureau. Retrieved February 16, 2005, at http://www.prb.org/

Rose, W. I., and Chesner, C. A., 1990, Worldwide dispersal of ash and gases from the earth's largest known eruption: Toba, Sumatra, 75 ka: *Palaeogeography, Palaeoclimatology, Palaeoecology,* v. 89, no. 3, p. 269–274.

Schwartz, P., 1991, *The art of the long view:* New York, Doubleday

Shanklin, Jonathan, 2004, *RAS Fact Sheet: The Ozone Layer.* Retrieved February 19, 2005, http://www.antarctica.ac.uk/met/jds/ozone/o3facts.html

Shanklin, Jonathan, 2008, Provisional mean total ozone at Halley (76° S), Antarctica: National Environment Research Council, British Antarctic Survey. Retrieved August 25, 2008, from http://www.antarctica.ac.uk/met/jds/ozone/data/ZOZ5699.DAT

Shelley, M., 1965, *Frankenstein, or the modern prometheus:* New York, New American Library.

Siegenthaler, U., and Oeschger, H., 1987, Biospheric CO2 emissions during the past 200 years reconstructed by deconvolution of ice-core data: *Tellus,* v. 39B, p. 361–364.

Simon, J. L., 1990, Population growth is not bad for humanity: *National Forum,* v. 70, no. 1, p. 12–16.

Smil, V., 1972, Energy and the environment: A delphi forecast: *Long Range Planning,* v. 5, no. 4, p. 27–32.

Stolarski, R., Bojkov, R., Bishop, L., Zerefos, C., Staehlin, J., and Zawodny, J., 1992, Measured trends is stratospheric ozone: *Science,* v. 256, p. 342–349.

Stommel, H., and Stommel, E., 1983, *Volcano weather: The story of 1816 the year without a summer:* Newport, RI, Seven Seas Press, 177 p.

Thompson, L., Mosley-Thompson, E., Dansgaard, W., and Grootes, P.M., 1986, The Little Ice Age as recorded in the stratigraphy of the tropical Quelccaya Ice Cap: *Science,* v. 234, p. 361–364.

UN Population Fund, 1999, *The day of 6 billion.* (See also counter for current population): Retrieved February 16, 2005, from http://www.unfpa.org/6billion/index.htm

APPENDIX I. UNITS AND CONVERSIONS

Prefixes for International System of Units

Multiples & Submultiples	Prefixes	Symbols
$1,000,000,000,000,000 = 10^{15}$	penta	P
$1,000,000,000,000 = 10^{12}$	tera	T
$1,000,000,000 = 10^9$	giga	G
$1,000,000 = 10^6$	mega	M
$1,000 = 10^3$	kilo	k
$100 = 10^2$	hecto	h
$10 = 10^1$	deka	da
$0.1 = 10^{-1}$	deci	d
$0.01 = 10^{-2}$	centi	c
$0.001 = 10^{-3}$	milli	m
$0.000001 = 10^{-6}$	micro	μ
$0.000000001 = 10^{-9}$	nano	n
$0.000000000001 = 10^{-12}$	pico	p
$0.000000000000001 = 10^{-15}$	femto	f

Units of Measure

Linear Measure

1 foot (ft)	= 12 inches (in)
1 mile (mi)	= 5,280 feet (ft)
1 chain (ch)	= 66 ft
1 rod (rd)	= 16.5 ft
1 fathom (fm)	= 6 ft
1 nautical mile	= 6,076.115 ft
1 kilometer (km)	= 1000 meters (m)
1 km	= 10^3 m
1 centimeter (cm)	= 0.01 m = 10^{-2} m
1 millimeter (mm)	= 0.001 m = 10^{-3} m
1 angstrom (Å)	= 0.0000000001 m = 10^{-10} m
1 micron (μ)	= 0.001 mm

Area Measure

1 square mile	= 640 acres
1 acre	= 43,560 sq ft
1 acre	= 4,840 sq yds = 160 sq rods
1 mile square	= 1 section = 640 acres
6 mile square	= 1 township = 36 sq miles
1 square meter	= 10,000 sq centimeters (cm)
100 square meters	= 1 are (a)
10,000 square meters	= 1 ha
100 ares	= 1 hectare (ha)
100 hectares	= 1 sq km

Conversions

English-Metric Conversions

1 inch	= 25.4 millimeters
1 foot	= 0.3048 meter
1 yard	= 0.9144 meter
1 mile	= 1.609 kilometers
1 sq inch	= 6.4516 sq centimeters
1 sq foot	= 0.0929 sq meter
1 sq yard	= 0.836 sq meter
1 sq mile	= 259 hectares
1 sq mile	= 2.59 sq kilometers
1 acre	= 0.4047 hectare
1 acre	= 4047 sq meters
1 cubic inch	= 16.39 cubic centimeters
1 cubic foot	= 0.0283 cubic meter
1 cubic yard	= 0.7646 cubic meter
1 quart (liq)	= 0.946 liter
1 gallon (U.S.)	= 0.003785 cubic meter
1 ounce (avdp)	= 28.35 grams
1 pound (avdp)	= 0.4536 kilogram
1 short ton	= 907.2 kilograms
1 horsepower	= 0.7457 kilowatt

Metric-English Conversions

1 millimeter	= 0.0394 inch
1 meter	= 3.281 feet
1 meter	= 1.094 yards
1 kilometer	= 0.6214 mile
1 sq centimeter	= 0.155 sq inch
1 sq meter	= 10.764 sq feet
1 sq meter	= 1.196 sq yards
1 hectare	= 2.471 acres
1 hectare	= 0.003861 sq mile
1 sq kilometer	= 0.3861 sq mile
1 cu centimeter	= 0.061 cu inch
1 cu meter	= 35.3 cu feet
1 cu meter	= 1.308 cu yards
1 liter	= 1.057 quarts
1 cu meter	= 264.2 gallons (U.S.)
1 gram	= 0.0353 ounce (avdp)
1 kilogram	= 2.205 pounds (avdp)
1 metric ton	= 2205 pounds (avdp)
1 kilowatt	= 1.341 horsepower

Volume and Cubic Measure

1 quart	= 2 pints = 57.75 cubic inches
4 quarts	= 1 gallon = 231 cubic inches
1 cubic foot	= 1728 cubic inches
1 cubic yard	= 27 cubic feet
1 barrel (oil)	= 42 gallons
1 barrel (proof spirits)	= 40 gallons
1 cubic foot	= 7.48 gallons
1 cubic inch	= 0.554 fluid ounce
1 gallon (U.S.)	= 128 U.S. fluid ounces = 0.833 British gal
1 liter	= 0.001 cubic meter = 1 cubic decimeter
1 liter	= 1000 milliliters
1 deciliter	= 100 milliliters
1 milliliter	= approximately 1 cubic centimeter (cc)
1 cubic meter (m^3)	= 1,000,000 cubic centimeters

Weights and Masses

1 short ton	= 2000 pounds = 907.2 kilograms
1 long ton	= 2240 pounds = 1016 kilograms
1 metric ton	= 2205 pounds = 1000 kilograms
1 pound (avoirdupois)	= 7000 grains
1 ounce (avoirdupois)	= 437.5 grains
1 gram	= 15.432 grains
1000 grams	= 1 kilogram

Additional Conversions

1 gallon of water = 8.3453 pounds of water
1 gallon per min = 8.0208 cubic ft per hour
1 acre-foot = 1233.46 m^3 = 325,829 gal
1 ft^3 per second = 0.0283 m^3 per second
1 ft^3 of fresh water = 62.4 lb = 28.3 kg
1 billion gallons per day (bgd) =
  3.785 million m^3 per day

360 degrees (°) = complete circle
1 degree (°) = 60 minutes (')
1 minute (') = 60 seconds (")

Temperature

To change from Fahrenheit (F) to Celsius (C)

$$°C = \frac{(°F - 32°)}{1.8}$$

To change from Celsius (C) to Fahrenheit (F)

$$°F = (°C \times 1.8) + 32°$$

Force

1 dyne (d) = the force that will produce an acceleration of 1 centimeter/second^2 when applied to a 1-gram mass.
1 newton (nt) = the force that will produce an acceleration of 1 meter/second^2 when applied to a 1-kilogram mass.
1 nt = 100,000 d = 1 × 10^5 d

Energy and Power

1 erg = the work done by a force of 1 dyne when its point of application moves through a distance of 1 centimeter in the direction of the force.
1 erg = 9.48 × 10^{-11} British thermal unit (Btu)
1 erg = 7.367 × 10^{-8} foot-pounds
1 erg = 2.778 × 10^{-14} kilowatt-hours
1 kilowatt-hour = 3413 Btu = 3.6 × 10^{13} ergs = 860,421 calories (cal)
1 Btu = 2.930 × 10^{-4} kilowatt-hours = 1.0548 × 10^{10} ergs = 252 calories (cal)
1 watt[*] = 3.413 Btu/hour
1 watt = 1.341 × 10^{-3} horsepower
1 watt = 1 joule per second
1 watt = 14.34 calories per minute
1 joule[*] = 1 × 10^7 ergs
1 joule = 1 newton-meter

[*] The watt and the joule are the internationally acceptable units for power (energy per unit time) and energy, respectively.

Heat

1 calorie (cal) = the amount of heat that will raise the temperature of 1 gram of water 1° Celsius (water at 4° Celsius).
1 calorie (gram) = 3.9685 × 10^{-3} Btu = 4.186 × 10^7 ergs

Pressure

1 millibar (mb) = 1000 dynes per cm^2 = 0.1 kilopascal (kPa)

1 atmosphere (atm) = 76 cm mercury = 14.70 lb/in^2 = 1013 millibars (mb)

APPENDIX II. GEOLOGIC TIME SCALE

1999 GEOLOGIC TIME SCALE

Geological Society of Amercia

CENOZOIC

PERIOD	EPOCH		AGE	PICKS (Ma)
QUATER-NARY	HOLOCENE			
	PLEISTOCENE	L	CALABRIAN	0.01
		E		1.8
NEOGENE	PLIOCENE	L	PIACENZIAN	
		E	ZANCLEAN	3.6
	MIOCENE	L	MESSINIAN	5.3
			TORTONIAN	7.1
			SERRAVALLIAN	11.2
		M	LANGHIAN	14.8
		E	BURDIGALIAN	16.4
			AQUITANIAN	20.5
PALEOGENE	OLIGOCENE	L	CHATTIAN	23.8
		E	RUPELIAN	28.5
	EOCENE	L	PRIABONIAN	33.7
			BARTONIAN	37.0
		M	LUTETIAN	41.3
		E	YPRESIAN	49.0
	PALEOCENE	L	THANETIAN	54.8
			SELANDIAN	57.9
		E	DANIAN	61.0
				65.0

(TERTIARY)

MESOZOIC

PERIOD	EPOCH	AGE	PICKS (Ma)	UNCERT. (m.y.)
CRETACEOUS	LATE	MAASTRICHTIAN	65	4.2
		CAMPANIAN	71.3	1
		SANTONIAN	83.5	1
		CONIACIAN	85.8	1
		TURONIAN	89.0	1
		CENOMANIAN	93.5	4
	EARLY	ALBIAN	99.0	1
		APTIAN	112	2
	(NEOCOMIAN)	BARREMIAN	121	3
		HAUTERIVIAN	127	3
		VALANGINIAN	132	4
		BERRIASIAN	137	4
JURASSIC	LATE	TITHONIAN	144	5
		KIMMERIDGIAN	151	6
		OXFORDIAN	154	7
	MIDDLE	CALLOVIAN	159	7
		BATHONIAN	164	8
		BAJOCIAN	169	8
		AALENIAN	176	8
	EARLY	TOARCIAN	180	8
		PLIENSBACHIAN	190	8
		SINEMURIAN	195	8
		HETTANGIAN	202	8
TRIASSIC	LATE	RHAETIAN	206	8
		NORIAN	210	8
		CARNIAN	221	9
	MIDDLE	LADINIAN	227	9
		ANISIAN	234	9
	EARLY	OLENEKIAN	242	9
		INDUAN	245	9
			248	10

PALEOZOIC

PERIOD	EPOCH	AGE	PICKS (Ma)
PERMIAN	L	TATARIAN	248
		UFIMIAN-KAZANIAN	252
		KUNGURIAN	256
	E	ARTINSKIAN	260
		SAKMARIAN	269
		ASSELIAN	282
CARBONIFEROUS (PENNSYLVANIAN)	L	GZELIAN	290
		KASIMOVIAN	296 S.
		MOSCOVIAN	303
	E	BASHKIRIAN	311 W.
CARBONIFEROUS (MISSISSIPPIAN)	L	SERPUKHOVIAN	323 N.
		VISEAN	327
	E	TOURNAISIAN	342
DEVONIAN	L	FAMENNIAN	354
		FRASNIAN	364
	M	GIVETIAN	370
		EIFELIAN	380
	E	EMSIAN	391
		PRAGHIAN	400
SILURIAN	L	LOCKHOVIAN	412
		PRIDOLIAN	417
		LUDLOVIAN	419
	E	WENLOCKIAN	423
		LLANDOVERIAN	428
ORDOVICIAN	L	ASHGILLIAN	443
		CARADOCIAN	449
	M	LLANDEILIAN	458
		LLANVIRNIAN	464
	E	ARENIGIAN	470
		TREMADOCIAN	485
CAMBRIAN*	D	SUNWAPTAN*	490
		STEPTOEAN*	495
	C	MARJUMAN*	500
	B	DELAMARAN*	506
		DYERAN*	512
	A	MONTEZUMAN*	516
			520
			543

PRECAMBRIAN

EON	ERA	BDY. AGES (Ma)
PROTEROZOIC	LATE	543
		900
	MIDDLE	
		1600
	EARLY	
		2500
ARCHEAN	LATE	
		3000
	MIDDLE	
		3400
	EARLY	
		3800?

APPENDIX III. POPULATION DATA

TABLE A Population, 1950 to 2025, and Rate of Growth

| World Regions | Rate of Population Growth | | | POPULATION (millions) | | | | | | | | | | | | |
| | 1970–75 | 1985–90 | 2000–05 | 1950 | | | 1975 | | | 2000 | | | 2025 | | |
				Total	Urban	Rural	Total	Urban	Rural	Total	Urban	Rural	Total	Urban	Rural
Eastern Africa	2.9	3.3	3.3	60	3	57	117	15	102	266	78	188	531	254	277
Middle Africa	2.5	2.8	2.9	29	4	25	48	14	34	96	50	46	183	125	58
Northern Africa	2.4	2.8	2.2	52	13	39	94	38	56	186	108	78	295	213	82
Southern Africa	2.3	2.6	2.4	17	6	11	29	13	16	54	33	21	91	67	24
Western Africa	3.0	3.2	3.2	64	6	58	123	25	98	275	101	174	542	298	244
Caribbean	2.0	1.7	1.5	17	6	11	28	14	14	42	27	15	59	44	15
Central America	3.2	2.5	1.9	37	15	22	80	46	34	150	107	43	223	180	43
Temperate South America	1.6	1.4	1.1	25	16	9	39	31	8	55	49	6	70	65	5
Tropical South America	2.5	2.2	1.7	86	31	55	175	107	68	304	239	65	436	375	61
Northern America	1.0	0.9	0.7	166	106	60	239	176	63	298	232	66	347	298	49
China	2.4	1.0	1.0	554	68	486	933	183	750	1,256	334	922	1,460	665	795
Japan	1.3	0.4	0.3	84	42	42	112	85	27	128	101	27	128	110	18
Other East Asia	2.2	1.7	1.1	33	9	24	57	30	27	87	68	19	108	93	15
Southeastern Asia	2.4	1.9	1.4	181	27	154	324	70	254	520	186	334	685	374	311
Southern Asia	2.4	2.0	1.4	472	75	397	846	180	666	1,386	468	918	1,816	948	868
Western Asia	2.9	2.8	2.3	42	10	32	85	41	44	168	110	58	270	209	61
Eastern Europe	0.5	0.5	0.4	88	37	51	106	60	46	121	86	35	131	106	25
North & West Europe	0.5	0.1	0.0	195	134	61	234	186	48	239	208	31	233	213	20
Southern Europe	1.0	0.5	0.3	109	48	61	134	80	54	153	111	42	163	133	30
Australia-New Zealand	1.7	1.1	1.0	10	8	2	17	14	3	22	20	2	28	26	2
Micronesia & Polynesia	2.6	2.4	1.9	2	0	2	5	1	4	8	2	6	12	5	7
USSR	1.0	0.9	0.7	180	71	109	253	152	101	315	234	81	367	306	61

Modified from United Nations Fund for Population Activities, *Population Images* (1987) after United Nations 1982 Assessment.

TABLE B Population, 1990 and 2025, and Average Growth Rate

| | Population | | Average Growth (%) |
	1990	2025	1990–95
World Total	5,292.2	8,504.2	1.7
More Developed*	1,206.6	1,353.9	0.5
Less Developed	4,085.6	7,150.3	2.1
Africa	642.1	1,596.9	3.0
Northern America	275.9	332.0	0.7
Latin America	448.1	757.4	1.9
Asia	3,112.7	4,912.5	1.8
China	1,139.1	1,512.6	1.4
Europe	498.4	515.2	0.2
USSR	288.6	352.1	0.7

*Northern America, Japan, Europe, Australia-New Zealand, USSR
Modified from United Nations Population Fund, 1991, *The State of World Population, 1991*: New York, United Nations, 48p.

MILE SCALE 1:62 500

UNITED STATES
DEPARTMENT OF THE INTERIOR
GEOLOGICAL SURVEY

TOPOGRAPHIC
MAP INFORMATION AND SYMBOLS
SEPTEMBER 1972

QUADRANGLE MAPS AND SERIES

Quadrangle maps cover four-sided areas bounded by parallels of latitude and meridians of longitude. Quadrangle size is given in minutes or degrees. The usual dimensions of quadrangles are: 7.5 by 7.5 minutes, 15 by 15 minutes, and 1 degree by 2 or 3 degrees.
Map series are groups of maps that conform to established specifications for size, scale, content, and other elements.

MAP SCALE DEPENDS ON QUADRANGLE SIZE

Map scale is the relationship between distance on a map and the corresponding distance on the ground.
Map scale is expressed as a numerical ratio or shown graphically by bar scales marked in feet, miles, and kilometers.

National Topographic Maps

Series	Scale	1 inch represents approximately	1 centimeter represents	Size (latitude × longitude)	Area (square miles)
Puerto Rico 7.5-minute	1:20,000	1,667 feet	200 meters	7.5 × 7.5 min.	71
7.5-minute	1:24,000	2,000 feet (exact)	240 meters	7.5 × 7.5 min.	49 to 70
7.5-minute	1:25,000	2,083 feet	250 meters	7.5 × 7.5 min.	49 to 70
7.5 × 15-minute	1:25,000	2,083 feet	250 meters	7.5 × 15 min.	98 to 140
USGS/DMA 15-minute	1:50,000	4,166 feet	500 meters	15 × 15 min.	197 to 282
15-minute	1:62,500	1 mile	625 meters	15 × 15 min.	197 to 282
Alaska 1:63,360	1:63,360	1 mile (exact)	633.6 meters	15 × 20 to 36 min.	207 to 281
County 1:50,000	1:50,000	4,166 feet	500 meters	County area	Varies
County 1:100,000	1:100,000	1.6 miles	1 kilometer	County area	Varies
30 × 60-minute	1:100,000	1.6 miles	1 kilometer	30 × 60 min.	1,568 to 2,240
U. S. 1:250,000	1:250,000	4 miles	2.5 kilometers	1° × 2° or 3°	4,580 to 8,669
State maps	1:500,000	8 miles	5 kilometers	State area	Varies
U. S. 1:1,000,000	1:1,000,000	16 miles	10 kilometers	4° × 6°	73,734 to 102,759
U. S. Sectional	1:2,000,000	32 miles	20 kilometers	State groups	Varies
Antarctica 1:250,000	1:250,000	4 miles	2.5 kilometers	1° × 3° to 15°	4,089 to 8,336
Antarctica 1:500,000	1:500,000	8 miles	5 kilometers	2° × 7.5°	28,174 to 30,462

CONTOUR LINES SHOW LAND SHAPES AND ELEVATION

The shape of the land, portrayed by contours, is the distinctive characteristic of topographic maps.
Contours are imaginary lines following the ground surface at a constant elevation above or below sea level.
Contour interval is the elevation difference represented by adjacent contour lines on maps.
Contour intervals depend on ground slope and map scale; they vary from 5 to 1,000 feet. Small contour intervals are used for flat areas; larger intervals are used for mountainous terrain.
Supplementary dotted contours, at less than the regular interval, are used in selected flat areas.
Index contours are heavier than others and most have elevation figures.
Relief shading, an overprint giving a three-dimensional impression, is used on selected maps.
Orthophotomaps, which depict terrain and other map features by color-enhanced photographic images, are available for selected areas.

COLORS DISTINGUISH KINDS OF MAP FEATURES

Black is used for manmade or cultural features, such as roads, buildings, names, and boundaries.
Blue is used for water or hydrographic features, such as lakes, rivers, canals, glaciers, and swamps.
Brown is used for relief or hypsographic features—land shapes portrayed by contour lines.
Green is used for woodland cover, with patterns to show scrub, vineyards, or orchards.
Red emphasizes important roads and is used to show public land subdivision lines, land grants, and fence and field lines.
Red tint indicates urban areas, in which only landmark buildings are shown.
Purple is used to show office revision from aerial photographs. The changes are not field checked.

INDEXES SHOW PUBLISHED TOPOGRAPHIC MAPS

Indexes for each State, Puerto Rico and the Virgin Islands of the United States, Guam, American Samoa, and Antarctica show available published maps. Index maps show quadrangle location, name, and survey date. Listed also are special maps and sheets, with prices, map dealers, Federal distribution centers, and map reference libraries, and instructions for ordering maps. Indexes and a booklet describing topographic maps are available free on request.

How to order maps and indexes

Mail orders: Free indexes may be ordered by State or index name from USGS map sales offices. Maps must be ordered by map name, State, and series/scale. Payment by money order or check payable to the Department of the Interior–USGS must accompany your order. A $1.00 postage and handling charge is applicable on orders less than $10.00. Your complete address, including ZIP code, is required. Mail your order and prepayment to: **USGS MAP SALES**
BOX 25286
DENVER, CO 80225

Residents of Alaska may order Alaska maps or an index for Alaska from:

USGS MAP SALES—ALASKA
101 12th AVENUE—BOX 12
FAIRBANKS, AK 99701

Commercial dealers: Names and addresses of dealers are listed in your local yellow pages.
Commercial dealers sell U.S. Geological Survey maps at their own prices.

MILE SCALE 1:24 000

FOOT SCALE 1:24 000

COLOR PLATES

Plate Number	Description
1	FIGURE 2.9 Topographic map symbols (USGS)
2	FIGURE 2.10 Bloomington, IN 7.5 minute quadrangle (USGS)
3	FFIGURE 2.11 Geologic map of west-central Great Lakes region of Canada and USA. Compiled by….
4	FIGURE 2.14 High altitude color photograph of Boston, MA (USGS)
5	FIGURE 2.16 Color stereopair aerial photos of Little Cottonwood Canyon, Salt Lake City, UT. North to left. See Figures 6.11 and 6.12. (USDA, SCS, 1992).
6	FIGURE II.3 Presidential disaster declarations: December 24, 1964 to March 3, 2007, (FEMA, 2008)
7	FIGURE II.4 Population density for counties: July 1, 2004 (U.S. Census Bureau, Population Estimates Program)
8	FIGURE 4.9 Mt. Rainier, WA 1: 500K Raven map (© 1987 Raven Maps and Images. Reprinted with Permission)
9	FIGURE 5.5 Geologic map of deposits and features of the 1980 eruptions of Mount St. Helens National Volcanic Monument (USDA, Forest Service, 1997)
10	FIGURE 6.11 Draper, UT 7.5 minute quadrangle (USGS) (see Figures 2.16 and 6.12)
11	FIGURE 6.12 Color oblique aerial photo, Little Cottonwood Canyon, Salt Lake City, UT. See Lower Bells Canyon Reservoir on Figures 2.16 and 6.11. North to the left. (USDA, SCS)
12	FIGURE 8.13 Calaveras Reservoir, CA (San Jose) 7.5 minute quadrangle (USGS)
13	FIGURE 8.23 Athens, OH 7.5 minute quadrangle (a. upper, 1975 & b. lower, 1995) (USGS)
14	FIGURE 8.29 Avalanche areas in the Aspen, 7.5 minute quadrangle, Pitkin County, CO. (USGS Map I-785-G, 1972)
15	FIGURE 10.5 Zanesville West, OH 7.5 minute quadrangle (USGS)
16	FIGURE 17.3 Geologic map of Waco West, TX quadrangle by Burket, 1963 (Burket, 1965)

NOTES

NOTES

NOTES

NOTES

NOTES

NOTES

NOTES

ADDITIONAL CREDITS

Figure 2.13   Avery, Thomas Eugene; Berlin, Graydon Lennis L.,1992. *Fundamentals of Remote Sensing & Airphoto Interpretation, 5th ed.,* Electronically reproduced by permission of Pearson Education, Inc., Upper Saddle River, New Jersey.

Figure 10.14   Keller, Edward A., 2002. *Introduction to Environmental Geology, 2nd ed.* Electronically reproduced by permission of Pearson Education, Inc., Upper Saddle River, New Jersey.

Figure 12.3   Reprinted by permission of Johnson Screens / a Weatherford Company.

Figure 14.3   Data from Painter, et al., 2001, *Sediment contamination in Lake Erie: A 25-year retrospective analysis: J. Great Lakes Research,* 27(4): 434-448; used with permission.

Figure 18.1   Thompson, L., and others, 1986. The Little Ice Age as recorded in the Stratigraphy of the Tropical Quelccaya Ice Cap: Science, v. 234, pp. 361–364. Adapted with permission from AAAS.

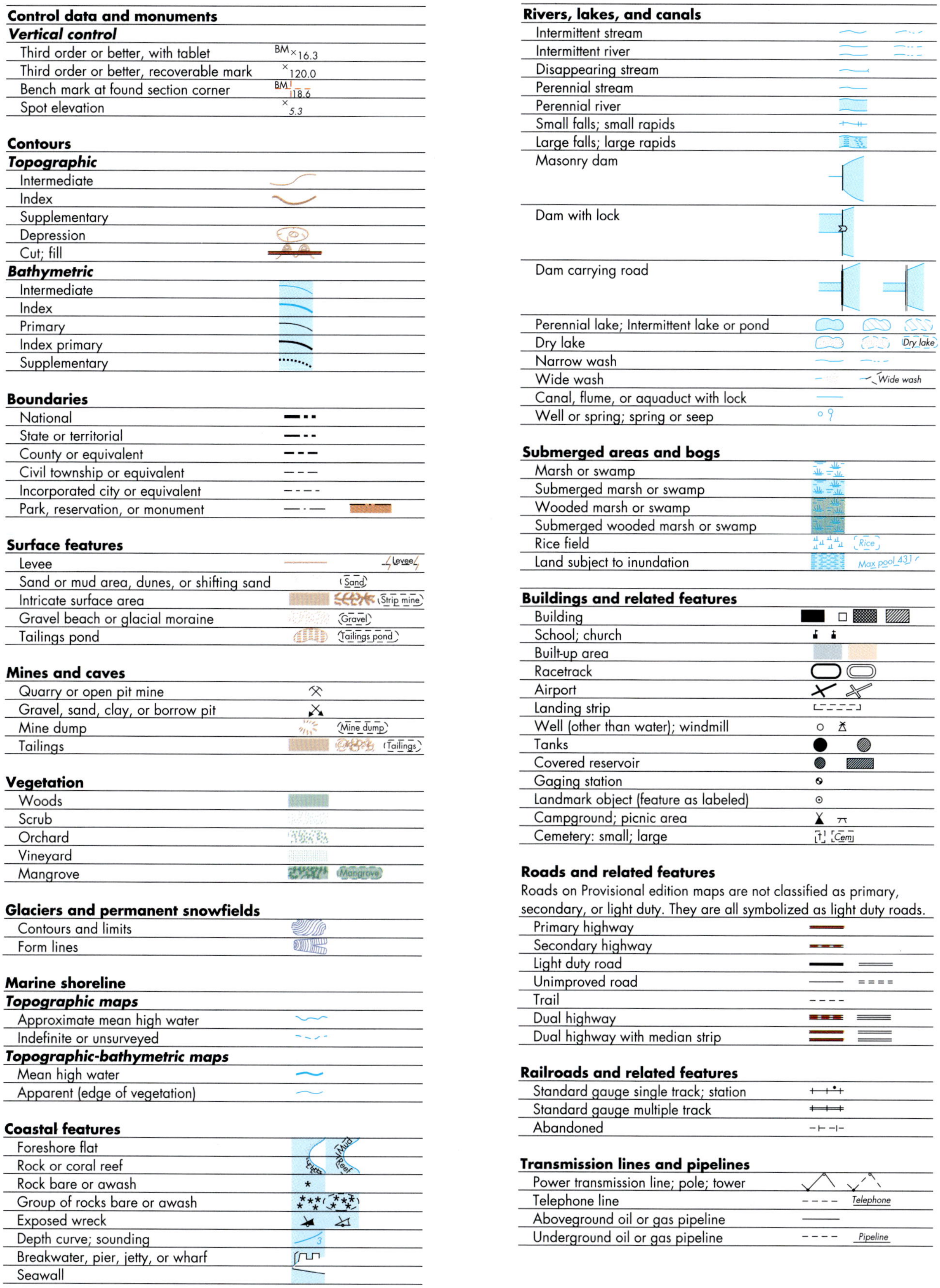

Control data and monuments

Vertical control

Third order or better, with tablet	BM × 16.3
Third order or better, recoverable mark	× 120.0
Bench mark at found section corner	BM 118.6
Spot elevation	× 5.3

Contours

Topographic

Intermediate	
Index	
Supplementary	
Depression	
Cut; fill	

Bathymetric

Intermediate	
Index	
Primary	
Index primary	
Supplementary	

Boundaries

National	
State or territorial	
County or equivalent	
Civil township or equivalent	
Incorporated city or equivalent	
Park, reservation, or monument	

Surface features

Levee	Levee
Sand or mud area, dunes, or shifting sand	Sand
Intricate surface area	Strip mine
Gravel beach or glacial moraine	Gravel
Tailings pond	Tailings pond

Mines and caves

Quarry or open pit mine	
Gravel, sand, clay, or borrow pit	
Mine dump	Mine dump
Tailings	Tailings

Vegetation

Woods	
Scrub	
Orchard	
Vineyard	
Mangrove	Mangrove

Glaciers and permanent snowfields

Contours and limits	
Form lines	

Marine shoreline

Topographic maps

Approximate mean high water	
Indefinite or unsurveyed	

Topographic-bathymetric maps

Mean high water	
Apparent (edge of vegetation)	

Coastal features

Foreshore flat	Mud
Rock or coral reef	Reef
Rock bare or awash	*
Group of rocks bare or awash	
Exposed wreck	
Depth curve; sounding	3
Breakwater, pier, jetty, or wharf	
Seawall	

Rivers, lakes, and canals

Intermittent stream	
Intermittent river	
Disappearing stream	
Perennial stream	
Perennial river	
Small falls; small rapids	
Large falls; large rapids	
Masonry dam	
Dam with lock	
Dam carrying road	
Perennial lake; Intermittent lake or pond	
Dry lake	Dry lake
Narrow wash	
Wide wash	Wide wash
Canal, flume, or aquaduct with lock	
Well or spring; spring or seep	

Submerged areas and bogs

Marsh or swamp	
Submerged marsh or swamp	
Wooded marsh or swamp	
Submerged wooded marsh or swamp	
Rice field	Rice
Land subject to inundation	Max pool 431

Buildings and related features

Building	
School; church	
Built-up area	
Racetrack	
Airport	
Landing strip	
Well (other than water); windmill	
Tanks	
Covered reservoir	
Gaging station	
Landmark object (feature as labeled)	
Campground; picnic area	
Cemetery: small; large	Cem

Roads and related features

Roads on Provisional edition maps are not classified as primary, secondary, or light duty. They are all symbolized as light duty roads.

Primary highway	
Secondary highway	
Light duty road	
Unimproved road	
Trail	
Dual highway	
Dual highway with median strip	

Railroads and related features

Standard gauge single track; station	
Standard gauge multiple track	
Abandoned	

Transmission lines and pipelines

Power transmission line; pole; tower	
Telephone line	Telephone
Aboveground oil or gas pipeline	
Underground oil or gas pipeline	Pipeline

FIGURE 2.9 Topographic map symbols (USGS)

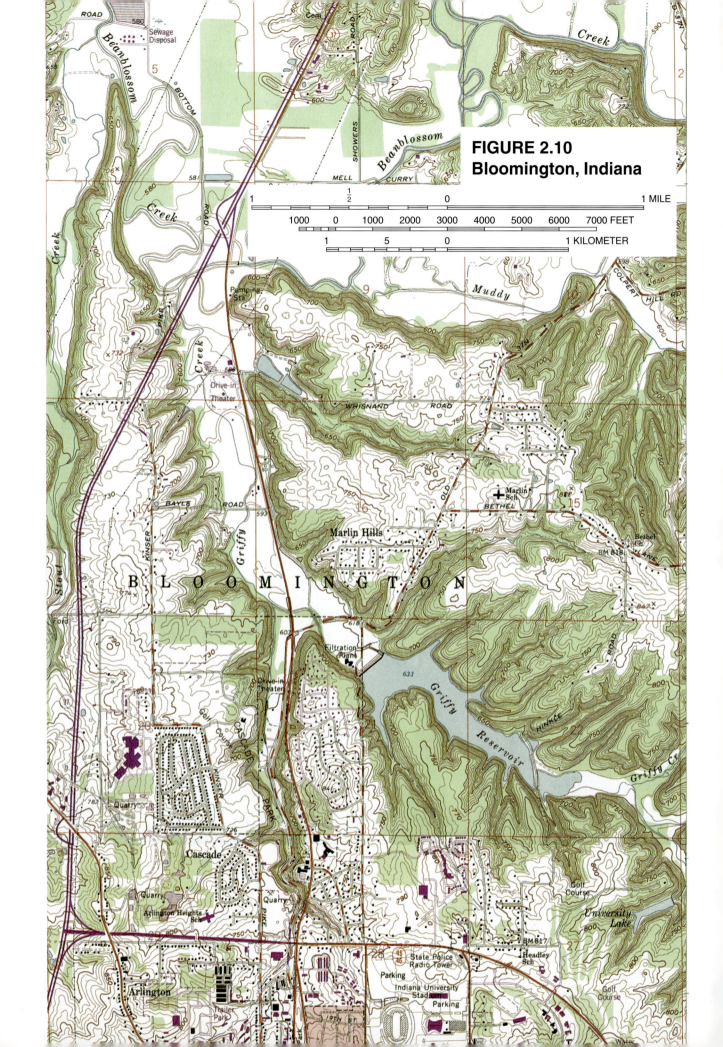

**FIGURE 2.10
Bloomington, Indiana**

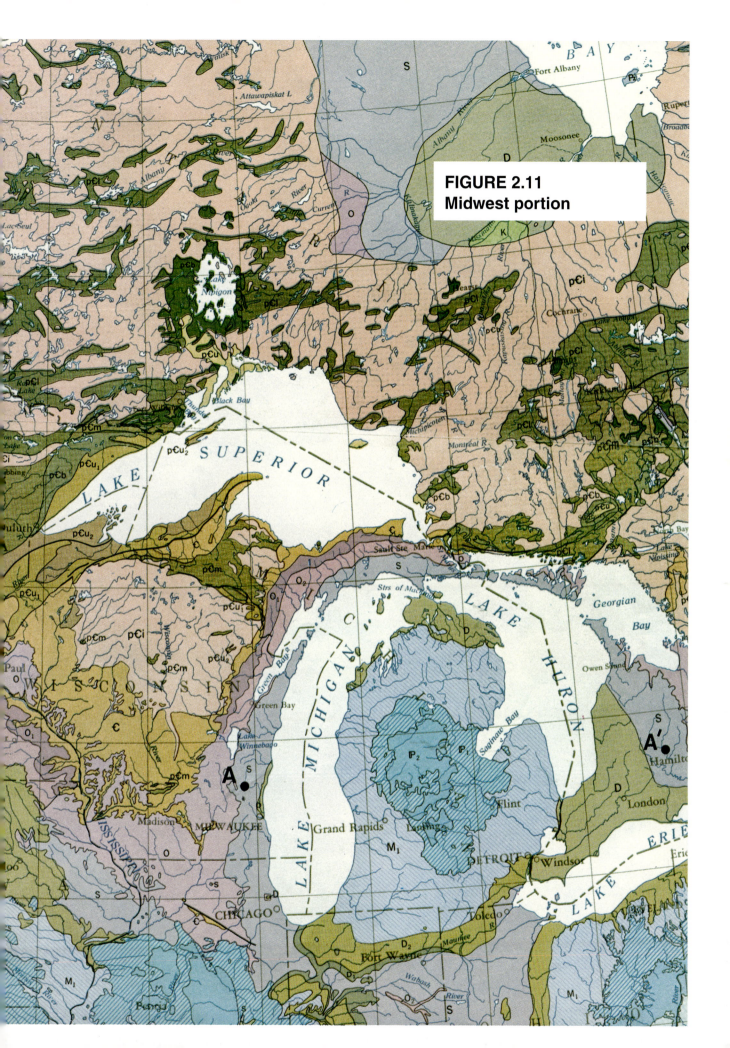

FIGURE 2.11
Midwest portion

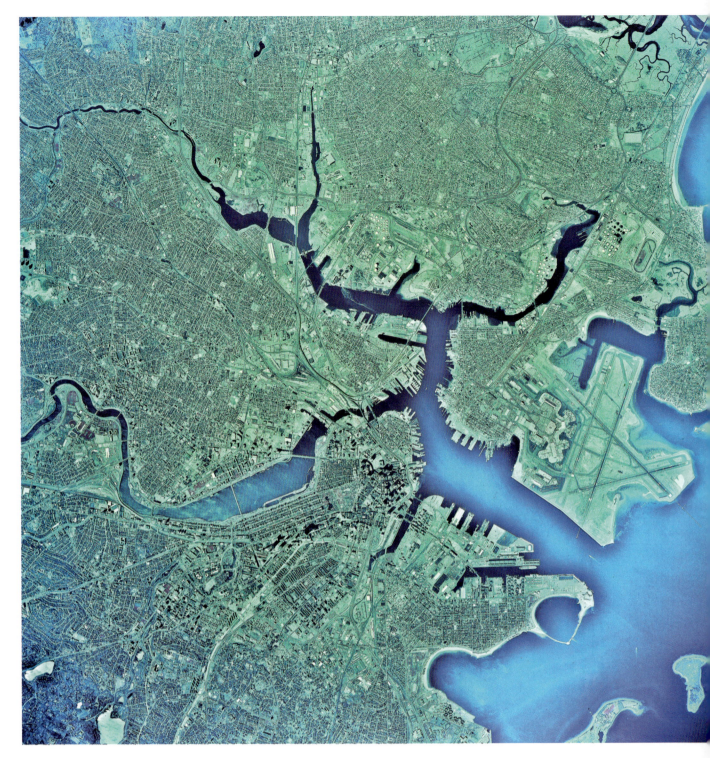

FIGURE 2.14   High altitude color map of Boston, MA (USGS)

FIGURE 2.16
Little Cottonwood Canyon
Salt Lake City, Utah

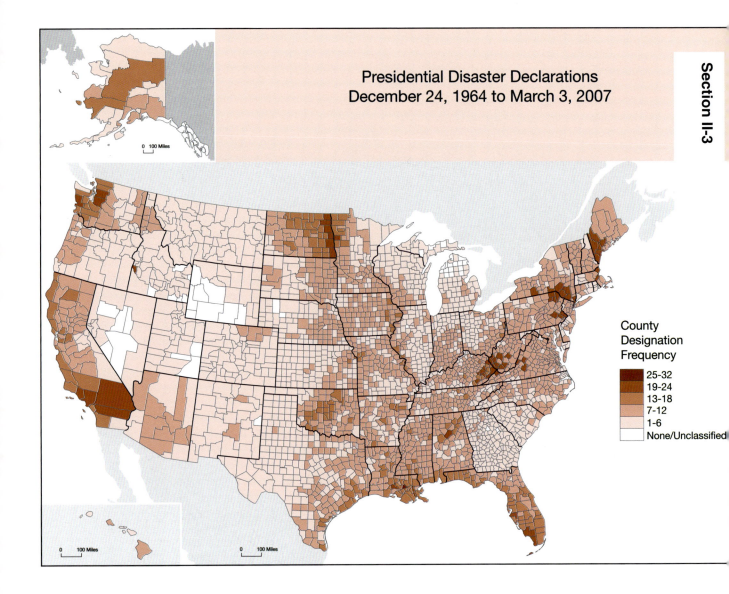

Presidential Disaster Declarations
December 24, 1964 to March 3, 2007

County
Designation
Frequency

25-32
19-24
13-18
7-12
1-6
None/Unclassified

0   100 Miles

0   100 Miles

0   100 Miles

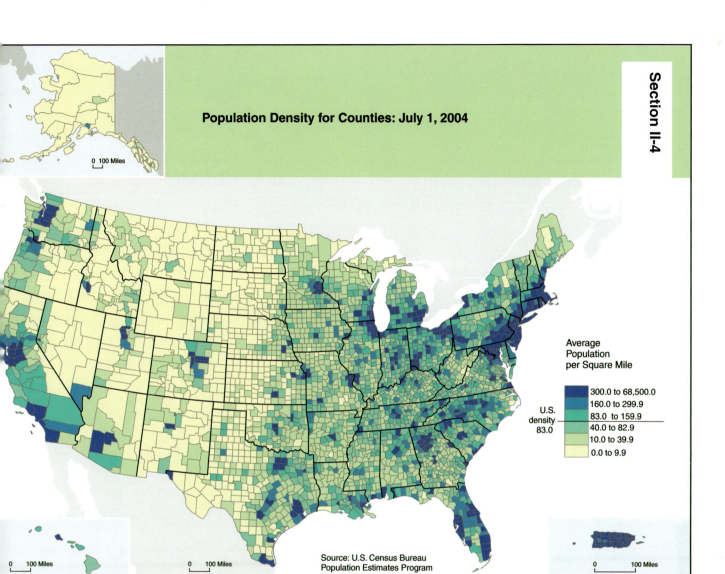

Population Density for Counties: July 1, 2004

0 100 Miles

Average
Population
per Square Mile

U.S.
density
83.0

	300.0 to 68,500.0
	160.0 to 299.9
	83.0 to 159.9
	40.0 to 82.9
	10.0 to 39.9
	0.0 to 9.9

0 100 Miles

0 100 Miles

Source: U.S. Census Bureau
Population Estimates Program

0 100 Miles

FIGURE 4.9
Mt. Rainier

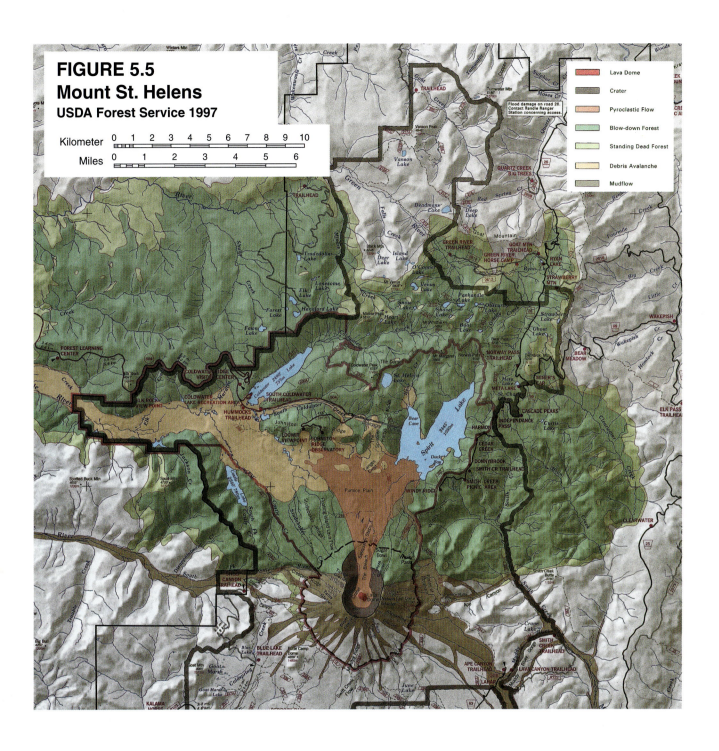

FIGURE 5.5
Mount St. Helens
USDA Forest Service 1997

Kilometer 0 1 2 3 4 5 6 7 8 9 10

Miles 0 1 2 3 4 5 6

Lava Dome
Crater
Pyroclastic Flow
Blow-down Forest
Standing Dead Forest
Debris Avalanche
Mudflow

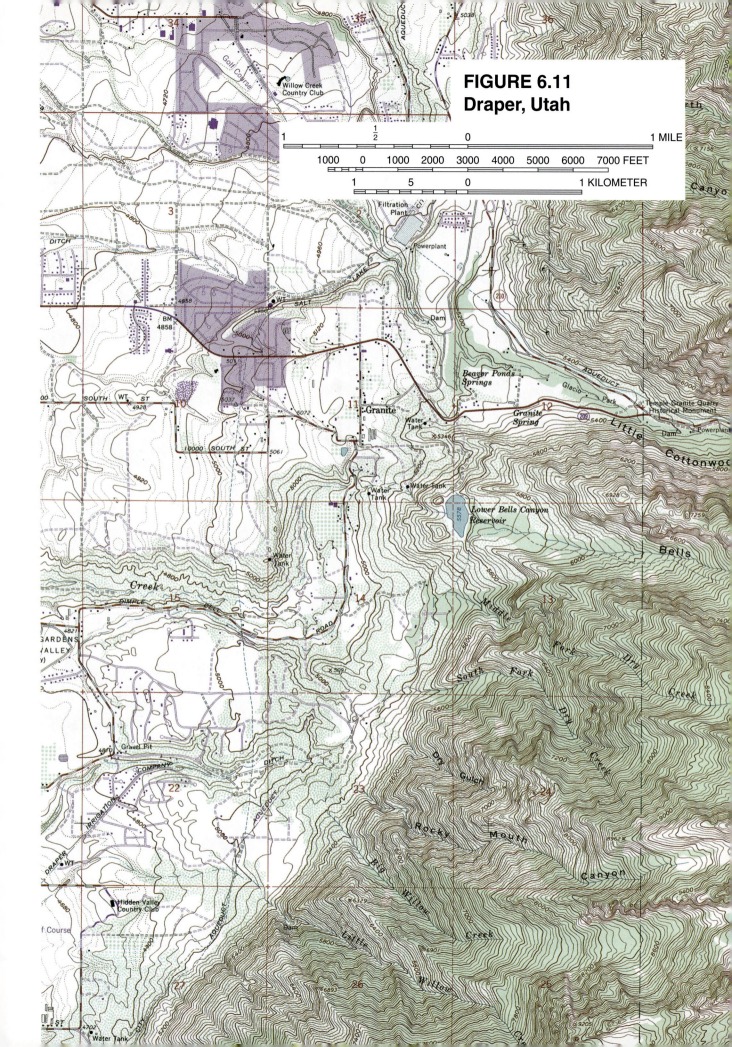

FIGURE 6.11
Draper, Utah

FIGURE 6.12
Salt Lake City, Utah

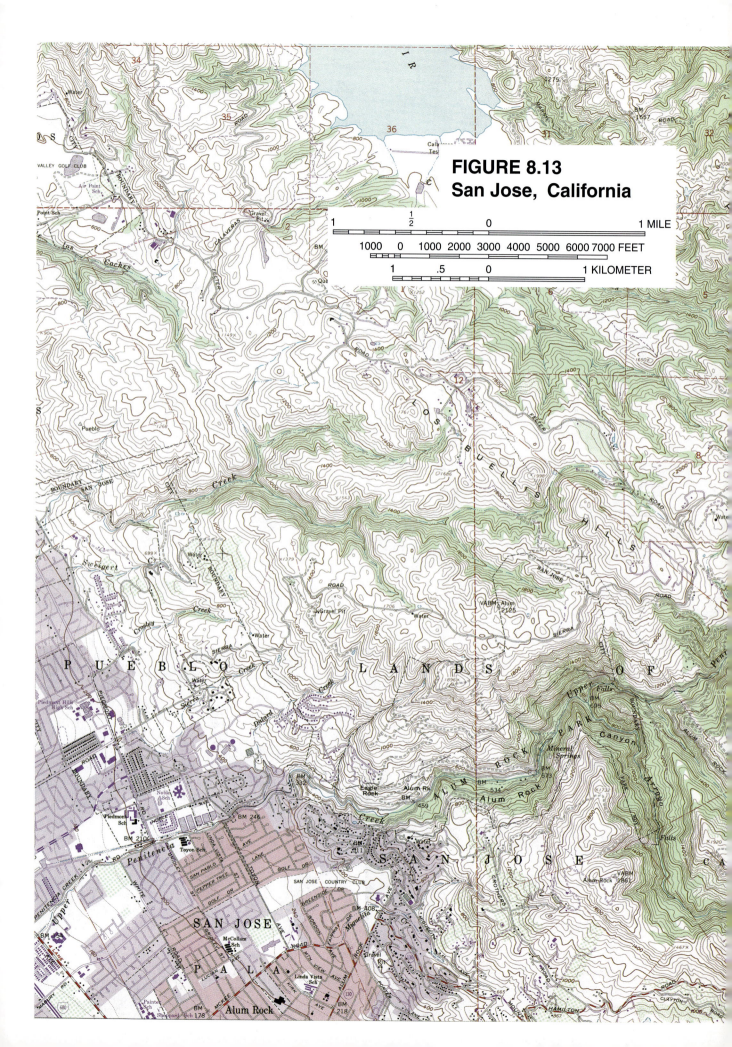

FIGURE 8.13
San Jose, California

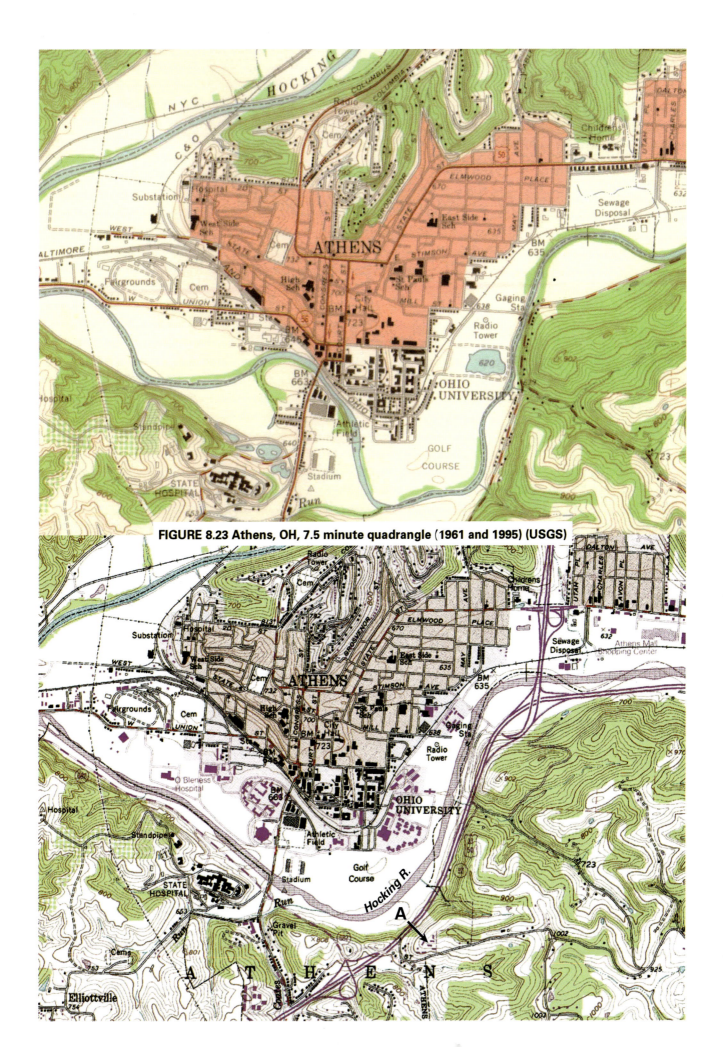

FIGURE 8.23 Athens, OH, 7.5 minute quadrangle (1961 and 1995) (USGS)

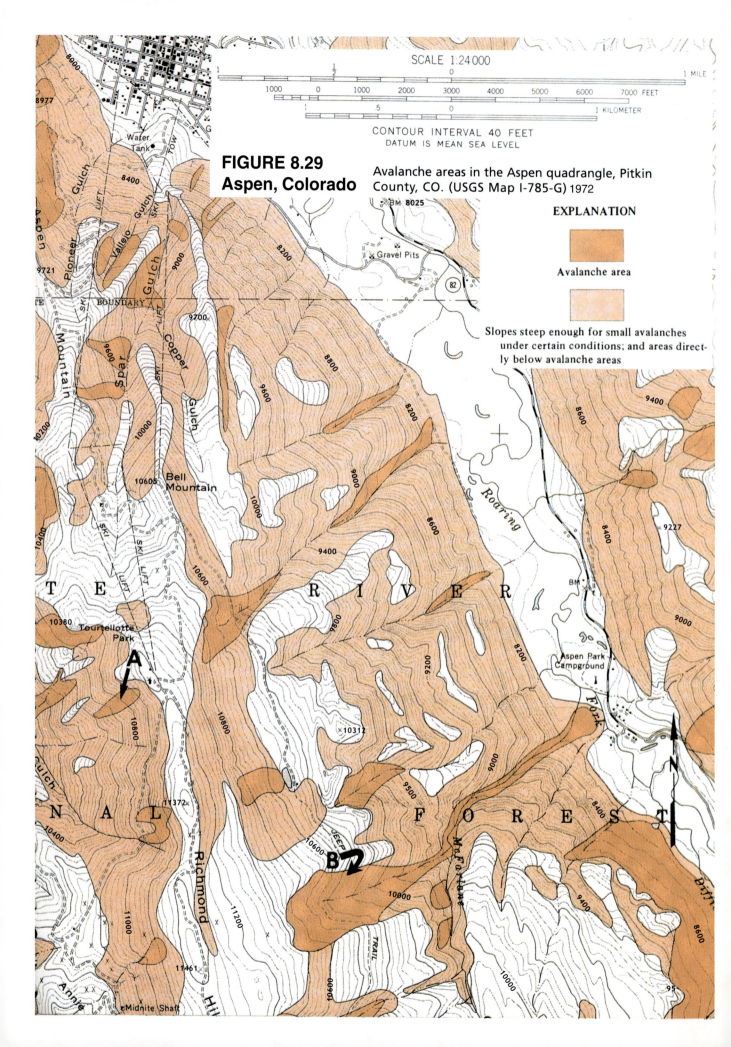

FIGURE 8.29
Aspen, Colorado

Avalanche areas in the Aspen quadrangle, Pitkin County, CO. (USGS Map I-785-G) 1972

SCALE 1:24 000

CONTOUR INTERVAL 40 FEET
DATUM IS MEAN SEA LEVEL

EXPLANATION

Avalanche area

Slopes steep enough for small avalanches under certain conditions; and areas directly below avalanche areas

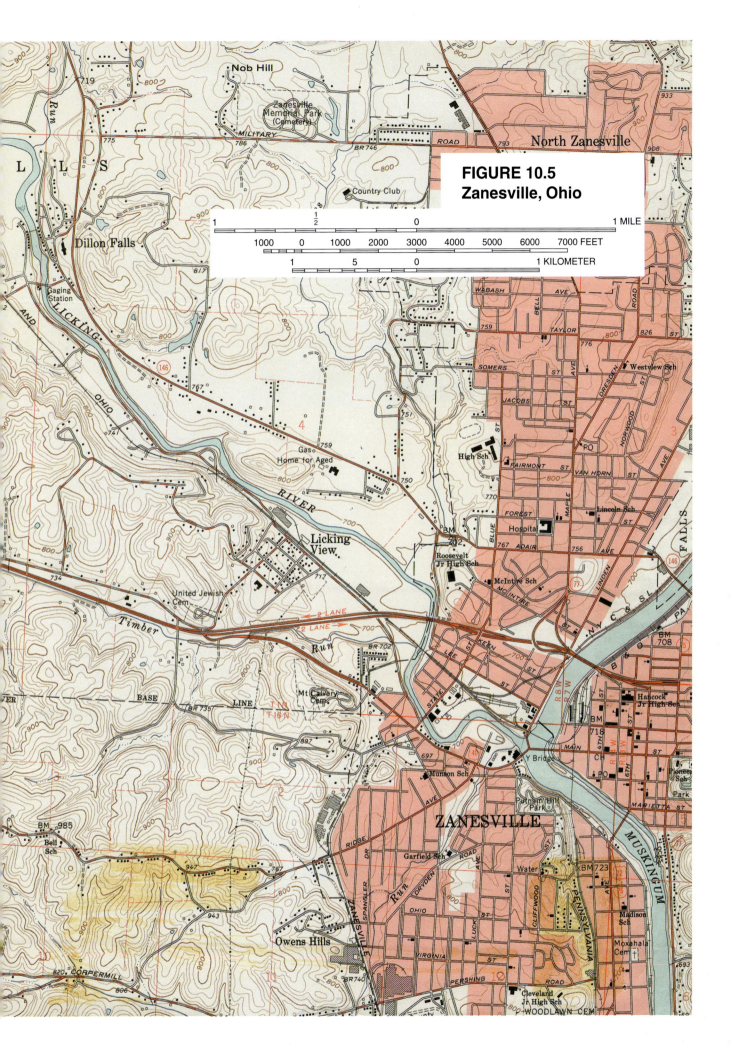

FIGURE 10.5
Zanesville, Ohio

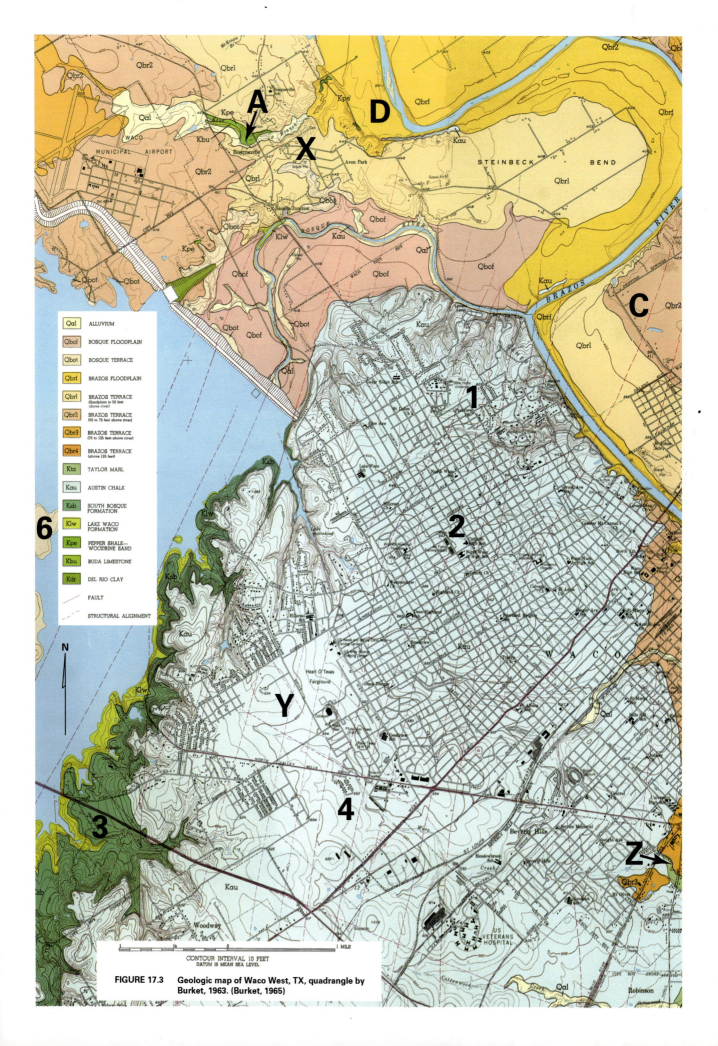

FIGURE 17.3   Geologic map of Waco West, TX, quadrangle by Burket, 1963. (Burket, 1965)

CONTOUR INTERVAL 10 FEET
DATUM IS MEAN SEA LEVEL

Qal	ALLUVIUM
Qbof	BOSQUE FLOODPLAIN
Qbot	BOSQUE TERRACE
Qbrf	BRAZOS FLOODPLAIN
Qbrl	BRAZOS TERRACE (floodplain to 50 feet above river)
Qbr2	BRAZOS TERRACE (50 to 75 feet above river)
Qbr3	BRAZOS TERRACE (75 to 125 feet above river)
Qbr4	BRAZOS TERRACE (above 125 feet)
Kta	TAYLOR MARL
Kau	AUSTIN CHALK
Ksb	SOUTH BOSQUE FORMATION
Klw	LAKE WACO FORMATION
Kpe	PEPPER SHALE— WOODBINE SAND
Kbu	BUDA LIMESTONE
Kdr	DEL RIO CLAY
	FAULT
	STRUCTURAL ALIGNMENT

N

1 MILE